OCCUPATIONAL ERGONOMICS

OCCUPATIONAL ERGONOMICS

A Practical Approach

THERESA STACK, LEE T. OSTROM, AND CHERYL A. WILHELMSEN

Published by John Wiley & Sons, Inc., Hoboken, New Jersey
Published simultaneously in Canada

Library of Congress Cataloging-in-Publication Data:

Names: Stack, Theresa, author. | Ostrom, Lee T., author. | Wilhelmsen, Cheryl A., 1933- author.
Title: Occupational ergonomics : a practical approach / by Theresa Stack, MS, CSP, CPE, Lee T. Ostrom, PhD, CSP, CPE, Cheryl Wilhelmsen, PhD.
Description: Hoboken, New Jersey : John Wiley & Sons, 2016. | Includes bibliographical references and index.
Identifiers: LCCN 2015050579| ISBN 9781118814215 (cloth) | ISBN 9781118814291 (epub)
Subjects: LCSH: Human engineering. | Industrial safety.
Classification: LCC T59.7 .S725 2016 | DDC 658.3/82–dc23 LC record available at http://lccn.loc.gov/2015050579

Printed in the United States of America

10 9 8 7 6

DEDICATION

This book is dedicated to extraordinary people of the US Military and their families, past and present, for their dedication and sacrifice in the name of freedom. In the words of Winston Churchill, Never was so much owed by so many to so few.

CONTENTS

LIST OF FIGURES

LIST OF TABLES

FOREWORD

Occupational ergonomics: A practical approach is a collaborative effort that merges engineering principles with practical applications to improve worker health, safety, and moral. It is packed with ergonomic evaluation tools and methods as well as well-crafted case studies to broaden the readers' understanding of reducing physical risk factor exposure and increasing productivity. The text is written from an educational perspective with exercises and questions following each chapter.

I became acquainted with the authors as the Navy Ergonomics Program Manager for 12 of my 32- year career. The expertise of the authors was needed to reduce the high rate of work-related musculoskeletal disorders experienced by Navy and civilian personnel.

I came to appreciate the creative and varied technical knowledge of the authors as we addressed the various and often unique processes of the Navy and Marine Corps facilities. The processes to be evaluated and improved ranged from laboratory work areas, to aircraft, hovercraft and helicopter repair to dolphin handling. Yes, dolphin handling. The Navy trains and uses dolphins to locate sea mines so they can be avoided or removed.

One of the many projects we worked on was at the Propeller Shop, where they maintain and repair propellers from the Landing Craft Air Cushioned (LCAC) Hovercrafts. The process consists of workers removing the 900-lb propeller from the hovercraft via an overhead crane and laying it onto a transportation fixture for repair and maintenance tasks.

The workers either stand hunched over or sit and reach overhead to work on the propellers with sanders, ratchets, rasps, putty knives, dead blow hammers, sledgehammers, and torque wrenches. Working for prolonged hours in these awkward and contorted postures resulted in 37% of the workers experiencing work-related pain or discomfort that did not improve when away from work, overnight, or over the weekend.

After evaluating the process, the authors worked with engineers to design and construct a fixture to hold the LCAC hovercraft propeller. The fixture allowed the propeller to rotate horizontally and vertically allowing the workers to adjust to a comfortable work height. This allowed the workers to sit or stand and be in a neutral posture while working, greatly reducing exposure to possible injury.

This is just one of many workplace improvements implemented at Navy and Marine Corps facilities that reduced physical risk factor exposure and increased productivity.

Occupational ergonomics: A practical approach contains all the important building blocks of occupational ergonomics – it revisits the fundamentals of workplace ergonomics and takes you step by step through more advanced workplace evaluations. The authors are experienced instructors and writers and have put together an excellent book using easy-to-follow examples gathered throughout their years of experience working in the field. I hope you find this book not only instructional but also an excellent desktop reference for occupational ergonomics.

CATHY ROTHWELL, P.E.

ACKNOWLEDGMENTS

The authors of *Occupational Ergonomics: A Practical Approach* would like to acknowledge and thank several people who were instrumental in this effort.

- Cathy Rothwell for providing unwavering support to the advancement of ergonomics within the United States Navy and the entire Naval Facilities Engineering Command Mishap Prevention and Hazard Abatement team.
- Ryan Haworth for the design and creation of the cover art work.
- Michael Stack for the bio-mechanics illustrations and other photographs.
- Kristopher Ostrom and Kateryna Savchenko for assistance with drawings, tables, and photographs.
- Malia Gonzalez and Keren Martin for allowing us to use portions of their student papers.
- Siobhan Wock, M.S. student in Industrial Hygiene at Montana Tech, for the keyword list and final editing.
- Raquel Wise for assisting with the development of the biomechanics chapter.
- Cherlyn Jewkes for her help with organizing the draft manuscript.
- Lidiia Ostrom for indexing the entire text.

ACKNOWLEDGMENTS

1

BOOK ORGANIZATION

LEARNING OBJECTIVES

Learning objectives are clearly stated to assist the student in knowing what will be discussed.

The goal of this module is d-ing, and leaning workstations. Common ly workspace envelopes are presented. At the end of this the skills to evaluate workstations at the United States.

INTRODUCTION

A workstation is location where a person performs one or more tasks that are required as part of his n have a profound impact on the person's form the required tasks. Reach and strength durance and visual capabilities are just a few of the factors that should be considered in work place design. The design guidelines to be discussed in this section include the following:

The banner on each page coincides with supplemental material.

- Accommodate people with a range of body sizes or anthropometric dimensions.
- Permit several working positions/postures to promote better blood flow and muscle movement.
- Design workstations from the working point of the hands. People work with their hands so we want their working height to be relative to their hand height.

Occupational Ergonomics: A Practical Approach, First Edition.
Theresa Stack, Lee T. Ostrom and Cheryl A. Wilhelmsen.

- Place tools, controls, and materials between the shoulder and waist height, where they have the greatest mechanical advantage.
- Provide higher work surfaces for precision work, and lower work surfaces for heavy work.
- Reduce compressive forces by rounding or padding work surface edges.
- Provide well-designed chairs in order to support the worker.

As each of the design guidelines are discussed, keep in mind that some of these principles may not be applicable to designs for individuals with special needs. Different tasks may also require different design guidelines; such as providing a brake pedal extension for a bus driver with shorter height.

CASE STUDY

See Figures 1.1 and 1.2.

Manual Materials Handling

Case studies are used in the chapters to relate the content to the student. Case studies can be used to prime in class discussion or exercises.

KEY POINTS

Key points and terms are identified.

- The desig[...] und impact on the person's ability to safely [...] tasks
- Reach ca[...] and visual capabilities are just a few of the factors that should be considered in workstation design.

REVIEW QUES[...]

Review questions for students for each chapter.

1. What anthropometric principle is us[...]t of the pull cord on a safety shower?
2. If a person is 20th percentile in height, will they be 20th percentile in weight? Why?
3. What anthropometric design principle is used as a last resort? Why?

EXERCISE

In class exercises can be perform[...]

In class exercises can be performed during the lectures.

In the exercise, the student groups wi[...]nbly task using ergonomic principles.

Figure 1.1 Worker moves 125-lb tire manually before the intervention (Photo courtesy of www.mohawklifts.com)

Figure 1.2 After the intervention, a tire dolly does much of the manual labor (Photo courtesy of www.mohawklifts.com)

2

THE BASICS OF ERGONOMICS

LEARNING OBJECTIVES

At the end of this module, students will be able to identify the basic principles of ergonomics, which will include a working definition of the term, and a brief history of ergonomics will be provided. Students will also be able to recognize the physical workplace risk factors and other contributors to the development of work-related musculoskeletal disorders (WMSDs) as well as potential resolutions to reduce or control workplace risk.

INTRODUCTION

Ergonomics is a field of study that involves the application of knowledge about physiological, psychological, and biomechanical capacities and limitations of the human (Butterworth, 1974). This knowledge is applied in the planning, design, and evaluation of work environments, jobs, tools, and equipment to enhance worker performance, safety, and health. Ergonomics is essentially fitting the workplace to the worker.

Ergonomics seeks to prevent WMSDs by applying principles to identify, evaluate, and control physical workplace risk factors.

History of Ergonomics

Wojciech Jastrzębowski, a Polish biologist, coined the word "ergonomics" as the science of work in 1857 in a philosophical narrative "based upon the truths drawn from

Occupational Ergonomics: A Practical Approach, First Edition.
Theresa Stack, Lee T. Ostrom and Cheryl A. Wilhelmsen.

the Science of Nature." The term ergonomics – **er·go·nom·ics \ˌûrg-go-'näm-iks** – is derived from two Greek words, Ergon meaning work and Nomos meaning principles or laws. Jasterzebowis understood the human and economic impacts of the industrial revolution during a time when a society of farmers traded in their hoes for 14-hour days in factories, growing iron and steel in lieu of wheat and potatoes. Factories brought people, process, and power together like never before.

A more commonly used definition of ergonomics today, as defined by one of the fathers of modern ergonomics Étienne Grandjean, is "fitting the work to the worker." Ergonomics is an applied science combining various disciplines that cater to the special needs of humans and is a goal-oriented science that seeks to reduce or eliminate injuries and disorders, increase productivity, and improve life quality.

Ergonomics, as defined by the Board of Certification for Professional Ergonomists (BCPE), "is a body of knowledge about human abilities, human limitations and human characteristics that are relevant to design. Ergonomic design is the application of this body of knowledge to the design of tools, machines, systems, tasks, jobs, and environments for safe, comfortable and effective human use" (Ergonomics).

The profession has two major branches with considerable overlap. One area of the discipline referred to as "industrial ergonomics," or "occupational biomechanics," concentrates on the physical aspects of work and human capabilities such as force, posture, and repetition. The second branch, referred to as "human factors," is oriented to the psychological aspects of work such as mental loading and decision making.

Bernardino Ramazzini was born in Carpi, Italy, in 1633. While he was still a medical student at Parma University, his attention was drawn to diseases suffered by workers. In 1682, when he was appointed chair of theory of medicine at the University of Modena, Ramazzini focused on workers' health problems in a systematic and scholarly way (Brauer, 2005). He visited workplaces, observed workers' activities, and discussed their illnesses with them. The medicine courses he taught were dedicated to the diseases of workers.

Primarily on the basis of this work, Ramazzini is called "the father of occupational medicine" (Report, 2012). Ramazzini systematized the existing knowledge and made a large personal contribution to the field by collecting his observations in **De Morbis Artificum Diatriba** (Diseases of Workers); the first edition was printed in Modena in 1700 and the second in Padua in 1713.

Each chapter of the **De Morbis Artificum Diatriba** contains a description of the disease associated with a particular work activity followed by a literature analysis, workplace description, questions for workers, disease description, remedies, and advice. The clinical picture was directly observed by Ramazzini, who questioned workers about their complaints. He regularly asked his patients about the kind of work they did and suggested that all physicians do the same expanding on the list by Hippocrates (Gilulian o Franco, 2009).

Ramazzini realized that not all workers' diseases were attributable to the working environment (chemical or physical agents).

The first and most potent is the harmful character of the material that they handle, for these emit noxious vapors and very fine particles … and induce particular diseases (in humans).

The second cause I ascribe to certain violent and irregular motions and unnatural posture of the body, by reason of which the natural structure of the vital machine is so impaired that serious disease gradually develop …

Ramazzini observed that a variety of common workers' diseases appeared to be due to prolonged, violent, and irregular motions and prolonged postures. Such cumulative trauma and repetitive motion injuries have recently been called the occupational epidemic of the 1990s (Giluliano Franco, 2009).

A good deal of evidence indicates that Greek civilization in the 5th century BC used ergonomic principles in the design of their tools, jobs, and workplaces. One outstanding example of this can be found in the description Hippocrates gave of how a surgeon's workplace should be designed and how the tools he uses should be arranged. Hippocrates also suggested to physicians that they inquire as to one's profession during examination.

In the 19th century, Frederick Winslow Taylor pioneered the "scientific management" method, which proposed a way to find the optimum method of carrying out a given task by maximizing human performance. Occupational ergonomics today seeks to decrease injury while enhancing performance. Taylor found, for example, you could triple the amount of coal workers were shoveling by incrementally reducing the size and weight of coal or ore in shovels until the fastest shoveling rate was reached, thus, literally matching the task and tools to the worker. Frank and Lillian Gilbreth expanded Taylor's methods in the early 1900s to develop the "time and motion study." They aimed to improve efficiency by eliminating unnecessary steps and actions. By applying this approach, the Gilbreths reduced the number of motions in bricklaying from 18 to 4.5, allowing bricklayers to increase their productivity from 120 to 350 bricks/h. The elimination of repetitive motions and extended reaching are methods of controlling WMSDs due to cumulative exposure to the same muscle group.

Prior to World War I (1914), the focus of aviation psychology or human factors was on the aviator himself, but the war shifted the focus onto the aircraft, in particular, the design of controls and displays, the size and shape of the aviator within the cockpit to reach and activate controls, and the effects of altitude and environmental factors on the pilot. The war witnessed the emergence of aeromedical research and the need for repeatable testing and measurement methods to ensure, as much as possible, a pilot's cognitive and physical capacities were maximized but not exceeded.

Studies on driver behavior started gaining momentum during this period as well as with Henry Ford (1920) providing millions of Americans with an automobile. Henry Ford was also concerned with efficiency of motion to decrease the cost of making an automobile while increasing its quality. Henry Ford stated, "The work must be brought to the man waist-high. No worker must ever have to stoop to attach a wheel, a bolt, a screw or anything else to the moving chassis (Ford)."

World War II (1940) marked the development of new and complex machines and weaponry, and these made new demands on operators' cognition and physical capacity. Now the design of equipment had to take into account human limitations and take advantage of human capabilities. It was observed that fully functional aircraft, flown by the best-trained pilots, still crashed. In 1943, Alphonse Chapanis, a lieutenant in the U.S. Army, showed that this so-called "pilot error" could be greatly reduced when more logical and differentiable controls replaced confusing designs in airplane cockpits. After the war, the Army Air Force published 19 volumes summarizing what had been established from research during the war. Research covered areas such as:

- Muscle force required to perform manual tasks
- Compressive low back disk force when lifting
- Cardiovascular response when performing heavy labor
- Perceived maximum load that can be carried, pushed, or pulled.

The beginning of the Cold War led to a major expansion of defense supported research laboratories in the areas of human factors and ergonomics. While most of the research following the war was military-sponsored, the scope of the research broadened from small equipment to entire workstations, and systems benefited the industrial sector as well as in defense. Concurrently, private opportunities started opening up in the civilian industry for improvements in workstation and task design. The focus shifted from research to participation through advice to engineers in the design of equipment, facilities, and processes.

The Human Factors Society, the main professional organization for human factors and ergonomics practitioners in the United States, was formed in 1957 with approximately 90 people attending the first annual meeting. The name was changed to the Human Factors and Ergonomics Society in 1992. Today the society has more than 4500 members (Society) and is the benchmark for an internationally recognized designation in the practice of ergonomics.

Starting in the mid-1960s, the discipline continued to grow and develop in previously established areas. Moreover, it expanded into computer hardware (1960s); computer software (1970s); nuclear power plants and weapon systems (1980s); the Internet and automation (1990s), and adaptive technology (2000s) just to name a few. Most recently, new areas of interest have emerged including neuro-ergonomics and nano-ergonomics.

A consistent theme influencing human factors and ergonomics has sought to grow over the years as an ever-expanding sphere and has emerged too, in order to keep pace with scientific advances. With the rapid advances in science and technology today, it's interesting to speculate on what newly discovered challenges human factors and ergonomics will be called upon to solve. Exoskeletons are being used to perform tasks and lighten the burden on the human today. As it was at its inception, human factors and ergonomics remain multidisciplinary professions.

Contributors to ergonomics and human factors concepts include industrial engineers, industrial psychologists, occupational medicine physicians, industrial hygienists, and safety engineers. Professions that use ergonomics/human factors information include architects, occupational therapists, physical therapists, industrial hygienists, designers, safety engineers, general engineering, occupational medicine professionals, and insurance loss control specialists.

WORK-RELATED MUSCULOSKELETAL DISORDERS

Ergonomics seeks to prevent Work-Related Musculoskeletal Disorders (WMSDs) by applying principles to identify, evaluate, and control physical workplace risk factors.

Musculoskeletal disorders (MSDs) are a class of disorders involving damage to muscles, tendons, ligaments, peripheral nerves, joints, cartilage (including vertebral discs), bones, and/or supporting blood vessels. WMSDs are MSDs aggravated by working conditions. WMSDs are not typically due to acute events but occur slowly over time due to repeated wear and tear or microtraumas to the tissue. For example, dental hygienists tend to develop hand-related tendon damage due to repeated gripping of small diameter tools while applying force.

> *Micro-trauma is a small, minor, limited area tissue damage or tear. Cumulative trauma occurs when rest or overnight sleep fails completely to heal the micro-trauma, and the residual trauma carries over to the next day. Damage continues to proliferate if the exposure or dose remains unchanged* (Labor, 1999).

WMSDs are also known as cumulative trauma disorders (CTDs), repetitive strain injuries (RSIs), repetitive motion trauma (RMT), or occupational overuse syndrome. Examples of WMSDs include epicondylitis (tennis elbow), tendinitis, DeQuervain's disease (tenosynovitis of the thumb), trigger finger, and Reynaud's syndrome (vibration white finger), Carpal tunnel syndrome (CTS) is a commonly known WMSD as is back strain. Occupational Safety and Health Association (OSHA) uses the term MSD.

> *OSHA defines MSD as a disorder of the muscles, nerves, tendons, ligaments, joints, cartilage or spinal discs that was not caused by a slip, trip, fall, motor vehicle accident or similar accident* (OSHA).

Researchers have identified specific physical workplace risk factors involved in the development of WMSDs. Exposure to these risk factors can result in

- decreased blood flow to muscles, nerves, and joints;
- nerve compression;
- tendon or tendon sheath damage;
- muscle, tendon, or ligament sprain or strain; and
- joint damage.

Prolonged exposure to the physical workplace risk factors can lead to permanent damage and a debilitating condition. Stages of WMSD development and specific WMSDs are discussed in Chapter 14.

When present for sufficient duration, frequency, or magnitude, physical workplace risk factors may contribute to the development of WMSDs. In addition, personal risk factors, such as physical conditioning, existing health problems, gender, age, work technique, hobbies, and organizational factors (e.g., job autonomy, quotas, deadlines) contribute to, but do not cause the development of WMSDs. Applying ergonomic principles to reduce a worker's exposure to the physical workplace risk factors decreases the chance of injury and illness. Figure 2.1 is an example of a task with a combination of physical workplace risk factors.

PHYSICAL WORKPLACE RISK FACTORS – OVERVIEW

Physical workplace risk factors are those aspects of a job or task that impose biomechanical stress on a worker. Researchers have identified specific physical workplace

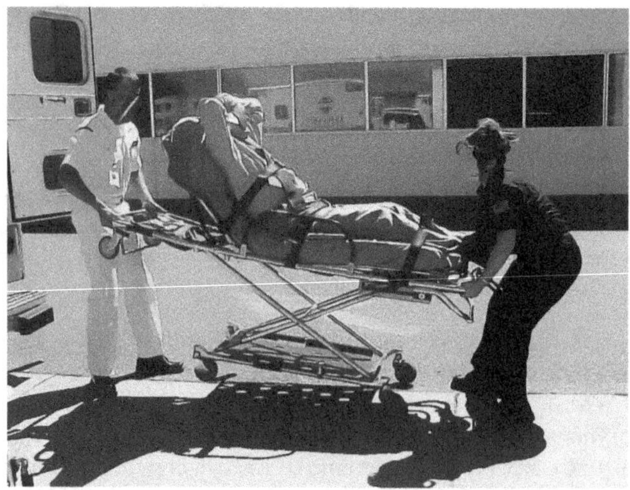

Figure 2.1 Awkward posture of the back, high spinal forces, and hand compression are found with emergency transport

risk factors that can cause or contribute to the development of WMSDs. Ergonomic principles are commonly used to mitigate the exposure. The following will be covered in detail.

- Postures – both awkward (nonneutral) and static
- Forces-including heavy, frequent, or awkward lifting
- Compression
- Repetition
- Vibration.

Force, repetition, and awkward postures, especially when occurring at high levels or in combination, are most often associated with the occurrence of WMSDs. While exposure to one risk factor may be enough to cause injury, typically physical workplace risk factors act in combination to cause injury. An example is the position of the wrist a waiter or waitress uses when carrying a tray (with hand bent back, wrist at almost a 90° angle). This position causes severe hand extension. When this extreme awkward posture is combined with the gravitational force created by a heavy tray and the position is repeated over a long shift, multiple risk factors are present that place the employee at risk of injury, Posture + Force + Duration + Repetition. Dialing down or decreasing any single risk factor will ultimately decrease the risk of injury. Other workplace conditions can contribute to but do not cause WMSDs. They can however cause other undesirable health conditions. These conditions can include the following:

- Duration
- Intensity
- Temperature

- Workplace
- Stress
- Organizational issues.

Personal risk factors can also contribute to the development of WMSDs, for example:

- Age
- Gender
- Hobbies
- Previous injuries
- Physical or medical conditions
- Smoking
- Fatigue.

Posture

The neutral posture is the optimal body position in order to minimize stress and provide the greatest strength and control as seen in Figure 2.2. The neutral posture is the body position in which there is the least amount of tension or pressure on the nerves, tendons, muscles, joints, and spinal discs. It is also the position in which muscles are at their resting length, neither contracted nor stretched. Muscles at this length can develop and maintain maximum force most efficiently. The neutral posture in the workplace can be recognized by the proper alignment of body landmarks. Neck, shoulders, and arms should be relaxed with elbows by the sides. The elbow should be open at an angle that is no less than 90°. The ears are roughly over the shoulders, shoulders over hips, hips over knees, and knees over ankles. The spine should have a slight S shape with the lower back slightly concaved.

The benefits of proper spinal alignment are discussed in Chapters 13 and 7. The seated neutral posture is discussed in Chapter 4 as well as the guide "Standing up on the Job" found in Appendix B.

Awkward Postures Awkward postures or nonneutral postures are those outside of the neutral posture. Awkward or unsupported postures can stretch the body's physical limits and can compress nerves and irritate tendons. Awkward postures are often significant contributors to MSDs because they increase the work and the muscle force required (OSHA).

Examples of awkward postures include the following:

- Raising hands above the head or elbows above the shoulders, a common awkward posture in manufacturing and manual material handling
- Kneeling or squatting, common in maintenance operations
- Working with back, neck, and/or wrist bent, for example, when using a microscope
- Sitting with feet unsupported that can cause blood to pool in the feet and flatten the natural curve in the lumbar spine. This problem is common among laboratory technicians who work at high benches.

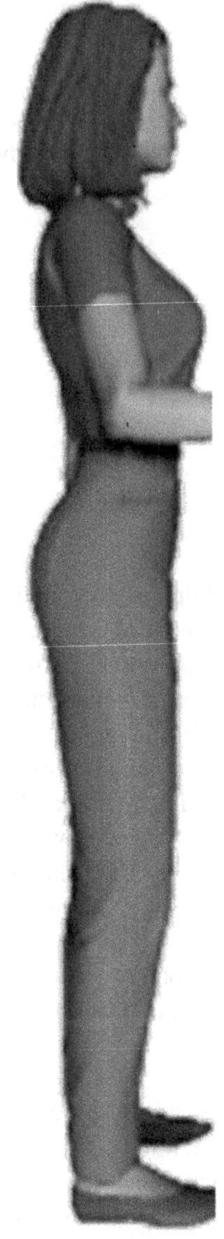

Fingers - are gently curved, in their natural resting position
Wrist - is in line with the forearm. It is neither bent up (extension) nor bent down (flexion)
Forearm - rests with the thumb up. It is not rotated to make the palm face down or up
Elbow - is in a neutral position when the angle between forearm and upper arm is close to a right angle (90°). Some extension (up to 110°) may be desirable
Upper arm -hangs straight down. It is not elevated to the side (abduction), pulled across in front of the body (adduction), raised to the front (flexion) nor raised towards the back (extension)
Shoulder - are in a resting position, neither hunched up nor pulled down, and not pulled forward or back
Neck - the head is balanced on the spinal column. It is not tilted forward, back or to either side. It is not rotated to the left or right
Back - the spine naturally assumes an S-shaped curve. The upper spine (thoracic region) is bent gently out; the lower spine (lumbar region) is bent gently in. The spine is not rotated (or twisted) to the left or right, and it is not bent to the left or right

Figure 2.2 Standing neutral postures (Adapted with permission from The Ergonomics Image Gallery)

Awkward postures are more fatiguing than neutral postures because the muscles, tendons, and ligaments are actively working to maintain the posture; the greater the posture deviations from neutral, the higher the stress on the human and resultant risk of injury.

In Figure 2.3, before, the worker had to bend and reach to gather parts for eyeglass assembly. After, an automatic storage system delivers parts at elbow height, reducing awkward postures and improving quality control through an automated inventory stem.

(a) (b)

Figure 2.3 (a) Workers are exposed to long reaches above the shoulders and below the knees. (b) Automated retrieval system delivers parts at elbow height reducing the repeated and frequent awkward postures

Static Postures Holding a posture for extended periods of time is known as a static posture or static muscle loading. Static postures prevent the flow of blood. These types of exertions put increased load or forces on the muscles or tendons, which contribute to fatigue (OSHA). Blood flow brings nutrients to the muscles and carries away waste products. Holding a muscle in contraction causes waste products to build up and leads to fatigue. Fatigue is considered a precursor to injury. For more information on fatigue, see the administrative chapter.

Examples of static postures include the following:

- Gripping tools that cannot be put down
- Holding arms out or up to perform tasks
- Standing in one place for prolonged periods.

Repetition

Repetition is a physical risk factor that occurs when the same motion or group of motions is performed over and over again. Different tasks may still utilize the same muscle groups and, therefore, not allow the muscles to rest leading to overuse. Repetition alone is not typically a problem, but when it occurs with other risk factors it magnifies the exposure.

Force

Force refers to the amount of physical effort that is required to accomplish a task or motion. Tasks or motions that require application of higher force place higher mechanical loads on muscles, tendons, and joints (OSHA) and can quickly lead to fatigue.

The force required to complete a movement increases when other risk factors are also involved. For example, more physical effort may be needed to perform a task when speed is increased or vibration present. Performing forceful exertions requires the application of muscle contraction that may cause them to fatigue quickly. The more force that must be applied, the more quickly the muscles fatigue. Excessive or prolonged exposure also leads to overuse of the muscles and may result in muscle strain or damage.

The power zone for lifting, with the greatest strength, endurance and control, and lowest risk of injury is holding the load close to the body between the knuckle and shoulder height.

Compression

Compression or contact stress is a concentrated force on a small surface area. Contact stress can reduce blood flow or cause tissue (e.g. tendon) irritation due to the constant pressure.

One of the most common sources of compression is a sharp or hard desk edge creating a compressive force on the forearm or elbows as we rest to stabilize the joint. Nerves in the forearm are close to the skin surface; compression of the forearm impedes nerve conduction.

Vibration

There are two types of vibration, single point and whole body. More information can be found in Chapter 10.

Single Point Vibration

Single point vibration is exposure to a single body part such as the upper extremity. This type of vibration is common with tool use.

The main outcome of prolonged exposure is a decrease in blood volume to the extremities. Vibrating tools can cause vascular spasms or a constriction of blood vessels in the fingers, which then appear white or pale. More information is provided in Chapter 9 of the book.

Whole Body Vibration

Whole body vibration is exposure to vibration through the entire body. This type of vibration can be found from vehicles such as forklifts, cranes, trucks, buses, ocean vessels, and aircrafts.

The main effect is usually to the spine, but studies indicate that high exposure can reduce circulation and cause disorientation and motion sickness.

High or prolonged exposure to whole body vibration can affect the skeletal muscles and digestive system and can cause lower back pain and pregnancy complications.

Contributing Factors

Contributing risk factors contribute to but do not cause WMSDs. Contributing factors, when personal in nature, can be outside of a safety professional's realm to mitigate without an active wellness program.

Duration and Magnitude

Duration is the time period in which an action continues or lasts. Continuous exposure to any risk factor may not allow sufficient recovery time for muscles, tendons, and nerves. Duration magnifies the risk factors as does intensity or magnitude.

Temperature Extremes

Temperature extreme are a contributing factor to the development of WMSDs. Working in cold environments places a greater aerobic demand on the worker, which means they fatigue faster. The cold also reduces dexterity and causes you to grip harder or apply more muscle force. Cold temperatures are especially problematic when present with vibration because both risk factors contract blood vessels.

Radiant heat from furnaces or direct exposure from the sun should be considered for warm temperature exposure. Workers move more slowly when hot, so simple tasks can take longer, thus increasing the duration of exposure to temperature extremes and other risk factors.

Note that personal protective equipment (PPE) can decrease evaporation from the skin thus reducing the body's ability to cool itself adding to thermal gain.

Inadequate Recovery

Working without rest can cause fatigue and contribute to injury. Muscles need time to rest to re-oxygenate and remove the waste products of muscle metabolism. There are many ways to rest a muscle group. Stretching, alternating tasks, and taking micro-breaks (brief pause) can aid in muscle recovery and readiness. Refer to the administrative controls, Chapter 5, more information on stretching programs and WMSDs, Chapter 13, for disease etiology.

Personal Risk Factors

Personal risk factors can also contribute to the development of workplace injury and illness.

- Age – as we age, the repair process in our body takes longer.
- Gender – due to anatomical and hormonal differences, certain WMSDs are more prevalent in women.
- Hobbies – knitting, crocheting, bowling, and computer gaming.

- Smoking – linked to back pain because smokers tend to heal more slowly due to the decrease in oxygen in the blood stream.
- Obesity and pregnancy – linked to carpal tunnel syndrome (CTS). Even a low level of exposure to workplace risk factors can create CTS in a pregnant worker. The symptoms usually disappear after the baby is born. Obesity impacts are lack of flexibility, fluid buildup, and increased pressure on the disks.
- Previous injury – puts a worker at risk of an MSD in the same place.
- Medications – can cause dehydration, swelling, decreased/increased metabolic rates, and a change in electrolyte levels.
- Fatigue – cause a reduction in performance due to a period of excessive activity followed by inadequate recovery time. Muscle fatigue is accompanied by a buildup of lactic acid in the working muscle.
- Physical conditions – poor fitness, particularly when combined with a body weight above the "ideal," is a prime cause of weariness and fatigue, which are commonly recognized to be factors that can contribute to the onset of musculoskeletal injuries.

WMSD SIGNS, SYMPTOMS, PREVENTION

Early identification of signs and symptoms of WMSDs can eliminate risk or reduce the severity of an injury. Reporting allows the occupational safety and health professional the opportunity correct issues before an employee is injured. Refer to Chapter 13 for more information.

Some signs and symptoms are as follows:

- Painful aching joints, muscles
- Pain, tingling or numbness
- Shooting or stabbing pains
- Swelling or inflammation
- Warmth
- Stiffness or difficulty moving
- Burning sensation
- Pain during the night
- Loss of strength and mobility.

Pain and discomfort are usually precursors to injury and should be considered warning signs or an indicator of a need for improvement.

SUMMARY

Understanding and applying ergonomics principles in the workplace will reduce the physical stress on the body and eliminate potentially serious, disabling WMSDs through matching jobs, tasks, and work environment to the worker. Ergonomics is also good economics. Other benefits include the following:

- Improved comfort, morale and job satisfaction, and worker health
- Improved productivity and reduced workers' compensation costs and employee turnover.

Ergonomics is essentially fitting the work to the worker and is effective in preventing MSDs that are caused or aggravated by the work environment. Physical ergonomics risk factors that can cause injury (when occurring in combination) are force, posture, duration, repetition, vibration, and compression. Other workplace risk factors (e.g., temperature and quotas) along with personal factors can contribute to but do not cause WMSDs.

KEY POINTS

- Ergonomics is not a new science; it was founded in the 1700s along with other occupational medicine disciplines. The ill effects have been documented since 1857 on work and health.
- Ergonomics seeks to prevent injuries before they occur.
- Early studies in ergonomics focused on productivity improvements.
- The US Military's involvement in flying technology in the 1940s, followed by the personal computer boom in the 1990s, brought ergonomics into the forefront of safety and health.
- Microtrauma is a small, minor, limited area tissue damage or tear. Cumulative trauma occurs when rest or overnight sleep fails to completely heal the microtrauma, and residual trauma carries over to the next day, adding to the total system trauma.
- Physical workplace risk factors that can cause WMSDs can be changed. Personal and contributing risk factors that do not cause WMSDs typically cannot be changed.
- Standing neutral posture:
  - Ears over shoulders
  - Shoulders over hips
  - Hips over knees
  - Knees over ankles
  - Elbows by the sides.

In Figure 2.4a, firefighters were required to lift an 86-lb smoke ejector fan from above the cab on the fire apparatus, walk across hoses, and lower it to others below. In Figure 2.4b, smoke ejector fan is mounted to bumper, greatly reducing awkward postures during lifting. The task resulted in physical workplace risk factor exposure, as well other safety risks. The fan was mounted to the rear bumper of the truck, virtually

Figure 2.4 (a) Firefighters carried a heavy fan over unstable footing and lowered it to workers below. (b) Mounting the fan on the bumper eliminated the hazards

eliminating the physical workplace risk factors by placing the lift to within the workers power zone.

REVIEW QUESTIONS

1. Define ergonomics.
2. Define WMSD.
3. List and briefly describe the physical workplace risk factors.
4. List at least three personal risk factors that contribute to but do not cause WMSDs.
5. List at least two workplace risk factors that contribute to but do not cause WMSDs.
6. What are the two differences between physical workplace risk factors and contributing factors for the development of WMSDs?
7. What is the difference between ergonomics and human factors?
8. Describe the contribution of Frederick W. Taylor to the advancement of ergonomics.
9. Describe the contribution of Dr. Ramazzini to the practice of occupational medicine.

EXERCISES

1. Students research a historic WMSD, such as musicians' nerve, and relate how it was caused in historic times to current times.

2. Students submit a personal biography highlighting their work or educational experience as well as exposure to occupational ergonomics. Students should list one goal they hope to meet by completing the course.
3. Students submit an assignment or verbally discuss an area in their work or home life, where they are already practicing ergonomics.

REFERENCES

Brauer, R. (2005). *Safety and Health for Engineers* 2nd edn. Hoboken: John Wiley & Sons Inc..

Butterworth (1974). *Applied Ergonomics Handbook.* London.

Ergonomics, B. o. (n.d.). About BCPE. Retrieved 2012, from Board of Certification in Professional Ergonomics: http://www.bcpe.org/.

Giluliano Franco, M. (2009). Bernardino Ramazzini: The Father of Occupational Medicine. Retrieved December 29, 2013, from National Center for Biotechnology Information: http://www.ncbi.nlm.nih.gov/pmc/articles/PMC1446786/.

Labor, U. S. (1999). Occupational Safety and Health Administration Federal Register Proposed Rule. Retrieved December 8, 2001, from Occupational Safety and Health Administration: https://www.osha.gov/pls/oshaweb/owadisp.show_document?p_table=federal_register&p_id=16305.

Report, T. E. (2012). Revisiting the Roots of Ergonomics. Retrieved January 4, 2013, from ErgoWed: https://ergoweb.com/revisiting-the-roots-of-ergonomics/.

ADDITIONAL SOURCES

Health, N. Y. (n.d.). Work Place Hazards. Retrieved December 26, 2013, from NYCOSH: http://nycosh.org/index.php?page=Hierarchy-of-Hazard-Controls.

Henry Ford Quotes. (n.d.). Retrieved March 5, 2015, from The Henry Ford: https://www.thehenryford.org/research/henryFordQuotes.aspx.

OSHA. (1999). *Federal Register V 64 No. 225 Proposed Rule.* OSHA.

Society, H. F. (n.d.). Membership. Retrieved December 26, 2013, from Human Factors and Ergonomics Society: https://www.hfes.org//Web/Membership/membership.html.

3

ANTHROPOMETRY

LEARNING OBJECTIVES

At the end of the chapter, students will have the ability to describe anthropometry, identify the best ergonomic design principle for a given situation, demonstrate how to use anthropometric data tables, and apply anthropometric principles to a workplace design.

INTRODUCTION

In basic terms, anthropometry is the measurement of the physical attributes of humans. Over time, the body dimensions of the human population have changed. In general, people have become taller and heavier than in the past. There is currently an obesity epidemic in the United States (US), and the result is that people are much heavier compared with the population around the 1930s and 1940s. This chapter is not concerned with how people in the US got to this point, rather how the tools people use must be changed to accommodate this heavier population.

In addition to these sorts of changes, the people who work within a population also change. In the early 1980s, a large number of female workers began working in heavy industries. At that time, safety equipment had not yet been adapted yet for smaller females. In one particular instance, a female with size 6 shoe was hired by a chemical company. Because size 6 female chemical boots were not available, the worker had to wear male size 7 chemical boots. This caused a big problem for the

Occupational Ergonomics: A Practical Approach, First Edition.
Theresa Stack, Lee T. Ostrom and Cheryl A. Wilhelmsen.

worker who had to walk and work in these boots. A female size 6 shoe is 8.9 in. in length. A male size 7 shoe is 9.7 in. in length. This is almost an inch difference, and makes a tremendous difference for the person wearing the shoes. This condition did not change until the mid-1980s. Now work boots of all sizes can be found.

It is obvious that ancient peoples used anthropometry of sorts to adapt tools and clothing to their needs. Even today, consumer goods such as clothing, appliances, cars, and tools are the biggest producers of anthropometric data. Though, in many instances, products adapted to one individual are still produced. Take a tailored article of clothing. In this case, the individual is measured and the product designed and manufactured from these data.

The savant, Alphonse Bertillon (born 1853), gave this name in 1883 to a system of identification depending on the unchanging character of certain measurements of parts of the human frame (Rhodes, 1956). He found by studying patient inquiry that several measures of physical features, along with dimensions of certain bones or bony structures in the body, remain fairly constant throughout adult life.

He concluded that when these measurements were made and recorded systematically every single individual would be found to be perfectly distinguishable from others. The system was soon adapted to police methods when crime fighters found value in being able to fix a person's identity. It prevented false impersonation and brought home, to any one charged with an offense, a person's responsibility for a wrongdoing. After its introduction in France in 1883 "Bertillonage," as it was called, became widely popular, and was credited with producing highly gratifying results. Many countries followed suit in the adoption of the method, integrating it within their justice systems.

However, it was almost a decade before England followed suit when in 1894 a special committee was sent to Paris for an investigation of the methods used and results obtained with them. It reported back favorably, especially on the use of measurements for primary classification, but also recommended the adoption, in part, of the system of "finger prints" as suggested by Francis Galton, in practice at that time in Bengal, India.

Anthropometry: Greek – Anthro, man; and pometry, measure, literally meaning "measurement of humans." **Physical anthropology** refers to the measurement of the human individual for understanding human physical variation.

Today, anthropometry plays an important role in industrial design, clothing design, architecture, and ergonomics. Changes in life styles, nutrition, and ethnic composition of populations lead to changes in the distribution of body dimensions (e.g., food consumption, exercise) and require regular updating of anthropometric data collections.

KEY POINTS

In ergonomics, anthropometry is used as the basis of setting up a workstation. The two primary objectives of the ergonomics process are to enhance performance and reduce fatigue. The ergonomics process is a multistep method to evaluate work, study how the body responds to these work demands, and use this information to design or improve work areas to best meet these two objectives. The design of a work area or equipment can have significant effects on worker fatigue, safety, and performance. In addition, with the expansion of machine technology, new and different equipment is

Figure 3.1 Daycare worker in a child's chair

continually introduced to the workplace each year. This expansion of technology in the workplace can both ameliorate workspace problems and create them. For a work area to flow efficiently and productively, both the equipment and the people must be operating smoothly. Any obstacle, difficult reach, congestion or confusion can impair work output and may, at times, compromise worker safety (Bradtmiller, n.d.).

In work design or modification, we need to answer the following questions:

- Does the person fit (body size) in a workplace (see Figure 3.1)?
- If one worker fits in a workplace, do all workers fit in a workplace?
- If all workers do not fit in a workplace, are some individuals more likely to experience fatigue, injury, or diminished work performance due to poor fit into a workplace?
- Does the design of the equipment have any adverse effects on the worker's safety and productivity?

It is essential to be aware of the potential effects, both positive and negative, that work design issues can have on both worker performance and fatigue. When designing or modifying equipment or a work area, the following factors need to be considered:

1. Safe clearances or heights – examples, doorways or walkways
2. Safe reach distances – examples, equipment controls
3. Code requirements
4. Safety features – examples, machine guards
5. Workstation design for work flows.

The basis for anthropometry is the careful measurement of human body dimensions of a set population. The dimensions that are measured include the following:

1. Height
2. Weight
3. Reach, both horizontal and overhead
4. Stoop
5. Grip strength
6. Circumferential measurements
7. Limb length.

In addition, gender, age, race, and nationality are variables that should be considered when evaluating standard anthropometric tables or creating specific population tables. Anthropometric data is compiled to make guidelines to make the work areas, equipment, tools, and product fit the size, reach, grip, clearance, and capacity of the working population. Worker populations contain individuals who are male and female, large and small, short and tall, young and old, and strong and weak. The goal of applying the principles of anthropometrics to the workplace as part of work area and system design is to enhance human performance, control fatigue, and prevent accidents.

ANTHROPOMETRIC TABLES

In the science of anthropometrics, measurements of the population's dimensions are obtained based on the population's size and strength capabilities and differences. From these measurements, a set of data is collected that reflects the studied population in terms of size and form.

This population can then be described in terms of a frequency distribution including the mean, median, standard deviation, and percentiles. The frequency distribution for each measurement of the population dimension is expressed in percentiles. The xth percentile indicates that x percent of the population has the same value or less than that value for a given measurement. The median or average value for a particular dimension is the 50th percentile. In addition, $100 - x$ of the population has a value higher than x.

Example anthropometric data tables are contained in this chapter. Table 3.1 contains male stature data. Table 3.2 contains female stature data. Table 3.3 contains male body mass index. Table 3.4 contains female body mass index. Others included are male and female weights (Tables 3.5 and 3.6), waist circumferences for males and females (Tables 3.7 and 3.8), and later in the chapter are recumbent length for small children and children's head circumference (Tables 3.10 and 3.11). These data are just a small sampling of the types of anthropometric data collected (CDC, 2012). Each of these data sets has multiple purposes. Stature data are used for designing ingress and egress ways, clearances, clothing, and seating design, for example. Body mass index data are used to assess obesity levels and waist size data are used in designing clothing, seating, and clearances. The question becomes, how long will these data be

TABLE 3.1 Height in Inches for Males Aged 20 and Over and Number of Examined Persons, Mean, Standard Error of the Mean, and Selected Percentiles, by Race and Ethnicity and Age

Race and Ethnicity	Number of Examined Persons	Mean	Standard Error of the Mean	Percentile								
				5th	10th	15th	25th	50th	75th	85th	90th	95th
All racial and ethnic groups[a]												
20 years and over	5647	69.3	0.08	64.3	65.4	66.2	67.3	69.3	71.2	72.3	73.0	74.1
20–29 years	895	69.4	0.13	64.4	65.5	66.5	67.4	69.4	71.5	72.6	73.1	74.1
30–39 years	948	69.5	0.14	64.4	65.4	66.3	67.5	69.5	71.6	72.6	73.3	74.2
40–49 years	934	69.6	0.17	64.9	66.0	66.6	67.6	69.6	71.5	72.6	73.5	74.4
50–59 years	938	69.5	0.13	64.5	65.8	66.3	67.7	69.7	71.3	72.3	72.8	74.4
60–69 years	932	68.8	0.10	63.9	64.8	65.7	66.9	69.0	70.7	71.9	72.6	73.6
70–79 years	646	68.2	0.13	63.6	64.6	65.2	66.3	68.2	70.0	70.9	71.9	72.7
80 years and over	354	67.2	0.15	62.5	63.7	64.4	65.3	67.3	69.2	69.9	70.4	71.5

(*continued*)

TABLE 3.1 *(Continued)*

Race and Ethnicity	Number of Examined Persons	Mean	Standard Error of the Mean	Percentile								
				5th	10th	15th	25th	50th	75th	85th	90th	95th
Non-Hispanic white												
20 years and over	2738	69.8	0.07	65.3	66.4	67.1	68.0	69.8	71.6	72.6	73.3	74.3
20–39 years	797	70.2	0.14	66.2	66.9	67.5	68.2	70.2	72.1	73.0	73.6	74.4
40–59 years	836	70.2	0.11	66.0	67.0	67.6	68.5	70.2	71.8	72.7	73.6	74.6
60 years and over	1105	68.7	0.09	64.2	64.9	65.7	66.9	68.9	70.6	71.6	72.4	73.4
Non-Hispanic black												
20 years and over	1091	69.5	0.10	65.1	65.9	66.5	67.6	69.4	71.2	72.2	73.0	74.0
20–39 years	356	69.7	0.15	65.5	66.2	66.9	67.9	69.4	71.4	72.3	73.3	74.0
40–59 years	373	69.5	0.21	65.0	66.1	66.6	67.7	69.6	71.4	72.3	73.2	74.3
60 years and over	362	68.6	0.17	64.2	64.9	65.7	66.7	68.9	70.3	71.5	72.1	73.1

Hispanic[b]												
20 years and over	1541	67.1	0.13	62.6	63.5	64.1	65.1	67.0	68.9	70.1	70.9	72.4
20–39 years	573	67.4	0.19	62.6	63.6	64.4	65.2	67.3	69.4	70.6	71.6	72.8
40–59 years	577	67.1	0.14	63.0	63.7	64.1	65.1	67.0	68.8	69.8	70.5	71.5
60 years and over	391	65.9	0.18	61.7	62.8	63.2	64.2	66.0	67.4	68.3	68.8	70.0
Mexican American												
20 years and over	990	66.9	0.15	62.5	63.3	63.9	65.0	66.7	68.5	69.8	70.8	72.2
20–39 years	386	67.1	0.24	62.5	63.3	64.0	65.0	66.9	69.1	70.3	71.5	72.8
40–59 years	371	66.7	0.14	62.9	63.5	64.1	65.1	66.5	68.2	69.4	70.0	71.2
60 years and over	233	65.8	0.21	61.9	62.6	63.2	64.2	65.9	67.2	68.2	68.5	69.8

Source: Adapted from Anthropometric Reference Data for Children and Adults: United States, 2007–2010; DHHS Publication No. (PHS) 2013-1602, 2012.
Source: CDC (2012).

[a]Persons of other races and ethnicities are included.

[b]Mexican Americans are included in the Hispanic group.

TABLE 3.2 Height in Inches for Females Aged 20 and Over and Number of Examined Persons, Mean, Standard Error of the Mean, and Selected Percentiles, by Race, Ethnicity, and Age

Race and Ethnicity	Number of Examined Persons	Mean	Standard Error of the Mean	Percentile								
				5th	10th	15th	25th	50th	75th	85th	90th	95th
All racial and ethnic groups[a]												
20 years and over	5971	63.8	0.05	59.3	60.3	60.9	61.9	63.8	65.7	66.6	67.3	68.4
20–29 years	980	64.2	0.09	59.9	60.6	61.3	62.2	64.1	66.0	67.0	67.6	68.9
30–39 years	1029	64.3	0.11	59.6	60.9	61.5	62.5	64.3	66.1	67.1	67.9	68.9
40–49 years	1060	64.2	0.09	59.8	60.8	61.4	62.4	64.0	66.0	66.9	67.6	68.7
50–59 years	873	63.9	0.12	59.6	60.4	61.0	62.0	64.0	65.6	66.6	67.0	67.9
60–69 years	952	63.6	0.10	59.3	60.2	61.0	62.0	63.8	65.3	66.2	66.8	38.6
70–79 years	679	62.6	0.13	58.4	59.3	59.8	60.7	62.8	64.3	65.4	66.1	66.9
80 years and over	398	61.4	0.14	56.9	57.9	58.8	59.8	61.5	62.9	63.8	64.3	65.5

Non-Hispanic white											
20 years and over	2764	64.2	0.06	59.9	60.8	61.5	62.5	64.2	66.0	66.8	67.5	68.7
20–39 years	824	64.9	0.10	60.6	61.7	65.2	63.2	64.9	66.5	67.6	68.4	69.5
40–59 years	861	64.5	0.11	60.5	61.2	61.9	62.8	64.4	66.1	66.9	67.6	68.7
60 years and over	1079	63.1	0.09	58.7	59.8	60.4	61.5	63.1	64.8	65.8	66.4	67.2
Non-Hispanic black												
20 years and over	1154	64.2	0.10	59.9	60.9	61.4	62.3	64.1	66.0	66.8	67.5	68.3
20–39 years	397	64.4	0.13	60.3	61.2	61.8	62.6	64.4	66.2	67.1	67.6	68.4
40–59 years	384	64.4	0.15	60.2	61.1	61.6	62.5	64.2	66.2	67.0	67.6	68.1
60 years and over	373	63.2	0.11	58.9	59.9	60.4	61.4	63.2	64.9	65.8	66.3	67.4

(*continued*)

TABLE 3.2 (*Continued*)

Race and Ethnicity	Number of Examined Persons	Mean	Standard Error of the Mean	Percentile								
				5th	10th	15th	25th	50th	75th	85th	90th	95th
Hispanic[b]												
20 years and over	1763	61.9	0.07	57.6	58.6	59.1	60.1	61.8	63.6	64.5	65.2	66.3
20–39 years	673	62.3	0.09	58.1	58.9	59.6	60.5	62.2	63.9	64.9	65.6	66.7
40–59 years	580	61.9	0.13	57.8	58.6	59.3	60.1	61.7	63.6	64.4	64.9	65.8
60 years and over	510	60.5	0.12	56.6	57.5	58.1	58.8	60.4	62.1	62.9	63.6	64.8
Mexican American												
20 years and over	1074	61.7	0.07	57.2	58.4	59.0	59.9	61.6	63.4	64.2	64.9	66.0
20–39 years	427	62.0	0.10	57.4	58.9	59.5	60.3	62.0	63.6	64.5	65.1	66.3
40–59 years	348	61.7	0.16	57.5	58.5	59.0	59.9	61.4	63.4	64.2	64.8	65.7
60 years and over	299	60.4	0.16	56.6	57.4	57.9	58.8	60.3	61.8	62.9	63.5	64.4

Source: Adapted from Anthropometric Reference Data for Children and Adults: United States, 2007–2010; DHHS Publication No. (PHS) 2013-1602, 2012.
Source: CDC (2012).
[a]Persons of other races and ethnicities are included.
[b]Mexican Americans are included in the Hispanic group.

TABLE 3.3 Body Mass Index for Males Aged 20 and Over and Number of Examined Persons, Mean, Standard Error of the Mean, and Selected Percentiles, by Race, Ethnicity, and Age

Race and Ethnicity	Number of Examined Persons	Mean	Standard Error of the Mean	Percentile								
				5th	10th	15th	25th	50th	75th	85th	90th	95th
All racial and ethnic groups[a]												
20 years and over	5635	28.6	0.13	20.7	22.2	23.2	24.7	27.8	31.5	33.9	35.8	39.2
20–29 years	894	26.8	0.24	19.4	20.7	21.4	22.9	25.6	29.9	32.3	33.8	36.5
30–39 years	948	29.0	0.22	21.0	22.4	23.3	24.9	28.0	32.0	34.1	36.2	40.5
40–49 years	933	29.0	0.29	21.2	22.9	24.0	25.4	28.2	31.7	34.4	36.1	39.6
50–59 years	934	29.2	0.31	21.5	22.9	23.9	25.5	28.2	32.0	34.5	37.1	39.9
60–69 years	930	29.5	0.25	21.3	22.7	23.8	25.3	28.8	32.5	34.7	37.0	40.0
70–79 years	643	28.8	0.26	21.4	22.9	23.8	25.6	28.3	31.3	33.5	35.4	37.8
80 years and over	353	27.2	0.22	20.7	21.8	22.8	24.4	27.0	29.6	31.3	32.7	34.5

(*continued*)

TABLE 3.3 (*Continued*)

Race and Ethnicity	Number of Examined Persons	Mean	Standard Error of the Mean	Percentile								
				5th	10th	15th	25th	50th	75th	85th	90th	95th
Non-Hispanic white												
20 years and over	2728	28.7	0.15	20.8	22.4	23.3	24.8	27.9	31.5	33.9	35.8	39.1
20–39 years	796	27.7	0.25	20.1	21.1	22.3	23.6	26.8	30.9	33.1	34.2	38.2
40–59 years	832	29.2	0.22	21.3	23.1	24.1	25.5	28.3	31.8	34.5	36.8	39.7
60 years and over	1100	29.2	0.18	21.4	23.2	24.2	25.5	28.6	31.9	34.0	35.9	38.8
Non-Hispanic black												
20 years and over	1090	29.0	0.23	19.7	21.4	22.4	23.7	28.0	32.7	35.5	38.1	41.9
20–39 years	356	28.7	0.39	19.6	21.1	21.8	23.1	27.3	32.6	35.8	38.1	42.7
40–59 years	372	29.4	0.38	19.7	22.1	22.9	24.8	28.3	33.1	35.8	38.4	41.1
60 years and over	362	28.8	0.32	19.8	21.6	22.6	24.1	28.2	31.6	34.3	36.9	40.8

Hispanic[b]												
20 years and over	1540	28.9	0.25	21.3	23.0	23.9	25.5	28.1	31.6	33.8	35.6	39.0
20–39 years	573	28.5	0.33	20.9	21.9	23.2	24.8	27.5	31.1	33.0	35.5	40.6
40–59 years	577	29.5	0.24	23.1	24.2	25.2	26.5	28.8	32.1	34.4	35.7	37.7
60 years and over	390	29.2	0.32	22.0	23.2	24.3	25.9	28.3	31.9	34.1	35.6	38.5
Mexican American												
20 years and over	990	29.0	0.30	21.4	23.2	24.0	25.7	28.1	31.6	33.7	35.6	39.5
20–39 years	386	28.8	0.42	20.9	22.2	23.5	25.1	27.8	31.4	33.0	35.3	41.1
40–59 years	371	29.5	0.28	23.1	24.3	25.3	26.4	28.7	32.0	34.3	35.8	37.8
60 years and over	233	29.2	0.44	22.1	23.2	24.2	25.9	28.3	31.8	33.9	35.3	39.0

Source: Adapted from Anthropometric Reference Data for Children and Adults: United States, 2007–2010; DHHS Publication No. (PHS) 2013-1602, 2012.
Source: CDC (2012).

Note: Body mass index (BMI) is calculated as follows: BMI = weight (kg)/height (m^2).

[a]Persons of other races and ethnicities are included.

[b]Mexican Americans are included in the Hispanic group.

TABLE 3.4 Body Mass Index Values for Females Aged 20 and Over and Number of Examined Persons, Mean, Standard Error of the Mean, and Selected Percentiles, by Race, Ethnicity, and Age

Race and Ethnicity	Number of Examined Persons	Mean	Standard Error of the Mean	Percentile								
				5th	10th	15th	25th	50th	75th	85th	90th	95th
All racial and ethnic groups[a]												
20 years and over	5841	28.7	0.12	19.5	20.7	21.7	23.3	27.3	32.5	36.1	38.2	42.0
20–29 years	906	27.5	0.42	18.8	19.9	20.6	21.7	25.3	31.5	36.0	38.0	43.9
30–39 years	982	28.7	0.33	19.4	20.6	21.6	23.4	27.2	32.8	36.0	38.1	41.6
40–49 years	1056	28.6	0.28	19.3	20.6	21.7	23.3	27.3	32.4	36.2	38.1	43.0
50–59 years	873	29.3	0.27	19.7	21.3	22.1	24.0	28.3	33.5	36.4	39.3	41.8
60–69 years	950	29.6	0.25	20.7	21.6	23.0	24.8	28.8	33.5	36.6	38.5	41.1
70–79 years	677	29.5	0.26	20.1	21.6	22.7	24.7	28.6	33.4	36.3	38.7	42.1
80 years and over	397	26.7	0.26	19.3	20.7	22.0	23.1	26.3	29.7	31.6	32.5	35.2

Non-Hispanic white												
20 years and over	2730	28.2	0.20	19.4	20.5	21.4	23.0	26.9	32.0	35.7	37.7	41.5
20–39 years	792	27.5	0.41	19.0	19.9	20.7	22.1	25.6	31.3	35.4	37.5	42.8
40–59 years	861	28.3	0.24	19.4	20.5	21.6	23.2	26.9	32.4	35.8	38.0	41.5
60 years and over	1077	28.7	0.20	20.1	21.4	22.5	24.0	27.8	32.2	35.5	37.6	40.7
Non-Hispanic black												
20 years and over	1126	32.0	0.31	20.5	22.1	24.0	25.9	30.8	36.5	40.2	42.8	47.2
20–39 years	372	31.4	0.46	19.8	21.6	22.8	25.0	30.3	35.9	40.0	42.1	47.5
40–59 years	383	33.1	0.49	21.2	23.0	24.5	26.7	31.2	37.7	41.3	44.8	48.0
60 years and over	371	31.1	0.33	20.8	22.9	24.5	26.6	30.3	34.8	38.0	40.0	44.2

(continued)

TABLE 3.4 (*Continued*)

Race and Ethnicity	Number of Examined Persons	Mean	Standard Error of the Mean	Percentile								
				5th	10th	15th	25th	50th	75th	85th	90th	95th
Hispanic[b]												
20 years and over	1707	29.5	0.19	20.8	21.9	22.9	24.9	28.5	33.0	36.0	38.0	40.9
20–39 years	619	28.8	0.23	20.3	21.3	22.1	23.6	26.9	32.7	36.1	38.2	40.9
40–59 years	579	30.2	0.34	21.3	23.0	24.2	26.2	29.6	33.2	35.9	37.8	40.7
60 years and over	509	29.9	0.17	21.1	22.8	23.9	25.7	29.4	33.2	36.0	37.7	41.7
Mexican American												
20 years and over	1031	29.8	0.17	21.1	22.2	23.3	25.2	29.1	33.6	36.1	38.1	40.9
20–39 years	386	29.2	0.29	20.9	21.8	22.3	24.0	27.6	33.5	36.5	38.3	40.9
40–59 years	347	30.6	0.37	21.6	23.5	25.0	26.7	30.1	33.8	35.9	36.7	40.8
60 years and over	298	30.0	0.21	21.1	23.0	24.1	26.1	29.6	33.1	36.0	37.4	41.7

Source: Adapted from Anthropometric Reference Data for Children and Adults: United States, 2007–2010; DHHS Publication No. (PHS) 2013-1602, 2012.
Source: CDC (2012).

[a]Persons of other races and ethnicities are included.

[b]Mexican Americans are included in the Hispanic group.

TABLE 3.5 Waist Circumference in Inches for Males Aged 20 and Over and Number of Examined Persons, Mean, Standard Error of the Mean, and Selected Percentiles, by Race, Ethnicity, and Age

Race and Ethnicity	Number of Examined Persons	Mean	Standard Error of the Mean	Percentile								
				5th	10th	15th	25th	50th	75th	85th	90th	95th
All racial and ethnic groups[a]												
20 years and over	5410	39.7	0.41	30.7	32.2	33.6	35.6	39.1	43.1	45.7	47.6	50.4
20–29 years	862	36.4	0.64	28.9	29.9	30.7	32.0	35.5	39.8	42.4	44.1	46.6
30–39 years	900	39.1	0.52	31.1	32.4	33.5	35.0	38.3	42.1	44.6	47.1	50.1
40–49 years	907	40.1	0.68	31.7	33.7	34.8	36.5	39.2	43.1	45.7	40.4	50.0
50–59 years	906	41.0	0.81	32.2	34.4	35.7	36.9	40.5	44.0	46.5	48.5	51.2
60–69 years	899	41.8	0.64	32.8	35.0	36.2	37.8	41.5	45.3	47.8	49.5	51.3
70–79 years	609	42.0	0.64	33.9	35.9	37.1	38.8	41.8	45.2	47.4	49.3	51.0
80 years and over	327	40.7	06.9	33.1	34.6	35.7	36.6	40.7	43.7	45.3	46.8	48.7

(continued)

TABLE 3.5 ***(Continued)***

Race and Ethnicity	Number of Examined Persons	Mean	Standard Error of the Mean	Percentile								
				5th	10th	15th	25th	50th	75th	85th	90th	95th
Non-Hispanic white												
20 years and over	2624	40.2	0.43	31.2	32.8	34.2	36.2	39.8	43.7	46.4	48.1	50.6
20–39 years	762	37.6	0.65	29.9	31.1	31.9	33.5	37.3	41.2	44.1	45.7	48.9
40–59 years	812	41.0	0.60	32.2	34.5	35.7	36.9	40.4	44.0	46.9	48.4	51.1
60 years and over	1050	40.9	0.41	34.3	35.8	36.9	38.4	41.8	45.3	47.4	49.3	51.0
Non-Hispanic black												
20 years and over	1033	38.6	0.62	28.7	30.1	31.1	33.3	37.9	42.7	45.6	47.4	51.9
20–39 years	335	36.9	0.86	28.1	29.1	29.8	31.2	35.7	41.0	44.4	46.7	50.8
40–59 years	351	39.6	0.91	29.9	31.6	32.8	35.2	38.0	43.3	45.7	47.4	52.4
60 years and over	347	40.7	0.90	30.8	32.8	33.9	36.5	40.2	44.2	47.1	49.0	52.7

Hispanic[b]												
20 years and over	1490	39.2	0.68	31.3	32.7	33.9	35.6	38.6	42.0	44.3	46.0	49.3
20–39 years	555	38.1	0.84	29.8	31.8	32.6	34.3	37.4	41.0	42.8	44.7	49.4
40–59 years	567	40.3	0.66	33.6	35.2	36.2	37.3	39.5	42.8	45.2	46.5	48.6
60 years and over	368	41.3	0.57	34.3	35.4	36.4	37.7	41.0	44.0	45.7	47.4	50.4
Mexican American												
20 years and over	957	39.3	0.80	31.5	33.1	34.4	36.0	38.8	42.0	44.2	45.8	49.4
20–39 years	371	38.4	41.01	30.2	32.0	33.0	34.8	37.6	41.3	42.9	44.5	49.6
40–59 years	367	40.3	0.82	33.8	35.5	36.4	37.2	39.4	42.8	45.2	46.2	48.5
60 years and over	219	41.2	0.75	34.3	35.6	36.7	37.6	40.9	43.9	45.5	46.7	50.1

Source: Adapted from Anthropometric Reference Data for Children and Adults: United States, 2007–2010; DHHS Publication No. (PHS) 2013-1602, 2012.
Source: CDC (2012).

[a]Persons of other races and ethnicities are included.

[b]Mexican Americans are included in the Hispanic group.

TABLE 3.6 Waist Circumference in Inches for Females Aged 20 and Over and Number of Examined Persons, Mean, Standard Error of the Mean, and Selected Percentiles, by Race, Ethnicity, and Age

Race and Ethnicity	Number of Examined Persons	Mean	Standard Error of the Mean	Percentile								
				5th	10th	15th	25th	50th	75th	85th	90th	95th
All racial and ethnic groups[a]												
20 years and over	5552	37.5	0.34	28.3	29.7	30.8	32.8	36.7	41.5	44.3	46.3	49.2
20–29 years	870	35.5	1.04	26.9	28.4	29.3	30.3	33.7	39.1	42.5	45.4	49.3
30–39 years	939	36.9	0.75	28.1	29.2	30.6	32.3	35.7	40.8	43.9	45.9	48.5
40–49 years	1024	37.2	0.62	28.3	29.7	30.8	32.6	36.1	40.9	44.0	46.1	48.9
50–59 years	839	38.4	0.55	28.9	30.5	32.0	33.8	37.6	42.3	44.9	47.1	49.8
60–69 years	903	39.3	0.57	30.6	32.4	33.4	34.8	39.1	43.0	45.4	47.2	50.0
70–79 years	632	39.3	0.62	30.3	32.0	33.6	35.2	38.8	42.9	45.6	47.1	49.6
80 years and over	345	36.4	0.59	29.4	30.7	31.8	33.3	36.5	39.9	42.2	43.5	44.9

Non-Hispanic white												
20 years and over	2602	37.3	0.49	28.3	29.5	30.7	32.6	36.4	41.3	44.3	46.2	49.0
20–39 years	762	35.9	1.00	27.7	28.7	29.4	30.8	34.3	39.5	42.6	45.4	48.5
40–59 years	829	37.5	0.58	28.4	29.8	30.9	32.7	36.4	41.5	44.5	46.5	49.2
60 years and over	1011	38.7	0.48	29.8	31.8	32.8	34.5	38.3	42.2	44.9	46.7	49.2
Non-Hispanic black												
20 years and over	1046	39.6	0.70	29.0	31.3	32.4	34.5	39.1	43.9	47.0	48.9	51.1
20–39 years	349	38.4	1.17	27.6	29.5	31.1	32.8	37.2	43.0	46.1	48.9	51.0
40–59 years	365	40.6	0.81	30.3	32.1	33.1	35.7	39.8	45.1	48.0	49.5	52.0
60 years and over	332	40.2	0.72	31.3	33.5	34.6	36.6	39.6	43.3	45.9	47.1	50.2

(*continued*)

TABLE 3.6 (*Continued*)

Race and Ethnicity	Number of Examined Persons	Mean	Standard Error of the Mean	Percentile								
				5th	10th	15th	25th	50th	75th	85th	90th	95th
Hispanic[b]												
20 years and over	1642	37.7	0.46	29.4	30.6	31.8	33.6	37.2	41.3	43.3	45.3	47.9
20–39 years	598	36.8	0.60	28.8	29.8	30.7	32.2	35.9	40.7	43.1	45.2	47.8
40–59 years	567	38.3	0.80	30.1	32.0	32.9	34.9	37.9	41.2	43.2	45.2	47.8
60 years and over	477	39.1	0.40	30.6	33.1	34.0	35.8	39.0	42.3	44.1	45.6	48.0
Mexican American												
20 years and over	998	38.1	0.47	29.7	31.0	32.1	34.0	37.7	41.8	43.9	45.5	48.3
20–39 years	372	37.4	0.74	29.2	30.1	31.1	32.6	36.2	41.6	43.9	45.6	48.5
40–59 years	342	38.8	0.89	30.5	32.3	33.3	35.4	38.4	41.9	43.3	45.4	47.5
60 years and over	284	39.4	0.50	31.4	33.4	34.4	36.1	39.3	42.4	44.1	45.4	48.3

Source: Adapted from Anthropometric Reference Data for Children and Adults: United States, 2007–2010; DHHS Publication No. (PHS) 2013-1602, 2012.
Source: CDC (2012).
Note: Pregnant women were excluded.
[a]Persons of other races and ethnicities are included.
[b]Mexican Americans are included in the Hispanic group.

TABLE 3.7 Weight in Pounds for Males Aged 20 and Over and Number of Examined Persons, Mean, Standard Error of the Mean, and Selected Percentiles, by Race, Ethnicity, and Age

Race and Ethnicity	Number of Examined Persons	Mean	Standard Error of the Mean	Percentile								
				5th	10th	15th	25th	50th	75th	85th	90th	95th
All racial and ethnic groups[a]												
20 years and over	5651	195.5	0.99	135.5	146.6	153.7	165.2	189.8	218.0	236.4	252.2	273.6
20–29 years	894	183.9	1.87	128.7	137.9	143.9	153.2	176.5	206.6	224.0	240.4	257.5
30–39 years	948	199.5	1.66	139.6	149.4	156.0	166.2	191.1	222.9	242.7	259.7	282.2
40–49 years	933	200.6	2.12	142.1	153.2	162.0	173.1	193.7	221.9	239.4	256.1	278.5
50–59 years	934	201.3	2.43	140.6	152.2	160.8	172.3	195.4	226.9	242.1	259.3	279.0
60–69 years	649	190.6	1.95	138.1	147.2	155.1	165.5	186.8	209.8	227.0	241.8	259.9
70–79 years	649	190.6	1.95	138.1	147.2	155.1	165.5	186.8	209.8	227.0	241.1	259.9
80 years and over	360	174.9	1.79	127.1	135.5	141.3	152.2	171.8	194.4	207.9	214.5	230.3

(continued)

TABLE 3.7 (*Continued*)

Race and Ethnicity	Number of Examined Persons	Mean	Standard Error of the Mean	Percentile								
				5th	10th	15th	25th	50th	75th	85th	90th	95th
Non-Hispanic white												
20 years and over	2738	199.2	0.92	140.4	151.8	159.1	170.3	194.0	221.7	239.6	254.9	273.6
20–39 years	796	194.7	1.71	135.7	145.3	153.0	162.8	188.1	217.2	238.0	250.3	270.6
40–59 years	832	204.9	1.56	146.3	159.0	165.4	177.3	198.5	227.4	246.7	259.5	283.7
60 years and over	1110	196.3	1.21	140.1	152.0	159.1	168.8	191.8	217.1	232.7	248.1	266.7
Non-Hispanic black												
20 years and over	1094	199.4	1.63	134.3	143.5	149.7	163.1	191.3	224.5	245.7	264.7	292.0
20–39 years	356	198.1	3.14	135.2	141.6	146.8	158.9	188.7	223.5	246.3	264.5	296.1
40–59 years	372	203.1	2.77	133.4	146.0	156.9	169.5	196.4	227.7	249.8	266.3	291.9
60 years and over	366	193.6	2.54	127.8	140.8	147.8	158.7	188.0	213.0	234.7	256.5	283.9

Hispanic[b]												
20 years and over	1541	186.1	1.95	133.2	143.4	148.5	157.5	180.0	205.4	221.0	236.4	266.7
20–39 years	573	185.1	2.66	129.2	140.4	146.1	154.9	176.1	204.6	221.4	241.2	271.7
40–59 years	577	189.4	1.77	141.6	150.3	155.0	165.2	184.2	207.7	223.3	232.4	258.0
60 years and over	391	180.8	2.50	129.1	138.8	146.5	154.7	177.1	200.5	212.9	223.7	241.2
Mexican American												
20 years and over	991	185.4	2.30	133.2	143.3	148.4	157.1	179.6	204.6	218.8	235.6	267.6
20–39 years	386	185.2	3.27	130.9	140.0	146.0	155.0	176.6	204.8	220.6	241.4	276.2
40–59 years	371	187.4	1.98	141.1	150.0	154.6	163.0	182.8	206.6	219.8	229.3	255.3
60 years and over	234	180.6	3.17	130.7	138.9	148.0	155.8	177.2	199.3	208.9	215.2	240.6

Source: Adapted from Anthropometric Reference Data for Children and Adults: United States, 2007–2010; DHHS Publication No. (PHS) 2013-1602, 2012.
Source: CDC (2012).

[a]Persons of other races and ethnicities are included.

[b]Mexican Americans are included in the Hispanic group.

TABLE 3.8 Weight in Pounds for Females Aged 20 and Over and Number of Examined Persons, Mean, Standard Error of the Mean, and Selected Percentiles, by Race, Ethnicity, and Age

Race and Ethnicity	Number of Examined Persons	Mean	Standard Error of the Mean	Percentile								
				5th	10th	15th	25th	50th	75th	85th	90th	95th
All racial and ethnic groups[a]												
20 years and over	5844	166.2	0.78	110.7	118.2	124.8	134.6	157.2	188.6	210.3	255.3	250.9
20–29 years	906	161.9	2.49	107.2	114.8	118.5	126.3	149.4	181.2	208.9	227.3	264.5
30–39 years	982	169.1	2.07	112.2	118.8	126.5	137.2	159.8	194.2	215.1	225.4	253.9
40–49 years	1056	168.0	1.64	111.9	120.8	126.5	134.9	158.4	189.0	212.0	228.7	253.2
50–59 years	873	170.0	1.82	112.7	123.3	128.8	138.5	161.4	193.8	213.3	230.2	255.3
60–69 years	951	170.5	1.44	116.5	126.1	132.2	140.4	165.7	192.5	211.0	226.3	241.4
70–79 years	679	164.9	1.52	109.9	118.0	125.7	136.9	159.4	187.1	201.7	218.4	240.6
80 years and over	397	143.1	1.60	100.2	109.8	114.1	123.1	140.0	158.5	172.9	182.7	192.6

Non-Hispanic white												
20 years and over	2730	165.4	1.13	111.6	118.6	125.0	134.9	156.6	187.4	209.3	223.7	246.8
20–39 years	792	164.7	2.55	110.4	117.7	123.0	132.0	154.0	186.6	212.3	225.4	257.5
40–59 years	861	167.7	1.45	112.3	121.3	126.7	136.2	158.5	189.7	212.2	227.0	254.6
Non-Hispanic black												
20 years and over	1128	187.9	1.72	118.9	129.5	136.5	151.0	177.5	215.5	237.4	255.1	287.8
20–39 years	372	186.2	2.57	113.8	125.6	134.4	145.1	176.2	216.6	241.8	258.9	294.5
40–59 years	383	194.7	2.68	125.4	133.9	144.2	156.4	183.4	221.5	243.5	257.3	290.3
60 years and over	373	177.8	2.30	116.4	125.9	135.0	149.1	171.4	200.3	221.0	233.9	253.0

(continued)

TABLE 3.8 ***(Continued)***

Race and Ethnicity	Number of Examined Persons	Mean	Standard Error of the Mean	Percentile								
				5th	10th	15th	25th	50th	75th	85th	90th	95th
Hispanic[b]												
20 years and over	1708	160.6	1.02	109.7	117.0	123.6	133.4	154.7	181.5	198.3	210.7	228.1
20–39 years	619	159.4	1.31	107.6	114.8	119.2	129.6	151.4	183.6	200.8	212.8	231.6
40–59 years	579	164.5	1.84	114.6	125.6	131.0	140.9	158.9	182.3	197.4	210.1	228.6
60 years and over	510	155.8	1.05	105.8	114.2	122.5	133.5	153.0	174.3	188.4	198.8	213.6
Mexican American												
20 years and over	1032	161.5	0.89	110.9	118.1	124.6	134.5	156.0	182.9	198.5	208.0	229.3
20–39 years	386	160.3	1.46	110.4	116.4	121.3	132.1	152.1	185.3	202.0	213.0	233.5
40–59 years	347	165.8	2.13	117.5	127.1	133.4	142.8	161.8	182.9	197.0	206.0	226.0
60 years and over	299	155.5	1.23	103.8	113.6	122.4	133.3	153.2	174.8	187.6	197.4	207.5

Source: Adapted from Anthropometric Reference Data for Children and Adults: United States, 2007–2010; DHHS Publication No. (PHS) 2013-1602, 2012.
Source: CDC (2012).
Note: Pregnant women were excluded.
[a]Persons of other races and ethnicities are included.
[b]Mexican Americans are included in the Hispanic group.

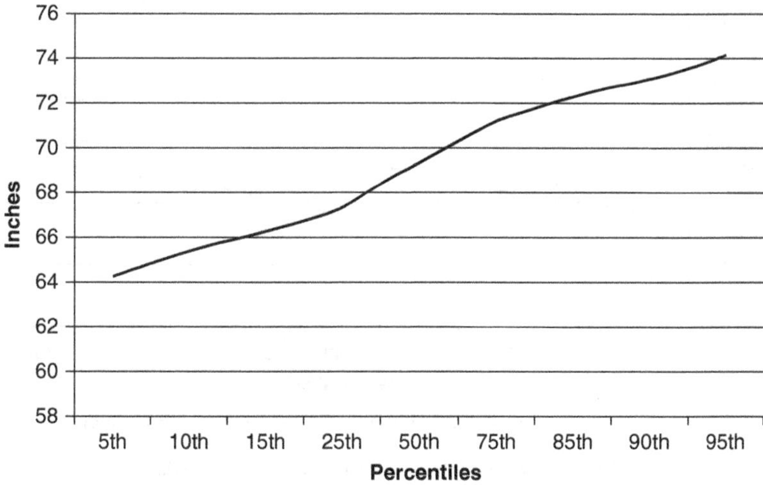

Figure 3.2 Graph of male stature height

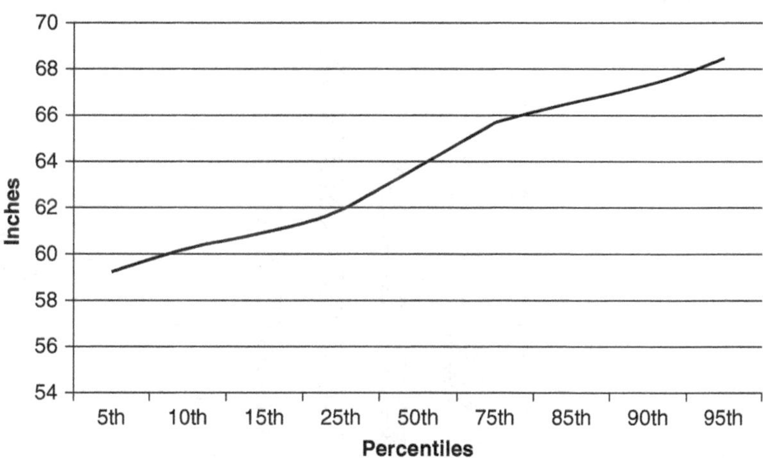

Figure 3.3 Graph of female stature data

relevant? Well, that is a hard question to answer. That is why there is almost always an ongoing effort to collect additional data.

Figure 3.2 shows a graph of the male stature data and Figure 3.3 shows a graph of the female stature data. Figure 3.4 shows the data for both genders. Generally speaking, anthropometric data follows a normal distribution as shown in Figure 3.5.

In ergonomic design, we do not design for the average person, or the 50th percentile, we design for the 95th percentile. In other words, 95% of the population can use the work area safely and efficiently, and 5% of the population may need to be accommodated. Conventionally, the 95th percentile has been chosen to determine clearance heights or lengths. That means 95% of the population will be able to pass through a door, while only 5% of the population may need to be accommodated. In addition, the 5th percentile female has been chosen to determine the functional

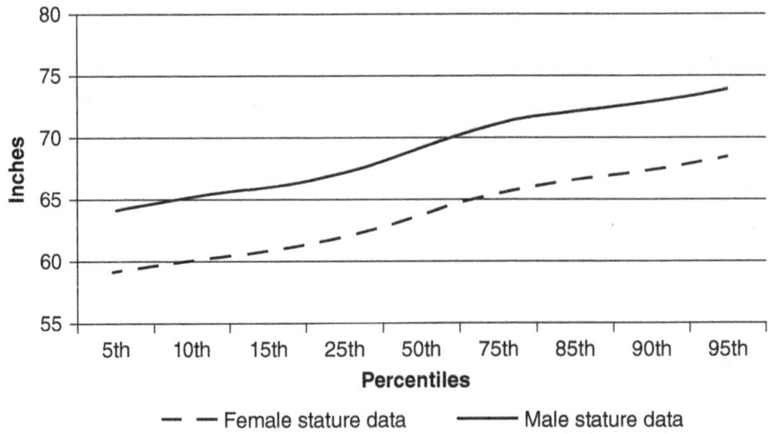

Figure 3.4 Graph of male and female stature data

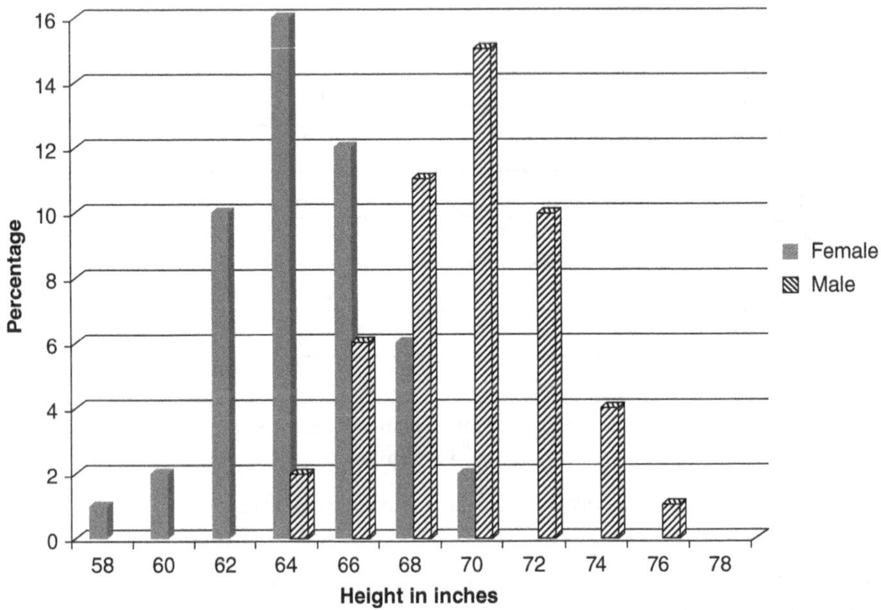

Figure 3.5 Distribution of male and female statures

reach distance, that means 95% of the population will be able to perform this reach, and only 5% of the population may need to be accommodated. Figure 3.6 shows the same graph of the distribution of male and female statures with the critical percentiles. Table 3.9 contains three (3) of the more commonly used anthropometric body dimensions (Helander, 2005).

In some cases, we have to adjust the value for a certain body dimension based on clothing or other considerations.

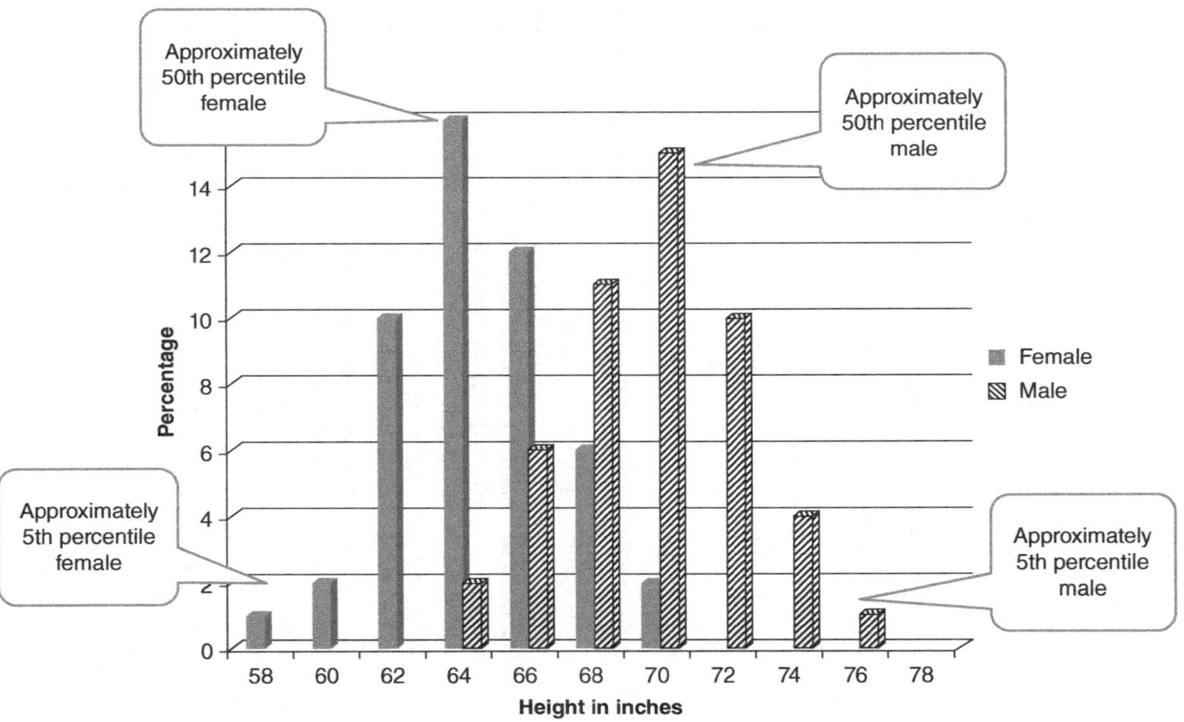

Figure 3.6 Statures with percentiles

TABLE 3.9 Critical Anthropometric Dimensions

Segment/ Body Dimension	Gender	5th Percentile	50th Percentile	95th Percentile
Stature	M	64.3	69.3	74.1
	F	59.3	63.8	68.4
Eye height	M	60.8	64.70	68.6
	F	57.3	60.3	65.3
Hip width (seated)	M	12.14	13.95	16
	F	12.29	14.34	17.22

Source: Adapted from Helander (2005).
Example Use of Anthropometric Data Tables.

Adjustments to accommodate clothing include the following:

- 2.5 cm (1.0 in.) for standing height (or seated knee height) to reflect the presence of a shoe heel
- 0.8 cm (0.3 in.) for breadths (due to the bulk of clothing)
- 3.0 cm (1.2 in.) for foot length (to accommodate for shoes being larger than feet).

Adjustments to accommodate for posture include

- 2.0 cm (0.8 in.) for standing height (due to slouching or lack of upright posture).

Table 3.10 contains recumbent lengths of children from 0 to 36 months in age, and Table 3.11 contains head circumference data. If we want to design a new crib, we would select the data we need from these tables, as well as other tables, and produce the design. We would use these data to design the length of the crib mattress and the width of the slates in the sides of the crib. Figure 3.7 is a conceptual design of the crib. Which percentiles of data would we select?

Well, one option would be the 50th percentile recumbent length for a newborn female (18.8 in.) and 95th percentile head circumference for a 6-month male child (18.1 in. or an estimated 5.74 in. diameter). In reality, a much more accurate head diameter would be needed. However, this would produce a crib that would be too short, and the slates would be spaced such that an infant could get his/her head through the slates and possibly become strangled. In fact, most all female and all but the largest male children could get their heads through the slates. The more reasonable approach would be to use the 5th percentile female head size, with a safety factor subtracted from the width for the slates. The Consumer Product Safety Commission recommends no more than 2.375-in. slate width or, as their website states, "If a soda can will fit through, the slats are too far apart" (CPSC, 2014). The 95th 3-year-old male recumbent length (42.2 in.) should be the starting point for the length of the crib, plus some comfort factor added on.

Average Person

Designing for the "average" person or the 50th percentile is a myth. If you designed a doorway for the "average" person, one-half of the population would not fit through

TABLE 3.10 Recumbent Length in Centimeters for Children from Birth Through Age 36 Months and Number of Examined Persons, Mean, Standard Error of the Mean, and Selected Percentiles, by Race, Ethnicity, and Age

Sex and Age[a]	Number of Examined Persons	Mean	Standard Error of the Mean	Percentile								
				5th	10th	15th	25th	50th	75th	85th	90th	95th
Male												
Birth to 2 months	82	22.4	0.36	20.0	20.2	20.9	21.6	22.6	23.1	23.3	23.7	24.0
3-5 months	110	25.1	0.35	23.2	23.5	24.1	24.2	25.0	25.9	26.5	26.8	*b*
6-8 months	103	27.2	0.41	25.5	26.0	26.2	26.5	27.2	27.7	28.3	28.6	28.9
9-11 months	108	29.1	0.32	27.4	27.7	27.9	28.3	28.8	29.8	30.2	30.7	30.8
1 year	316	31.9	0.37	29.0	29.4	29.8	30.7	31.9	33.2	33.6	34.2	35.2
2 years	313	36.3	0.36	33.2	33.7	34.2	35.0	36.4	37.4	38.1	38.6	39.4
3 years	197	39.3	0.37	36.2	37.3	37.6	38.2	39.3	40.2	41.3	42.0	42.2[c]
Female												
Birth to 2 months	81	22.0	0.55	18.8	20.3	20.7	21.5	22.2[d]	22.9	23.1	23.1	*b*
3-5 months	104	24.6	0.26	22.6	23.3	23.5	24.0	24.6[d]	25.5	25.7	26.0	26.1
6-8 months	103	26.6	0.29	24.9	25.3	25.5	25.8	26.6	27.2	27.6	27.7	28.1
9-11 months	119	28.1	0.29	*b*	26.6	26.9	27.4	28.1	28.8	29.4	29.6	29.8
1 year	296	31.5	0.36	28.5	29.0	29.5	30.3	31.6	32.8	33.3	33.8	34.7
2 years	261	35.7	0.36	33.3	33.6	34.1	34.4	35.5	36.9	37.5	37.9	38.5
3 years	183	39.3	0.39	36.5	36.9	37.2	38.0	39.4	40.5	41.1	41.6	42.6

50th Percentile Female

95th Percentile Male

Source: Adapted from Anthropometric Reference Data for Children and Adults: United States, 2007–2010; DHHS Publication No. (PHS) 2013-1602, 2012.
Source: CDC (2012).
[a]Refers to the age at time of examination.
[b]Missing data.
[c]95th percentile male.
[d]50th percentile female.
[e]Standard error not calculated by SUDAAN.

TABLE 3.11 Head Circumference in Centimeters for Infants from Birth Through Age 6 Months and Number of Examined Persons, Mean, Standard Error of the Mean, and Selected Percentiles, by Race, Ethnicity, and Age

Sex and Age[a]	Number of Examined Persons	Mean	Standard Error of the Mean	Percentile								
				5th	10th	15th	25th	50th	75th	85th	90th	95th
Male												
Birth to 2 months	82	15.3	0.23	14.0	14.3	14.6	14.8	15.5	15.7	15.8	16.0	16.3
3-5 months	111	16.7	0.15	15.7	15.8	15.9	16.4	16.7	17.0	17.2	17.4	*b*
6 months	38	17.4	0.34	*b*	16.7	16.9	17.0	17.5	17.7	17.9	18.0	18.1[c]
Female												
Birth to 2 months	82	15.1	0.24	13.6	14.3	[illegible]	[illegible]	[illegible]	[illegible]	[illegible]	[illegible]	[illegible]6.1
3-5 months	104	16.4	0.12	15.5	15.6	15.7	15.9	[illegible]	[illegible]	16.9	17.0	17.1
6 months	36	17.2	0.29	*b*	16.4	16.5	16.6	17.1	17.6	17.9	17.9	*b*

95th Percentile Male

[a]Refers to the age at time of examination.
[b]Missing data.
[c]95th percentile male.
[d]Standard error not calculated by SUDAAN.
[e]Estimate does not meet standards of reliability or precision based on less than 12 degrees of freedom.

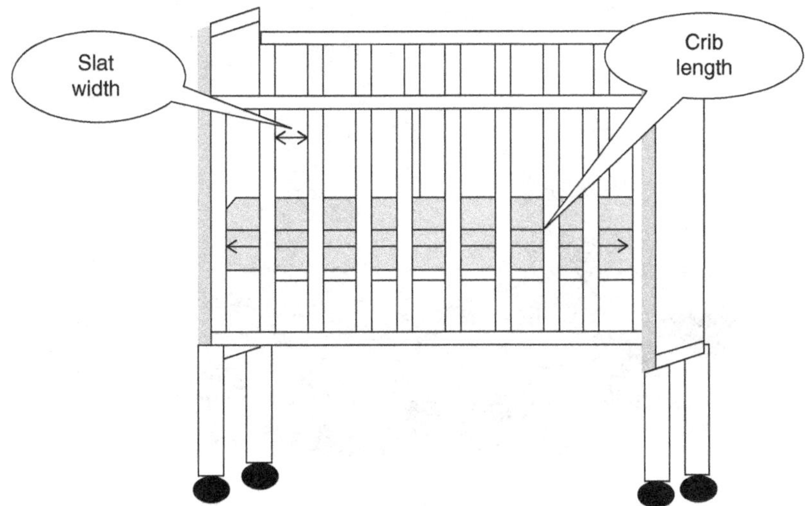

Figure 3.7 Crib design

the door. Design considerations for doorways, for example, must be for the 99th percentile male. Reach considerations must be given to the smallest person or the 5th percentile female. There are numerous guidelines and anthropometric tables to assist in designing tasks and equipment, which address all aspects from stature and forward functional reach to eye height for particular tasks.

ANTHROPOMETRY IN DESIGN

All of us have experienced problems with equipment, workspaces, or even our homes not being designed with our body dimensions in mind. The following example illustrates ergonomic issues when a design does not accommodate a wide range of body sizes. In the first example, the task is to remove the ladder from the top of the van (Figure 3.8). The larger individual in the photo can do the task, but the smaller individual cannot. The solution to this issue is to design a device that will lower the ladder to the side of the vehicle so a full range of individuals can remove it at their power zone (Figure 3.9).

Poor Ergonomic Design

Ergonomists use the following steps to design equipment based on anthropometric design principles:

1. Define the population – who are we designing for?
2. Determine critical body dimensions – what allows use?
3. Select the percentage of the population to be accommodated (or excluded) – how many do we need to accommodate?

(a)

(b)

Figure 3.8 (a) Poor design with the ladder on top of the van. (b) Unsafe way of getting ladder

4. Select the anthropometry principle:
   Range, extreme, average
5. Locate data tables.
6. Adjust for clothing, posture.
7. Test.*

Next, we need to understand what it means to apply anthropometric data for a certain design principle. There are three ergonomic design principles based on anthropometry: design for a range, design for the extreme, and design for the average. These are explained below.

Figure 3.9 Better ergonomic design

Design for a Range

Principle: Allow for adjustments in position, size, intensity, and duration of the product or system, to accommodate unexpected circumstances and maximize use (preferred option).

Design: Common to use from the 5th percentile female to the 95th percentile male; can result in accommodation of 95% of 50/50 mixed population group because of overlap in male and female body dimensions.

Examples of use: Car seats, desk height, keyboard support, footrests, purchase in different sizes (chairs, shoes, and tools).

Some ways of using this principle in design are as follows:

- Designing six-way adjustability into car seats
- Providing adjustable height computer workstations
- Providing safe platforms for smaller workers to stand on when working at a higher workstation
- Providing work fixtures to aid workers reaching for equipment.

Design for the Extreme

Principle: Accommodate largest percent of the population group where adjustability is costly or not feasible – Maximum Levels: 95th–100th percentile

Clearance, Load Tolerance, Girth (e.g., doorways, size of escape hatches, entry ways, strength of ladders) – Minimum Levels: 1st–5th percentile

Reach, Strength (e.g., distance of control button from operator, force required to operate control lever or button).

Practical Design

Use 1st–5th or 95th–100th percentiles of population group as extremes, typically the smallest female and largest male.

Examples of uses: Egress ways, control configurations, and safety showers.

Design for the Average

Principle: Design for the 50th percentile

Acceptable for short-term use

Accommodates small population group.

Design: 50th percentile only

Used as a last resort – may exclude 50% of the population

There is no "average" person (e.g., average height may not mean average arm length).

Examples of use: Self-serve checkout counter, water fountain.

Rule of Thumb

Design so the tall can fit and the small can reach.

CASE STUDY

Introduction

The following is a description of an aviation accident that occurred on January 8, 2003. The description of the event, analysis, and conclusions come directly from the National Transportation Safety Board (NTSB) report (Beechcraft, 2004). Note that not all of the information is provided, only that relating to the anthropometric aspects of the accident. In general, the text has not been changed from the original report. However, in some cases, the text has been changed for readability. Also, the notes have been provided at the end of the sections.

History of Flight

On January 8, 2003, about 0847:28 eastern standard time, Air Midwest (doing business as US Airways Express) flight 5481, a Raytheon (Beechcraft) 1900D, N233YV, crashed shortly after takeoff from runway 18R at Charlotte-Douglas International Airport (CLT), Charlotte, North Carolina. The 2 flight crewmembers and 19 passengers aboard the airplane were killed, persons on the ground received minor injuries, and the airplane was destroyed by impact forces and a postcrash fire. Flight 5481 was a regularly scheduled passenger flight to Greenville-Spartanburg International Airport (GSP), Greer, South Carolina, and was operating under the provisions of 14 *Code of Federal Regulations* (CFR) Part 121 on an instrument flight rules flight plan. Visual meteorological conditions prevailed at the time of the accident.

The accident airplane had been flown from the Tri-State/Milton J. Ferguson Field, Huntington, West Virginia (HTS), to CLT on January 7, 2003 (the day before the accident). Air Midwest records indicated that the accident pilots flew the accident airplane on six flight legs that day. The first officer (the nonflying pilot) of the flight from HTS to CLT told the accident first officer, when handing off the airplane, that everything was normal, and it was a good flying airplane. The accident pilots began their trip sequence about 1340 and ended their trip sequence at CLT about 2045. Another flight crew met the accident airplane for a trip that night from CLT to Lynchburg Regional Airport/Preston Glenn Field (LYH), Lynchburg, Virginia. That flight crew flew the accident airplane back to CLT the next morning (January 8th), arriving at 0715. According to postaccident interviews, neither the captain nor the first officer of those two flight legs noticed anything unusual about the airplane.

On January 8, 2003, the accident flight crew was scheduled to fly two flight legs on a 1-day trip sequence, CLT to GSP and GSP to Raleigh–Durham International Airport (RDU), Raleigh–Durham, North Carolina, and then to travel on duty as passengers from RDU to CLT. An Air Midwest pilot saw the captain in the gate area about 0745 and the first officer about 0800. The dispatch release for flight 5481 showed that a maximum of 32 bags were allowed on the flight. One of the two ramp agents working flight 5481 stated, in a postaccident interview, that 23 bags had been checked and that 8 bags were carried on the airplane. The ramp agent stated that two of the checked bags were heavy, with an estimated weight of between 70 and 80 lb. The ramp agent also stated that he told the captain that some of the bags were heavy, although they were not marked as such. According to the ramp agent, the captain indicated that the bags were fine because a child would be on board, which would allow for the extra baggage weight. The ramp agent estimated that the forward cargo compartment was about 98% full by volume. Cockpit voice recorder (CVR) information early in the recording indicated that the flight crew was completing the preflight paperwork regarding the airplane's weight and balance. Air Midwest records indicated that flight 5481 departed the gate on time about 0830.

The captain was the flying pilot, and the first officer was the nonflying pilot. Flight data recorder (FDR) data indicated that, beginning about 0835:16, the flight crew performed a control check of the elevators. The pitch control position parameter, which measures the position of the control column, recorded values from 15° airplane nose up (ANU) to 16.5° airplane nose down (AND). These values corresponded to elevator positions from full ANU to 7° AND. About 0837:20, the CVR recorded the first officer contacting the CLT Air Traffic Control Tower (ATCT) ground controller and informing him that flight 5481 was ready to taxi. The ground controller instructed the flight crew to taxi to runway 18R. About 0846:18, the tower (local) controller cleared flight 5481 for takeoff and instructed the flight crew to turn right to a heading of 230° after takeoff. About 0846:35, the captain asked the first officer to set the takeoff power, and the first officer stated that the power had been set. About 0846:48, the airplane's airspeed was above 102 knots, and the elevator position was 7° AND. About 3 s later, the elevator position was 1° AND, and the pitch attitude of the airplane began to increase. After 0846:53, the pitch trim started moving AND and about 3 s later, the captain called for the landing gear to be retracted. About 0846:57, the elevator position returned to 7° AND, and, about 2 s later, the CVR recorded the sound of the landing gear retracting.

About 0847:02, the first officer stated, "wuh," and the captain stated, "oh." About 0847:03, the captain stated, "help me," At that point, the airplane was about 90 ft above ground level, and FDR data showed that the airplane's pitch attitude was 20° ANU and airspeed was 139 knots. About 0847:04, the CVR recorded the captain asking, "you got it?" and FDR data indicated that the flight crew was forcefully commanding AND.

During the next 8 s, the CVR recorded multiple statements and sounds from both flight crewmembers associated with their efforts to push the airplane's nose down. Also, about 0847:09, the CVR recorded a change in engine/propeller noise and, about 1 s later, the beginning of a sound similar to the stall warning horn. About 0847:13, the FDR recorded a maximum pitch attitude of 54° ANU. About 0847:16, the captain radioed the ATCT and stated, "we have an emergency for Air [Midwest] fifty four eighty one," and the CVR recorded the end of the sound similar to the stall warning horn. About 0847:18, the airplane's pitch attitude decreased through 0°, and the elevator position began to move ANU. By 0847:19, the airplane was about 1150 ft above ground level, and the FDR recorded a maximum left roll of 127° and a minimum airspeed of 31 knots. About 1 s later, the FDR recorded a pitch attitude of 42° AND. About 0847:21, the captain stated, "pull the power back," the elevator position reached full ANU, and the airplane's pitch altitude was 39° AND. At 0847:21.7, the CVR recorded the beginning of a sound similar to the stall warning horn, which continued to the end of the recording. About 0847:22, the airplane's roll altitude stabilized at about 20° left wing down; the pitch attitude began to increase; and the elevator position moved in the AND direction, reaching about 8° ANU. About 1 s later, the elevator position began moving in the ANU direction. About 0847:24, the airplane rolled right through wings level, and the pitch altitude increased to about 5° AND. About 0847:26, the FDR recorded a maximum right roll of 68° and a maximum vertical acceleration of 1.9 Gs. About the same time, the captain stated, "oh my god ahh," and the first officer stated something similar to, "uh uh god ahh [expletive]." The CVR recording ended at 0847:28.1. The FDR's last recorded pitch altitude was 47° AND; roll altitude was 66° to the right and pitch control position was 19.2° ANU, which corresponded to an elevator position of full ANU. The airplane struck a US Airways maintenance hangar on CLT property and came to rest about 1650 ft east of the runway 18R centerline and about 7600 ft beyond the runway 18R threshold. ATCT controllers heard an emergency locator transmitter signal beginning about 0847:29. The accident occurred at 35°12′25″ north latitude and 80°56′46.85″ west longitude during daylight hours.

Analysis

On the day of the accident, the first officer was seen conducting a walk-around inspection of the airplane. He did not report anything unusual about the airplane, including its elevator control system. The accident flight crew filled out the Air Midwest Beechcraft 1900D Load Manifest form for the flight using the average weight values for passengers and baggage in Air Midwest's weight and balance program at the time of the accident. The load manifest indicated a taxi fuel burn of 220 lb, even though Air Midwest assumes a taxi fuel burn of 110 lb. These figures resulted in a calculated airplane weight of 17,028 lb and a center of gravity (CG) position of 37.8% mean aerodynamic chord (MAC), which were within the Beech 1900D certified weight and CG limits of 17,120 lb and 40% MAC, respectively. The two ramp agents assigned to the accident flight handled the baggage according to company procedures and interacted appropriately with the flight crew. In a postaccident interview, one of the ramp agents reported that he told the captain that 2 of the 31 bags aboard the airplane had an estimated weight of between 70 and 80 lb. However, the bags did not have a heavy bag tag attached to them. (Gate agents use these tags to indicate an overweight bag, that is, a bag that weighs between 70 and 100 lb.) Also, the bags were not identified as overweight on the OF-11E form (the US Airways Express Load Report that is used to account for all passengers, baggage, and cargo loaded on a US Airways Express flight). As a result, the flight crew was not required to account for the extra weight of the reportedly heavy bags on the load manifest form.

Even if the flight crew had (1) accounted for the two reportedly heavy bags or (2) accounted for the two heavy bags, estimated an additional 110 lb of fuel at takeoff,[127] and recorded the 12-year-old passenger's weight as 80 lb rather than 175 lb,[128] the Air Midwest weight and balance program would still have indicated that flight 5481 was within the Beech 1900D certified weight and CG limits. However, for the second scenario, flight 5481's calculated weight would have exceeded the Beech 1900D weight limit if the 12-year-old passenger's weight had remained 175 lb.

Table 3.12 shows calculations for Air Midwest flight 5481 using the weight and balance program in effect at the time of the accident.

Findings

1. The captain and the first officer were properly certified and qualified under Federal regulations. No evidence indicated any preexisting medical or behavioral conditions that might have adversely affected their performance during the accident flight. Flight crew fatigue was not a factor in this accident.
2. The accident airplane was properly certified and equipped in accordance with Federal regulations. Except for the elevator control system, no evidence indicated that the airplane was improperly maintained. The recovered components showed no evidence of any preexisting structural, engine, or systems failures.

TABLE 3.12 Load Calculations

Load Manifest		Load Manifest Plus Two Bags		Load Manifest Plus Two Bags and 110 lb of Fuel But with 80 lb for the Child Passenger	
Weight in Pounds	CG in Percent of Mac	Weight in Pounds	CG in Percent of Mac	Weight in Pounds	CG in Percent of Mac
17,028	37.8	17,078	38.8	17,093	38.8

Source: Adapted from National Transportation Safety Board, Loss of Pitch Control During Takeoff Air Midwest Flight 5481, Raytheon (Beechcraft) 1900D, N233YV, AAR-04-0, 2/26/2004.

3. Weather was not a factor in this accident. The air traffic controllers that handled the accident flight were properly trained and provided appropriate air traffic control services. The emergency response for this accident was timely and effective. The accident was not survivable for the airplane occupants because they were subjected to impact forces that exceeded the limits of human tolerance.
4. The accident airplane entered the detail six maintenance check with an elevator control system that was rigged to achieve full elevator travel in the downward direction.
5. The accident airplane's elevator control system was incorrectly rigged during the detail six maintenance check, and the incorrect rigging restricted the airplane's elevator travel to 7° AND, or about one-half of the downward travel specified by the airplane manufacturer.
6. The changes in the elevator control system resulting from the incorrect rigging were not conspicuous to the flight crew.
7. The Raytheon Aerospace quality assurance inspector did not provide adequate on-the-job training and supervision to the Structural Modifications and Repair Technicians mechanic who examined and incorrectly adjusted the elevator control system on the accident airplane.
8. Because the Raytheon Aerospace quality assurance inspector and the Structural Modifications and Repair Technicians mechanic did not diligently follow the elevator control system rigging procedure as written; they missed a critical step that would have likely detected the misrigging and thus prevented the accident.
9. A complete functional check at the end of maintenance for critical flight systems or their components would help to ensure their safe operation, but no such check is currently required.
10. Flight 5481 had an excessive aft center of gravity, which, combined with the reduced downward elevator travel resulting from the incorrect elevator rigging, rendered the airplane uncontrollable in the pitch axis.
11. Air Midwest's weight and balance program at the time of the accident was not correct and resulted in substantially inaccurate weight and balance calculations for flight 5481.
12. Air Midwest's revised weight and balance program is also unacceptable because it may result in an inaccurate calculation of an airplane's center of gravity position.
13. Air Midwest did not adequately oversee the work performed by Raytheon Aerospace and Structural Modifications and Repair Technicians personnel at its Huntington, West Virginia, maintenance station and did not ensure that the accident airplane was returned to service in an airworthy condition.
14. When an inspector provides on-the-job training for a required inspection item (RII) maintenance task and then inspects that same task, the independent nature of the RII inspection is compromised.

15. Air carriers that use contractors to perform RII maintenance tasks and inspections need to provide substantial and direct oversight during each work shift to ensure that this work is being properly conducted.
16. Air Midwest did not have maintenance training policies and procedures in place to ensure that each of its maintenance stations had an effective on-the-job training program.
17. It is important that air carrier on-the-job training programs are developed in accordance with detailed guidance that emphasizes effective training practices.
18. Air Midwest did not ensure that its maintenance training was conducted and documented in accordance with the company's maintenance training program, which degraded the quality of training and inspection activities at the Huntington, West Virginia, maintenance station.
19. Air Midwest's Continuing Analysis and Surveillance System (CASS) program was not being effectively implemented because it did not adequately identify deficiencies in the air carrier's maintenance program, including some that were found by the Federal Aviation Administration (FAA) before the flight 5481 accident.
20. Accurate and usable work cards developed jointly by air carriers and aircraft manufacturers would improve the performance of maintenance for critical flight systems.
21. The FAA's failure to aggressively pursue the serious deficiencies in Air Midwest's maintenance training program that were previously and consistently identified permitted the practices that prevailed at the Huntington, West Virginia, maintenance station and during the accident airplane's detail six maintenance check.
22. Updated CASS guidance would help FAA aviation safety inspectors ensure that CASS programs are being effectively implemented at 14 *CFR* Part 121 air carriers.
23. Because proper aircraft maintenance is crucial to safety, air carrier maintenance training programs should be subject to the same standard that exists for other air carrier training programs (i.e., FAA approval).
24. The lessons learned by the FAA through its human factors research program need to be used to develop mandatory programs to prevent human error in aviation maintenance.
25. The use of average weights does not necessarily ensure that an aircraft will be loaded within its weight and center of gravity envelope.
26. The FAA's average weight assumptions in Advisory Circular 120-27C, "Aircraft Weight and Balance Control," were not correct.
27. Periodic sampling of passenger and baggage weights would determine whether air carrier average weight programs were accurately representing passenger and baggage loads.
28. Current safety margins in air carrier average weight and balance programs do not ensure that aircrafts will be loaded within their manufacturer-certified and FAA-approved weight and center of gravity envelope.

29. Technology may enable air carriers to accurately determine weight and effectively control balance while maintaining operational efficiency.
30. Beech 1900 mechanics would benefit from using Airliner Maintenance Manuals with more specific instructions for critical flight system procedures.
31. Because the CVR can be one of the most valuable tools used for accident investigation, reliable daily test procedures are needed to safeguard CVR data.

Probable Cause

The NTSB determines the probable cause of this accident was the airplane's loss of pitch control during takeoff. The loss of pitch control resulted from the incorrect rigging of the elevator control system compounded by the airplane's aft center of gravity, which was substantially aft of the certified aft limit. Contributing to the cause of the accident were (1) Air Midwest's lack of oversight of the work being performed at the Huntington, West Virginia, maintenance station; (2) Air Midwest's maintenance procedures and documentation; (3) Air Midwest's weight and balance program at the time of the accident; (4) the Raytheon Aerospace quality assurance inspector's failure to detect the incorrect rigging of the elevator control system; (5) the Federal Aviation Administration's (FAA) average weight assumptions in its weight and balance program guidance at the time of the accident; and (6) the FAA's lack of oversight of Air Midwest's maintenance program and its weight and balance program.

OTHER BACKGROUND INFORMATION

Weight and Balance

This section discusses the accident airplane's calculated weight and balance, which was determined using Air Midwest's FAA-approved weight and balance program, and the loading conditions that existed on the day of the accident.

The accident airplane's last weighing on September 8, 2002, determined that the airplane's empty weight was 10,293 lb. The airplane's balance was determined by the location of the CG, which is usually described as a given number of inches aft of the reference datum. At the time of the airplane's weighing, the CG was determined to be located 282.1 in. aft of the reference datum, which corresponds to a CG location of 14.4% MAC.

According to the Air Midwest Beech 1900D (Beechcraft, 2004) load manifest form for flight 5481, the operating empty weight was 10,673 lb, the passenger weight was 3325 lb, the weight in the coat closet was 10 lb, the AFT1 cargo compartment weight was 775 lb, the AFT2 cargo compartment weight was 45 lb, the zero fuel weight was 14,818 lb, the fuel weight at takeoff was 2200 lb, and the gross takeoff weight was 17,018 lb. The flight crew made a 10-lb addition error when summing the weights that comprise the zero fuel weight. As a result, the calculated zero fuel weight was actually 14,828 lb, and the calculated gross takeoff weight was actually 17,028 lb. The Beech 1900D maximum gross takeoff weight is 17,120 lb. The load manifest form also indicated that the calculated CG index for the accident flight was 81 (37.8% MAC). The Air Midwest CG takeoff limits range from indexes of about

TABLE 3.13 Weight and Balance for the Last 10 Flights

Date	Gross Takeoff Weight (lb)	CG (Percent of MAC)	Number of Passengers	Cargo (lb)
01-08-03	17,028	37.8	19	820
01-08-03	16,278	25.9	15	470
01-07-03	15,118	19.6	6	195
01-07-03	13,303	17.3	2	70
01-07-03	14,528	19.0	3	120
01-07-03	12,618	12.6	0	45
01-07-03	14,653	20.8	7	345
01-07-03	14,278	23.7	9	320
01-07-03	14,413	24.2	5	455
01-07-03	13,318	13.6	0	45

Source: Adapted from National Transportation Safety Board, Loss of Pitch Control During Takeoff Air Midwest Flight 5481, Raytheon (Beechcraft) 1900D, N233YV, AAR-04-0, 2/26/2004.

23–85 (16.7–39.2% MAC) when a Beech 1900D airplane is at a gross takeoff weight of 17,028 lb. The Beech 1900D aft CG limit is 40% MAC.

Table 3.13 shows information reported on Beech 1900D load manifest forms for flights flown by the accident airplane after the January 6, 2003 D6 maintenance check.

The table shows that the accident flight was calculated to be the most aft loaded of all of the postmaintenance flights.

OTHER RELATED ACCIDENTS

Ryan Air Service Flight 103, Homer, Alaska

On November 23, 1987, Ryan Air Service flight 103, a Beech 1900C, 107 N401RA, crashed short of the runway during arrival at the Homer, Alaska airport. Flight 103 was a scheduled 14 CFR Part 135 flight operating from Kodiak, Alaska, to Anchorage, Alaska, with intermediate stops in Homer and Kenai, Alaska. The 2 flight crewmembers and 16 passengers were killed, and 3 passengers were seriously injured. The accident investigation revealed that the airplane was loaded with about 600 lb more cargo than the first officer had requested. The airplane was 400–500 lb over the airplane's maximum takeoff weight and 100–200 lb over its maximum landing weight. In addition, the CG position was 12–16% MAC aft of the allowable aft limit, and the flight crew did not comply with company and FAA procedures in computing the CG position. Even with an extreme aft CG, the airplane was able to take off and establish cruise flight. Evidence indicated that the flight crew lost control of the airplane as its flaps were lowered for landing.

The Safety Board determined that the probable cause of this accident was the failure of the flight crew to properly supervise the loading of the airplane, which resulted in the CG being displaced to such an aft location that airplane control was lost when the flaps were lowered for landing.

Two data points in the figure show the CG range for the flight 103 airplane.

ValuJet Airlines Flight 592, Everglades, Near Miami, Florida

On May 11, 1996, ValuJet Airlines flight 592, a Douglas DC-9-32, N904VJ, crashed into the Everglades, near Miami, Florida, about 10 min after takeoff from Miami International Airport. The 2 pilots, 3 flight attendants, and all 105 passengers were killed. Flight 592 was operating under 14 CFR 121 with a scheduled destination of the William B. Hartsfield International Airport, Atlanta, Georgia. A fire erupted in the airplane's class D cargo compartment. The fire was initiated by the actuation of one or more oxygen generators being improperly carried as cargo. The oxygen generators were prepared and packaged for carriage aboard flight 592 by SabreTech, a 14 CFR Part 145 repair station in Miami that performed heavy maintenance for ValuJet Airlines.

The Safety Board's investigation of the FAA's oversight of ValuJet Airlines revealed that inspectors from the Aircraft Maintenance Division within the Office of Flight Standards had recommended recertification of the airline 3 months before the accident. Specifically, in a February 14, 1996, summary report, the inspectors indicated, consideration should be given to an immediate FAR 121 recertification of this airline because of safety-related issues, such as the absence of adequate policies and procedures for maintenance personnel. The inspectors also indicated that the overall surveillance of ValuJet Airlines should be increased with special attention directed toward "manuals and procedures, structural inspections, the adequacy of the maintenance program, and shops and facilities."

The Safety Board determined that the probable causes of this accident were (1) the failure of SabreTech to properly prepare, package, and identify unexpended chemical oxygen generators before presenting them to ValuJet for carriage; (2) the failure of ValuJet to oversee its contract maintenance program to ensure compliance with maintenance, maintenance training, and hazardous materials requirements and practices; and (3) the failure of the FAA to require smoke detection and fire suppression systems in class D cargo compartments. Contributing to the accident was the failure of the FAA to adequately monitor ValuJet's heavy maintenance programs and responsibilities, including ValuJet's oversight of its contractors, and SabreTech's repair station certificate; the failure of the FAA to adequately respond to prior chemical oxygen generator fires with programs to address the potential hazards; and ValuJet's failure to ensure that both ValuJet and contract maintenance facility employees were aware of the carrier's "no-carry" hazardous materials policy and had received appropriate hazardous materials training.

During its investigation of the ValuJet Airlines flight 592 accident, the Safety Board determined that the SabreTech mechanics had many shortcomings, including their failure to install safety caps, improper maintenance entries, use of improper tags, and inadequate communications between the maintenance shop floor and stores department. In its final report on the accident, the Board indicated that these shortcomings resulted from human failures that might have been avoided if more attention were given to human factors issues in the maintenance environment. As a result, the Board issued Safety Recommendation A-97-70 on September 9, 1997. Safety Recommendation A-97-70 asked the FAA to include, in its development and approval of air carrier maintenance procedures and programs, explicit consideration of human factors issues, including training, procedures development, redundancy, supervision, and the work environment, to improve the performance of personnel and their adherence to procedures.

On October 2, 2000, the FAA stated that it had reviewed the information contained in its report, "Human Factors in Aviation Maintenance and Inspection, Strategic Program Plan," and that it was amending AC 120-16C, "Continuous Airworthiness Maintenance Programs," to include information from the report. The FAA indicated that the revisions to the AC would also expand on CASS programs. On April 24, 2001, the Safety Board stated that it was difficult to determine whether the revisions to AC 120-16C would address the issues in this recommendation. On July 21, 2003, the FAA stated that Chapter 10 of AC 120-16D included human factors as part of initial training. On February 23, 2004, the Safety Board stated that, although AC 120-16D addressed many of the human factors issues related to training, procedures development, redundancy, supervision, and work environment, the AC would be significantly strengthened if the FAA added specific references to its available human factors information related to aviation maintenance operations, such as the Human Factors Guide for Aviation Maintenance and Inspection. The Board also stated that it continued to investigate major accidents in which incorrect maintenance led to a loss of control of the airplane and that human factors in aviation maintenance was an important safety issue. Pending the inclusion in AC 120-16D of references to FAA-published guidance on human factors in aviation maintenance, Safety Recommendation A-97-70 was classified "Open Acceptable Response."

In its final report on the ValuJet Airlines flight 592 accident, the Safety Board determined that the FAA's surveillance of ValuJet before the accident did not include any significant oversight of the air carrier's heavy maintenance contractors, including SabreTech. The Board further determined that the FAA's limited oversight of ValuJet's maintenance contractors was not sufficient to detect potential problems. The ValuJet PMI was not required to conduct surveillance of the air carrier's contract maintenance facilities. Thus, the Board concluded that the lack of an explicit requirement for a PMI of a 14 CFR Part 121 air carrier to regularly inspect repair stations that are performing heavy maintenance for the carrier is a significant deficiency in the FAA's oversight of the carrier's total maintenance program. As a result, the Board issued Safety Recommendation A-97-74 on September 9, 1997. Safety Recommendation A-97-74 asked the FAA to ensure that Part 121 air carriers' maintenance functions receive the same level of FAA surveillance, regardless of whether those functions are performed in-house or by a contract maintenance facility.

On April 22, 1998, the FAA stated that it issued Flight Standards Handbook Bulletin for Airworthiness 96-05C, "Air Carrier Operations Specifications Authorization to Make Arrangements With Other Organizations to Perform Substantial Maintenance." On December 15, 1997, the FAA indicated that the bulletin described detailed procedures to ensure that surveillance of each 14 CFR Part 121 air carrier's maintenance function entails the performance of the maintenance, the adequacy of the maintenance organization, the competency of maintenance personnel, and the adequacy of maintenance facilities and equipment, regardless of whether those functions are performed in-house or by a contract maintenance facility.

On July 23, 1999, the Safety Board stated that the FAA's actions met the intent of Safety Recommendation A-97-74 and classified it "Closed Acceptable Action."

Fine Airlines Flight 101, Miami, Florida

On August 7, 1997, Fine Airlines flight 101, a Douglas DC-8-61, N27UA, crashed after takeoff from Miami International Airport. The 3 flight crewmembers and 1 passenger on board the airplane were killed, 1 person on the ground was killed, and the airplane was destroyed by impact forces and a postcrash fire. The cargo flight, which had a scheduled destination of Santo Domingo, Dominican Republic, was operated under 14 CFR Part 121 as a supplemental air carrier.

The accident airplane was loaded incorrectly, which resulted in an aft CG. Also, an incorrect stabilizer trim setting precipitated an extreme pitch-up at rotation. The Safety Board determined that the probable cause of this accident was the failure of Fine Air to exercise operational control over the cargo loading process and the failure of Aeromar (a cargo shipper) to load the airplane as specified by Fine Air. Contributing to the accident was the failure of the FAA to adequately monitor Fine Air's operational control responsibilities for cargo loading and to ensure that known cargo-related deficiencies were corrected at Fine Air.

In its final report on the Fine Airlines accident, the Safety Board discussed the Sum Total Aft and Nose (STAN) system, which is an electronic system installed on some cargo airplanes that allows flight crews to verify an airplane's weight and balance before departure. According to the report, the STAN system uses pressure transducers to convert main gear and nose gear shock strut air pressure to an electronic signal. The system then provides flight crews with a digital readout in the cockpit (on the flight engineer's instrument panel) of the airplane's gross weight and CG values. The Safety Board's final report on the Fine Airlines accident concluded that, if the flight crew had an independent method in the cockpit for verifying the airplane's actual weight and balance and gross weight, it might have alerted them to loading anomalies and prevented the accident. As a result, on July 10, 1998, the Safety Board issued Safety Recommendation A-98-49 to the FAA. Safety Recommendation A-98-49 asked the FAA to evaluate the benefit of the STAN and similar systems and require, if warranted, the installation of a system that displays airplane weight and balance and gross weight in the cockpit of transport-category cargo airplanes. On December 30, 1998, the FAA stated that it had completed an evaluation of the reliability of onboard weight and balance systems. The FAA found that some operators had reliability and accuracy concerns with such systems because of factors such as wind, ramp slope, oleo stiction, low hydraulic pressure, and asymmetrical gear loads. The FAA stated that the results of its evaluation did not support imposing a requirement to install a system that displays airplane weight and balance and gross weight in the cockpit of transport-category cargo airplanes. On January 11, 2000, the Safety Board stated that, on the basis of the FAA's evaluation and subsequent determination that onboard weight and balance systems do not yet meet the quality standards for a mandatory system, Safety Recommendation A-98-49 was classified "Closed-Acceptable Action."

Emery Airlines Flight 17, Rancho Cordova, California

On February 16, 2000, Emery Airlines flight 17, a McDonnell Douglas DC-8-71F, N8079U, crashed in an automobile salvage yard shortly after takeoff while attempting to return to Sacramento Mather Airport, Rancho Cordova, California, for an emergency landing. Flight 17 was a scheduled 14 CFR Part 121 cargo flight from Sacramento to James M. Cox Dayton International Airport, Dayton, Ohio. The 2 pilots and the flight engineer were killed, and the airplane was destroyed.

The Safety Board's investigation of this accident determined that the bolt attaching the accident airplane's right elevator control tab crank fitting to the pushrod was improperly secured and inspected during either the airplane's most recent D inspection (heavy maintenance accomplished every 12 years) or subsequent maintenance. Tennessee Technical Services (TTS), an Emery Airlines maintenance contractor, performed the accident airplane's last D inspection between August 27 and November 17, 1999. Eight days after the D inspection was completed, a pilot reported increased control column forces. Emery maintenance personnel found that the left and the right elevator dampers were reversed, and the maintenance logbook indicated that the maintenance personnel moved the dampers to their correct positions. Emery maintenance personnel could have come in contact with the bolt at the control tab crank fitting while troubleshooting the reported problem. The Safety Board determined that the probable cause of the accident was a loss of pitch control resulting from the disconnection of the right elevator control tab. The disconnection was caused by the failure to properly secure and inspect the attachment bolt.

Safety Recommendation A-03-31

In its final report on the Emery Airlines flight 17 accident, the Safety Board discussed the accident airplane's last B-2 maintenance inspection (the second of four segmented inspections generally accomplished at 136-h intervals) on January 21 and 22, 2000. The B-2 inspection includes a visual check of the elevators and tabs for general condition, corrosion, leakage, and security of attachment. The DC-8 elevator assembly design requires the elevator control tab inboard fairing to be removed for maintenance personnel to inspect the inboard hinge fitting and the control tab crank fitting to pushrod attachment.

During postaccident interviews, Emery Airlines maintenance personnel stated that, when performing the accident airplane's last B-2 maintenance, they did not remove the elevator control tab inboard fairing or inspect the crank fitting to pushrod attachment. During public hearing testimony, witnesses from Emery Airlines indicated that its B-2 inspection was intended to be a general visual inspection that was to be accomplished without removing access or inspection panels or fairings. However, witnesses from TTS stated that removal of the control tab fairing was necessary to satisfactorily perform the tasks described on the Emery Airlines B-2 work card, even though that step was not specifically listed on the work card. The Safety Board noted that several air carriers have tried to clarify the intended scope of maintenance tasks by including, on their work cards, an enumeration of the actions that are necessary for the proper accomplishment of the associated work task. The Board stated that the inclusion of this additional detail on work cards, although not required by the FAA, should result in more consistent accomplishment of maintenance tasks. As a result, the Board issued Safety Recommendation A-03-31 on August 18, 2003. Safety Recommendation A-03-31 asked the FAA to Require all 14 CFR Part 121 air carrier operators to revise their task documents and/or work cards to describe explicitly the process to be followed in accomplishing maintenance tasks.

On January 12, 2004, the FAA stated that Safety Recommendation A-03-31 was limited to DC-8 operators only. The FAA also stated that the Boeing Company issued temporary revisions to the DC-8 AMM on May 8, 2002, and that these revisions explicitly described the maintenance task process to be followed. On January 23, 2004, the Safety Board classified Safety Recommendation A-03-31 "Open-Response Received."

Colgan Air Flight 9446, Yarmouth, Massachusetts

On August 26, 2003, Colgan Air (doing business as US Airways Express) flight 9446, a Beech 1900D, N240CJ, crashed into water near Yarmouth, Massachusetts. The 2 flight crewmembers were killed, and the airplane was substantially damaged. The repositioning flight, which was conducted under 14 CFR Part 91, departed Barnstable Municipal Airport, Hyannis, Massachusetts, for Albany International Airport (ALB), Albany, New York.

Shortly after takeoff, the flight crewmembers declared an emergency and reported a trim problem. The airplane had reached an altitude of about 1100 ft msl. The flight crew requested to land on a specific runway, and the controller cleared the flight to land on any runway. No further transmissions were received from the flight crew. FDR data indicated that the airplane's airspeed continued to increase to about 250 knots and that the airplane's last recorded altitude was about 300 ft msl.

The accident airplane's FDR pitch trim control position parameter had been placed on the minimum equipment list (deferred maintenance) for the flight because the parameter was not calibrated. The Safety Board's airplane performance study for this accident determined that the recorded pitch trim control positions did not reflect the actual pitch trim control positions. The difference between the recorded and actual pitch trim control positions was about 2.1° ANU. According to the airplane performance study, the airplane began the flight with a pitch trim control position of about 0.5° ANU. Shortly after takeoff, the pitch trim control position moved to about 0.8° AND and remained there for about 10 s. The pitch trim control position then moved to about 5° AND and remained there for the rest of the flight. Calculations showed that the airplane would have required about 200 lb of aft (pulling) control force to maintain level flight in the out-of-trim condition.

The accident flight was the first flight after maintenance had been performed on the airplane. The maintenance work included replacement of both elevator trim tab actuators (because of a failed freeplay check) in accordance with Beech 1900D AMM section 27-30-06, "Elevator Trim Tab Actuator, Removal and Installation and the Actuator Cable Replacement." The procedure required that the elevator be removed before the actuators were replaced. The Safety Board's investigation of this accident determined that the mechanics skipped this procedural step and replaced the actuators with the elevators installed. The mechanics thought that the forward elevator trim tab cable had become jammed or kinked during the replacement of the trim tab actuator. The mechanics then tried to replace the cable according to the procedure in Beech 1900D AMM section 27-30-04, "Elevator Trim Tab Cables, Removal and Installation." A postaccident examination of a section of the forward elevator trim cable revealed evidence consistent with a misrouted cable. The Beech 1900D AMM and Colgan Air work cards did not include a trim system check at the end of the elevator trim tab cable procedure. Although the mechanics stated that they checked the trim system, evidence was consistent with the trim system operating in a direction opposite from the command of the trim wheel. The Safety Board's investigation of this accident revealed that the illustration of the forward elevator trim tab cable drum appeared backward in section 27-30-04 in the Beech 1900D AMM. On October 22, 2003, Raytheon Aircraft Company revised its Beech 1900D AMM elevator trim tab cable rigging procedure to show the correct illustration for the forward elevator trim tab cable drum (Beechcraft, 2004).

Federal Aviation Administration Airworthiness Directive 2003-20-10

On October 15, 2003, the FAA issued AD 2003-20-10, which applied to all Beech model 1900, 1900C, and 1900D airplanes. The FAA reviewed Raytheon Aircraft's current maintenance procedures for the elevator trim system and determined that the figures in the applicable maintenance manuals depicted the elevator trim cable drum at 180° from the installed position and showed the open, keyed side of the drum instead of the flat side of the drum. The FAA's review of the maintenance procedure also identified the need to add a step to visually confirm that the trim wheel position and the trim tab position were consistent. According to the FAA, such a check would detect and correct any problems with the elevator trim system installation before problems occur during operation. AD 2003-20-10 warned that an incorrectly installed elevator trim system component, if not detected and corrected, could result in difficulties in controlling the airplane or a total loss of pitch control. As a result, the AD required operators of Beech 1900 series airplanes to replace the incorrect figure in the elevator trim system maintenance procedures with the corrected figure, incorporate a temporary revision to the applicable maintenance manual that describes the elevator trim operational check, and perform an elevator trim operational check each time maintenance is accomplished on the elevator trim system.

CommutAir Flight 8718, Albany, New York

On October 16, 2003, CommutAir (doing business as Continental Connection) flight 8718, a Beech 1900D, N850CA, aborted takeoff from ALB because of an elevator control system discrepancy. The pilot stated that, during the takeoff roll, the control column would not move aft when the airspeed reached V1. The intended destination for the positioning flight, which was conducted under 14 CFR Part 91, was Westchester County Airport, White Plains, New York. The 2 flight crewmembers were not injured, and the airplane was not damaged. Maintenance was performed on the airplane 1 day before the incident, and the incident flight was the first postmaintenance flight. Maintenance records showed that a worn detent pin was replaced on the right thrust lever assembly. The mechanic who replaced the detent pin stated that he looked in the Beech 1900D AMM for a procedure to replace the pin or a procedure to access the thrust lever assembly. The mechanic indicated that the Beech 1900D AMM did not contain either procedure, so he looked in a CommutAir manual that contained a list of the manufacturer's field service kits, found a reference to a field service kit for "thrust lever, replaceable detent pin," and obtained the kit. The field service kit installation instructions included a step to remove the thrust control assembly from the center pedestal, but the instructions did not provide information about how to access the thrust lever assembly so that it could be removed from the pedestal. The instructions also did not provide any reference to technical documents that contained instructions for the removal and installation of the thrust lever assembly. The mechanic stated that he accessed the thrust lever assembly by removing the elevator trim wheel and an access panel from the left side of the cockpit pedestal. No write-up was generated for the trim wheel removal, and no markings or tags were placed on the trim wheel to ensure proper reinstallation. The investigation of this incident determined that, when the elevator trim wheel was reinstalled, the mechanic did not properly align the elevator trim tab position indicator with the elevator trim tab position. As a result, the trim wheel was reinstalled incorrectly. The mechanic did not perform a functional test of the elevator trim control system, as required by AD 2003-20-10. The Safety Board and Raytheon Aircraft Company performed a functional check of the elevator trim control system. The check verified that the elevator trim tab position indicator did not accurately reflect the elevator trim tab position. Specifically, when the elevator trim tab position indicator was set at 3 units of ANU trim, the elevator trim tab was deflected 4.6° AND from its neutral position. (The elevator trim tab's full AND position is 5.5° from neutral.) The functional check determined that the elevator trim tab position indicator pointer was off by about 6 units, which equates to about 8.8° of elevator trim tab, or about 37% of the trim tab's full range of travel. After the cause of the elevator trim control system discrepancy was determined, a CommutAir mechanic properly aligned the elevator trim tab position indicator to the elevator trim tab position. Elevator and elevator trim control system functional tests were accomplished to verify that the systems were operating according to the requirements described in the Beech 1900D AMM. A high-speed taxi test and a flight test were accomplished to verify the functionality of the elevator and elevator trim control systems.

Notes

- Unless otherwise indicated, all times in this report are Eastern Standard Time based on a 24-h clock.
- Raytheon Aircraft Company acquired Beech Aircraft Corporation in February 1980.
- The ramp agents working this flight were employees of Piedmont Airlines, which runs US Airways Express ground operations at CLT.
- In calculating the weight and balance of the airplane, the flight crew used Air Midwest's standard adult weight figure (175 lb) for this child, who was 12 years of age.
- A cargo net separates the forward (AFT1) cargo compartment from the aft (AFT2) cargo compartment. A ramp agent stated that the cargo net was in place before the accident flight.
- An elevator is an aerodynamic control surface hinged to the back of the horizontal stabilizer. An elevator moves up and down to control the airplane's wing angle of attack, pitch, and climb. Normal elevator travel for the Beech 1900D is from 20° to 21° ANU to 14° to 15° AND, and the elevator neutral position is

0°. The elevator control check in the Beech 1900D involves moving the control column from the full forward position to the full aft position.

- In a properly rigged elevator control system, the FDR pitch control position parameter accurately reflects the elevator position. For the accident flight, however, the recorded pitch control positions did not reflect the actual elevator positions. The recorded pitch control positions were about 9° more AND than the actual elevator positions.
- According to Air Midwest's Beech 1900D Performance Manual, the rotation speed during takeoff is 105 knots.
- The three previous flight crews who flew the accident airplane also did not report anything unusual about the elevator control system. In fact, the first officer of the flight from Huntington, West Virginia (HTS), to Charlotte, North Carolina, on January 7, 2003, stated that "everything was normal" and "it was a good flying airplane."
- The flight crew made a 10-lb addition error when summing the weights that comprise the zero fuel weight. This addition error was not a factor in the accident.
- The Air Midwest Flight Operations Procedures Manual at the time of the accident stated that an average weight of 175 lb could be used for each adult passenger during the winter and that an average of 80 lb could be used for children between the ages of 2 and 12 years.
- The reference datum is an imaginary vertical plane, arbitrarily fixed somewhere along the longitudinal axis of the airplane, from which all horizontal distances are measured for weight and balance purposes.
- According to the FAA's *Aircraft Weight and Balance Handbook*, the MAC is the chord of an imaginary airfoil that has all of the aerodynamic characteristics of the actual airfoil. The chord is drawn through the geographic center of the plan area of the wing. The location of the CG with respect to the MAC is important because it predicts the handling characteristics of the aircraft.
- The Beech 1900C is generally similar in size and capacity to the 1900D. The airplanes have numerous similar components and share a common FAA type certificate.
- For more information, see NTSB, *Ryan Air Service, Inc., Flight 103, Beech 1900C, N401RA, Homer, Alaska, November 23, 1987*, Aircraft Accident Report NTSB/AAR-88/11 (Washington, DC: NTSB, 1988).
- For more information, see NTSB, *In-flight Fire and Impact With Terrain, ValuJet Airlines Flight 592, DC-9-32, N904VJ, Everglades, Near Miami, Florida, May 11, 1996,* Aircraft Accident Report NTSB/AAR-97/06 (Washington, DC: NTSB, 1997).
- For more information, see NTSB, *Uncontrolled Impact With Terrain, Fine Airlines Flight 101, Douglas DC-8-61, N27UA, Miami, Florida, August 7, 1997*, Aircraft Accident Report NTSB/AAR-98/02 (Washington, DC: NTSB, 1998).
- For more information, see NTSB, *Loss of Pitch Control on Takeoff, Emery Worldwide Airlines, Inc., Flight 17, McDonnell Douglas DC-8-71F, N8079U,*

Rancho Cordova, California, February 16, 2000, Aircraft Accident Report NTSB/AAR-03/02 (Washington, DC: NTSB, 2003).

- Additional information about this accident, NYC03MA183, can be found on the Safety Board's Website.
- The FAA indicated that, although the figures in the manuals were incorrectly depicted, the step-by-step instructions in the procedure, if followed correctly, would result in the proper installation and action of the elevator trim system. If only the figures were used, a reversing of the action of the elevator manual trim system could result.
- Raytheon Aircraft Company addressed these issues in its Safety Communiqué number 234, dated September 2003.
- According to 14 CFR Part 1, "V1 means the maximum speed in the takeoff at which the pilot must take the first action (e.g., apply brakes, reduce thrust, deploy speed brakes) to stop the airplane within the accelerate-stop distance. V1 also means the minimum speed in the takeoff, following a failure of the critical engine at VEF [the speed at which the critical engine is assumed to fail during takeoff], at which the pilot can continue the takeoff and achieve the required height above the takeoff surface within the takeoff distance."
- When the elevator trim tab position indicator is at 0°, the elevator trim tab position should also be at 0°.
- The quality assurance inspector also did not perform a functional test of the elevator trim control system. The inspector stated that he did not know that the mechanic's work involved the trim system.
- Additional information about this incident, NYC04IA010, can be found on the Safety Board's website.

REVIEW QUESTIONS

1. Find your height and weight for your gender and age on the anthropometric data tables and see if the percentiles are the same or different. Do you think most peoples' percentiles would be the same or different?
2. Which anthropometric data table and what percentile would one use to design a safety shower and why?
3. What types of anthropometric data would you collect to aid you in the design of a comfortable airliner seat?
4. Do you feel current airliner seats are comfortable? Why/why not?
5. Why is anthropometric data important?
6. What anthropometric data would you collect for use in the design of a machine guard to prevent someone from having their hand caught in the machinery?
7. Which principle *best* accommodates the widest range of workers?
8. Which principle is used as the last resort?

REFERENCES

Beechcraft, R. (2004). *Loss of Pitch Control Takeoff Air Midwest Flight.* Charlotte: National Transportation Safety Board.

Bradtmiller, B. P. (n.d.). Applied Anthropometry Improves Fit. Retrieved April 2, 2014, from https://ergoweb.com/complex-anthropometry-made-simple.

CDC. (2012). Anthropometric Reference Data for Children and Adults: United States, 2007–2010, Hyattsville, Maryland, October 2012 DHHS Publication No. (PHS) 2013–1602.

CPSC. (2014). Safety Standards for Full-Size Baby Cribs and Non-Full-Size Baby Cribs. http://www.cpsc.gov/en/Regulations-Laws--Standards/Rulemaking/Final-and-Proposed-Rules/Full-Size-Cribs/.

Helander, M. (2005). *A Guide to Human Factors and Ergonomics* 2nd edn. CRC Press.

Rhodes, H., *Alphonse Bertillon: Father of Scientific Detection*, In H. Rhodes (ed.) New York: Abelard-Schuman, p. 27, 1956.

ADDITIONAL SOURCES

http://www.cpsc.gov/en/Newsroom/News-Releases/1995/CPSC-Warns-Consumers-That-Used-Cribs-Can-Be-Deadly/.

National Health Statistics Reports. Anthropometric Reference Data for Children and Adults: United States, 2003–2006. See more at: http://www.talladaptations.com/2014/04/definining-tall-how-tall-is-tall.html#sthash.gpl1evHW.dpuf.

Sheldon, W.H. with the collaboration of S.S. Stevens and W.B. Tucker. (1940). *The Varieties of Human Physique*. Harper and Brothers, New York.

4

OFFICE ERGONOMICS

LEARNING OBJECTIVE

At the end of the chapter, students will have the ability to describe what ergonomics means, identify the best ergonomic posture for a workstation, demonstrate how to set up a productive computer office workstation, and identify risk factors associated with a computer workstation.

INTRODUCTION

Ergonomics, the science of designing workplaces, equipment, and jobs to fit the capabilities and limitations of workers, has shown that poor workplace design and bad work habits are counterproductive and costly.

One major problem with computer workstations is they keep an operator at that workstation, in virtually the same posture, all day long.

Our bodies are designed to be an upright, dynamic system. Chronic sitting stretches and contracts muscles inappropriately and causes stress to the body. A poorly designed workspace causes neck and back pain, decreasing productivity and job satisfaction.

The goal is to prevent musculoskeletal disorders by surveying the workplace, taking preventative steps, relying on employee input, and addressing the problems early. Figure 4.1 depicts what a neutral posture should be for those working at a desk on a computer.

This chapter will refer to the Work-Related Musculoskeletal chapter in the book when discussing the various disorders caused by poor office ergonomics.

Occupational Ergonomics: A Practical Approach, First Edition.
Theresa Stack, Lee T. Ostrom and Cheryl A. Wilhelmsen.

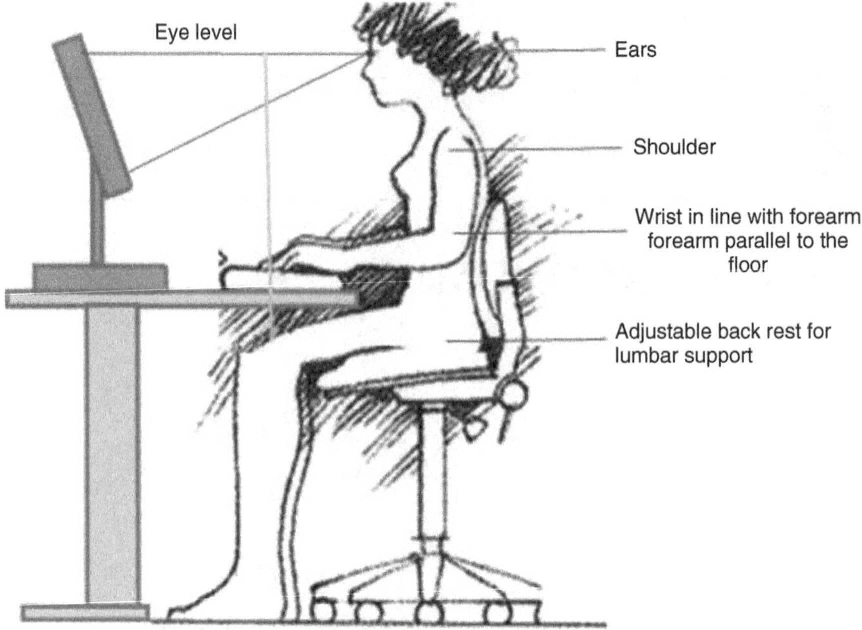

Figure 4.1 Neutral posture

Statistics have found that workers annually have ergonomic work-related musculoskeletal injuries, resulting in millions of dollars in total injury costs per year. Carpal tunnel comprises 0.7% and tendonitis comprises 0.2% of those injuries, while sprains, strains, and tears comprise 36.7%, and soreness and pain comprise 17.4%. The number of musculoskeletal disorders with days away from work in 2013 within private industry totaled 307,640, state government 18,410, and local government 54,560. Nursing assistants, laborers and freight, stock, and material movers, as well as heavy and tractor-trailer truck drivers, are among the highest incidence rates of injuries and illnesses due to musculoskeletal disorders (Demographics, 2014).

PRINCIPLES OF OFFICE ERGONOMICS

There are many principles to consider when assessing an office environment. Ergonomist need to consider the posture required by the workplace. This is dependent on the tasks or activities associated with the workplace. Another factor to consider is the environment of the workplace.

The goal is to arrange the environment to help the worker maintain a neutral posture, in order to minimize the incidents of work-related musculoskeletal disorders (WMSDs). WMSDs are discussed in Chapter 13.

All of the components within the workspace (e.g., chair, desk, keyboard) must be balanced. Figure 4.2 provides an example of an ideal typing position.

Figure 4.2 Example of good sitting posture (Permission from Marci Merrick)

General Considerations

The key points of interface between the office worker and the office environment are as follows:

- The eyes and the visual display terminal (VDT) screen
- The hands and the keyboard
- The back, the legs, and the chair
- The feet and the floor.

We will talk about each of these key points and some suggestions on how to improve the comfort and efficiency for the office worker a little later in this chapter.

There is no optimum posture while sitting, the next posture is the best posture. Instead, the goal is to have the ability to shift your body by raising and lowering the position of the chair. You need to move around, changing how you sit as your body tells you to move. Ensure a person can attain a posture in which the ears line up over the shoulders and hips. The hip joint and knee joint should be at a 90–110° angle. Avoid slouching postures.

A poor sitting position may lead to the following:

- Neck and shoulder fatigue
- Low back pain
- Numbness and discomfort in the legs.

Ergonomic Stressors Associated with Computer Use

Let us think about the physical and contributing risk factors and which apply to computer use.

First, the physical risk factors associated with computer use are as follows:

- Awkward postures (hands, arms, wrists, neck, shoulder, and legs)
- Static muscle loading
- Repetition
- Contact stress (thighs, arms, and palms)
- Duration.

The contributing risk factors associated with computer use are as follows:

- Inadequate recovery or rest
- Extreme temperatures – hot or cold
- Stress on the job
- Fatigue.

After looking at the risk factors, we can identify some of them in the office area. Steady typing involves heavy repetition with little opportunity for muscle recovery. Inadequate recovery leads to fatigue.

An improperly setup workstation can lead a computer user to exhibit poor postures. If these postures are sustained, there may be static muscle loading. Computer tasks can often be visually or mentally demanding, which are contributing risk factors. Stress and constant attention requirements are also contributing factors. A computer workstation can cause contact stress for the computer user on the thighs (from the chair), arms (from the chair or workstation edge), and palms (from the workstation or the keyboard tray) (Figure 4.3).

The phone is another area of possible ergonomics stress. If phone use is frequent or sustained, a headset might be an option. Normal telephone operation often involves bending the neck for long periods. Avoid cradling your phone, as this tends to cause neck pain and headaches. Figure 4.4 illustrates poor neck and shoulder posture. Hands-free headsets are available and greatly reduce this hazard by allowing the head to remain straight. Figure 4.5 illustrates an example of a hands-free headset that eliminates risk factors.

People are concerned with sitting in an office on a computer all day. Some of their concerns are eyestrain or computer vision syndrome. Eye fatigue or eye strain is among the most common problems experienced. The eyes are focused using muscles that can become overworked. Intensive visual tasks can lead to soreness, temporary blurring, headaches, dryness, and redness. According to the 2015 Digital Eye Strain Report released by the Vision Council, nearly one-third of adults (30%) spend more than half their waking hours (9+) using a digital device. About 72% of adults are unaware of the potential dangers of blue light to eyes. Blue light is light that appears while but is exposing the eye to hidden spikes in intensity at wavelengths within the blue portion of the spectrum. Light-emitting diode (LED) lights and compact fluorescent lamps can emit a high level of blue light, typically the wavelength starting at

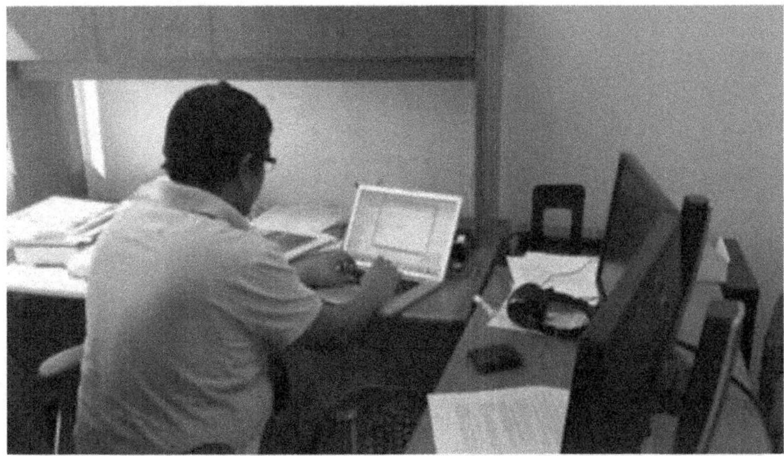

Figure 4.3 Bad workstation creates bad posture – twisted posture (Permission from Raghunath Kanakali)

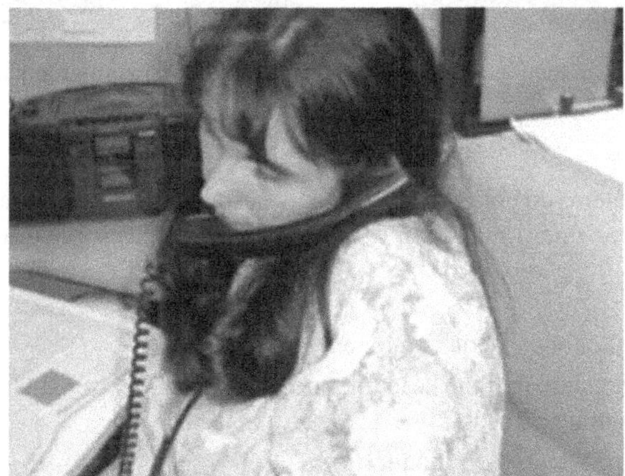

Figure 4.4 Poor neck and posture while cradling the phone (Permission from Cheryl Wilhelmsen)

400 nm, according to the vision council Hindsight is 20/20/20 report of 2015 (Council, 2015).

The symptoms most commonly associated with these digital devices are as follows:

- Eye strain 32.8%
- Neck/shoulder/back pain 32.6%
- Headache 24%
- Blurred vision 23.3%
- Dry eyes 22.8% (Council, 2015).

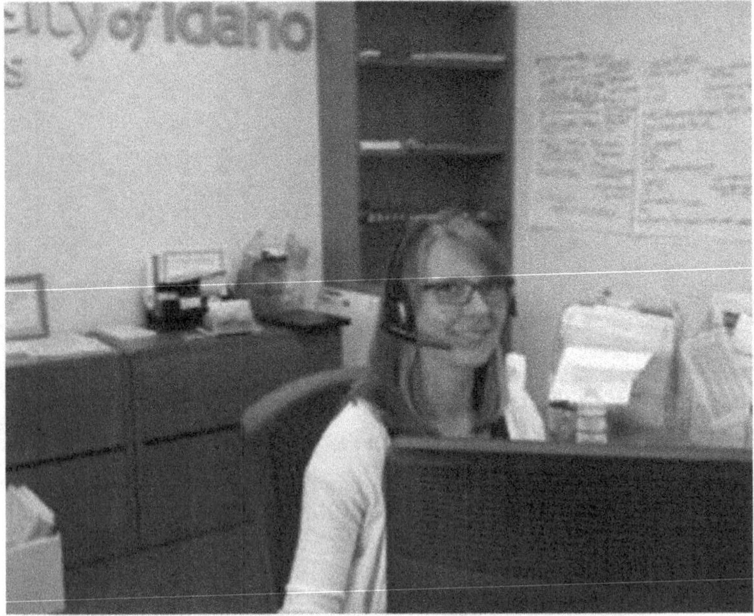

Figure 4.5 Example of a hands-free headset (Permission from Jennifer Flynn)

If these problems do not go away, one may need computer glasses and should visit eye doctor. Some things may be able to do to help alleviate some of these issues are as follows:

- Increase the font size on computer.
  - People tend to pull the head forward to view smaller print that can strain both their vision and neck.
- Correctly position their monitor(s).
- Minimize screen glare.
- Wear the right glasses.
- Take a break.
  - Giving eyes a rest can help reduce eyestrain and fatigue. OSHA recommends workers take a 10-min break for every hour spent on the computer. Focus eyes on objects at farther ranges or simply close them and rest them (OSHA).

Neck and back pain or discomfort in the upper back or shoulders is another common problem associated with working with computer stations. The Administrative Controls chapter provides some great stretching exercises and tips to help reduce neck and back fatigue. Again, the computer workstation needs to be properly set up for each individual.

The impact of digital devices varies with each generation. Kids born between 1997 and 2014 spend more than 3 h a day using digital devices, Millennials born

1981–1996 reported nearly 7 in 10 symptoms of digital eye strain, Gen-X born 1965–1980 reported 63% or 6 in 10 report symptoms of digital eye strain, where the Boomers born 1946–1964 reported 81% own a TV and 26% or one in every four boomers spend at least 9 h on digital devices a day (Council, 2015).

Digital devices are great, they help us connect with others and share and capture information, but users tend to fail to notice how many hours are spent on these devices, which can affect vision health starting at a very young age.

Office Ergonomics: Evaluating

There are several indicators that an evaluation is necessary:

- Person is observed sitting in an awkward posture.
- Elbows are not bent at a 90° angle.
- Wrists are deviated.
- Person is leaning forward.
- Person looks uncomfortable.
- Person's feet are not supported.
- Person has experienced a musculoskeletal injury or illness.
- Person complains of aches and pains.
- Person complains that the workstation is not comfortable.

If any or all of the above are observed, an ergonomic assessment is encouraged.

When conducting an evaluation of an office computer workstation, it is important to start with a good introduction. Employee cooperation is much improved if one conducts evaluations upon request.

It is important to record job factors. What is the employee's job title and what are their responsibilities. In order to identify possible causes or ergonomic stress, they need to know what the person does on a regular basis. We can record tasks as a percentage of the day or as a range of hours (e.g., filing, talking on the phone, photocopying).

If a workstation is shared, it needs to be adjustable and you should explain to the user how to set up the workstation for their use. If the workstation is not shared, you should be able to set it up for the user. Figure 4.6 is a drawing of office furniture location within easy reach of the user.

One should also record how much of the person's time is spent using the keyboard and mouse. Keyboard use of greater than 4 h/day should be thoroughly evaluated. Some people only use a mouse once or twice a day, in which case the mouse can be located almost anywhere. If the mouse is used throughout the day, it is important to evaluate its location and use. What percentage of time the worker spends on each task, the environment and arrangement of the office equipment, and the postures used during these tasks are key factors in arranging the workers' office area. Again, the goal is to minimize the incidents of WMSDs. Figure 4.7 illustrates an example of a device that helps to adjust the workstation.

Filing, writing, or adding machines are other possible areas of ergonomic concern. If the employee uses these or other tools with high frequency, their use should be evaluated.

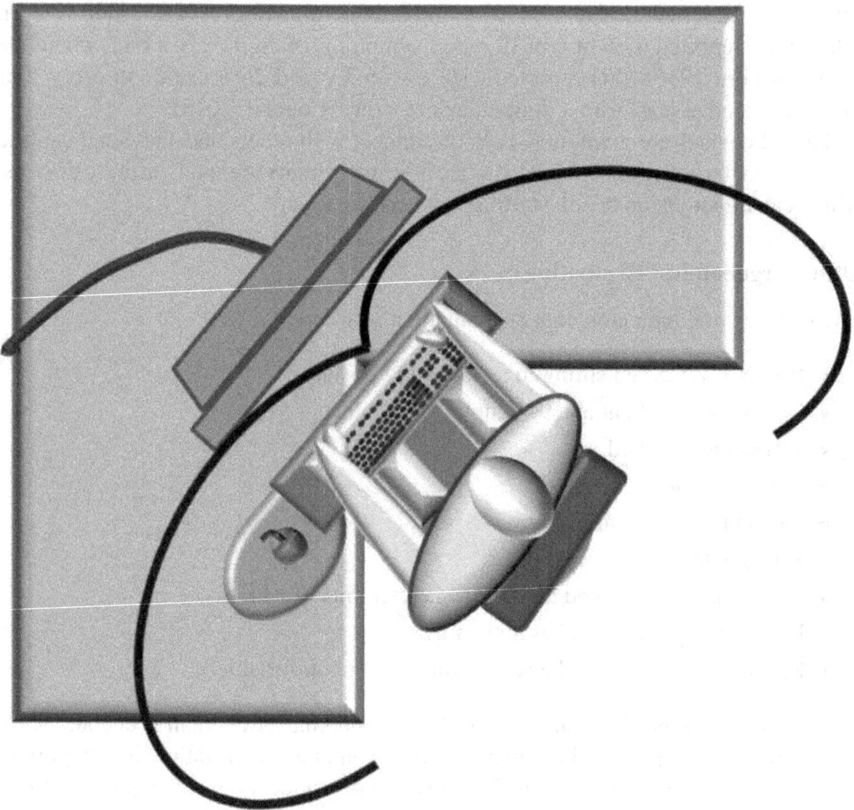

Figure 4.6 Easy reach workstation

WORKSTATIONS

You should record what type of workstation a person is using. One can create a quick drawing with items such as the monitor, keyboard, and phone to help you in making recommendations. Correct adjustment is one of the best methods of preventing employees from developing WMSDs. A fixed-height desk may require an adjustable-height keyboard, whereas an adjustable-height desk may just need proper adjustment. The shape of the desk may determine the type of keyboard tray recommended.

Reach

- The most critical and frequently used items should be placed directly in front of the user and accessible without flexing the body forward or fully extending the arms (i.e., the mouse and keyboard).
- All less frequently used items, (i.e., the telephone or note pads) should be within a comfortable reach zone. For example, with the arms fully extended but not flexing the body forward.

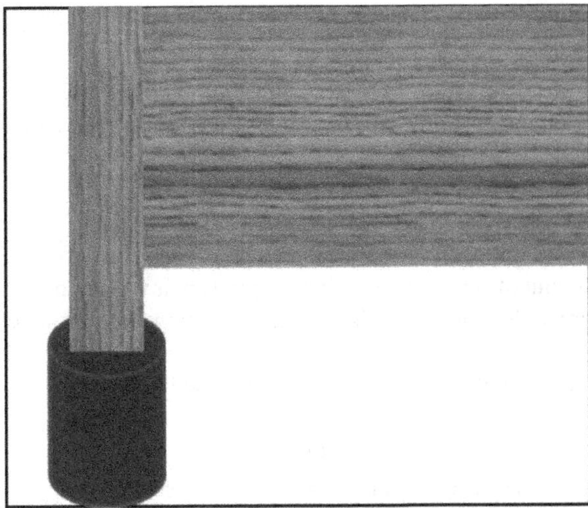

Figure 4.7 Use a deskalator to raise a low surface (Original art by Lee Ostrom)

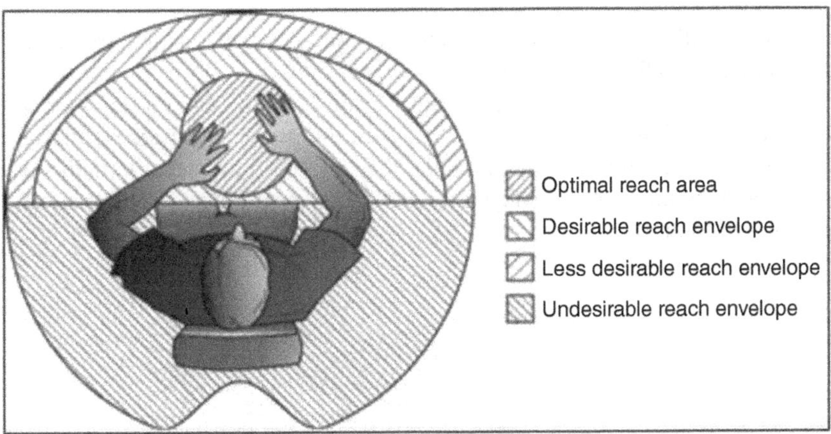

Figure 4.8 Reach envelops while seated

Figure 4.8 illustrates the optimal reach area and the reach envelope that is both desirable and undesirable. If the workstation is adjustable, then you want to adjust the chair height so that the person's feet are resting comfortably on the floor. Thighs should be parallel to the floor with the hips at the same height, or slightly higher than the knees.

Other things to consider are as follows:

- The most commonly used items are accessible.
- Always analyze office layout to see if items can be moved closer to the user.
- Analyze what tasks are being performed and the workflow.

- Remind employees to stand when accessing books stored above the desk.
- Housekeeping.
    - A lack of organization or an abundance of personal items are delicate issues that should be handled with tact suggestions for improvement.

Chairs

Chairs are one of the most important tools in an office. They affect posture, circulation, and the amount of strain on the spine. A good chair will support the back without forcing a posture. Selecting a chair is a bit like selecting a pair of shoes, one size does not fit all.

The following are indications that a chair may be inadequate:

- The person is sitting on the edge of the chair.
- There is excess space around the person on the seat pan.
- The person completely fills up the seat pan.
- The chair is not adjustable.
- The person yells, "I hate my chair!"

Figure 4.9 illustrates an adjustable chair where feet are flat on the floor and hips and knees are parallel to the floor.

One should always have a person's feet resting comfortably on the floor. If the workstation is adjustable, adjust the chair until the person's feet are resting on the floor. If the workstation is not adjustable, they will need a footrest to help support the feet and the knees at 90° when necessary. Figures 4.10 and 4.11 illustrate some examples of footrests.

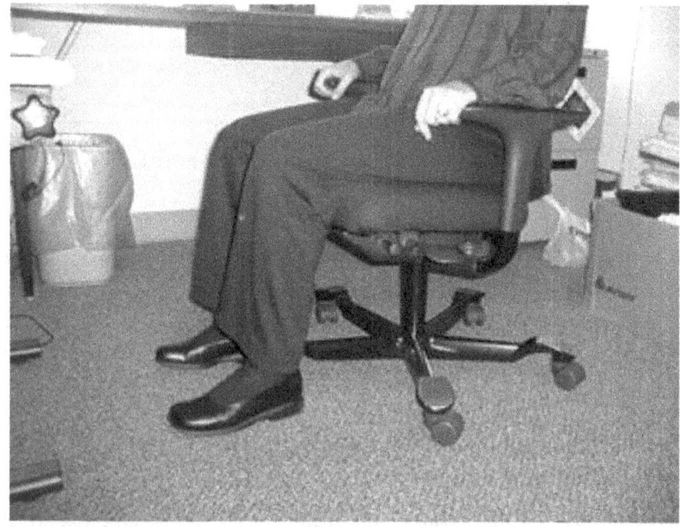

Figure 4.9 Adjustable chair

Figure 4.10 Use a footrest, when feet do not touch the floor (Original artwork by Lee Ostrom)

Figure 4.11 Use an F-ring, when feet do not touch the floor (Original artwork by Lee Ostrom)

There should be 2 in. of thigh clearance under the desk and 2 in. of space between the front edge of the chair and the back of the knee (popliteal area). If there is too much space, the thighs are not fully supported, and the chair may be too small for the user. If the back of the knees hit the front of the chair, the chair may be too big for the user. To remedy this, adjust the backrest of the chair forward or back (or some chairs have a seat pan that adjusts forward and back).

Popliteal height is "the vertical distance from the floor to the side of the knee." The worker sits with the thighs parallel, the feet in line with the thighs, and the knees flexed 90° (National Safety Council, 2012).

- Thighs should be parallel with the floor.
- Hips should be at the same height or slightly higher than the knees.

Remember, we always start the alignment of the worker with the feet flat on the floor. This helps align the rest of the body. The lumbar support should be positioned to support the lumbar area of the back (located in the small of the back above the waistband). The user's arms should be comfortably at his/her sides with the elbows at a 90–110° angle. The wrists should be completely flat to encourage blood flow and nerve conduction.

If the chair has armrests, it should be adjusted so that the arms are at a 90–110° angle without the user having to hunch over with shoulders or back (which may indicate armrests which are too low). Armrests should not prevent the user from getting close enough to the work surface. Oftentimes, armrests can be removed from the chair if they are not needed or interfere with normal tasks. Armrests are useful for resting, particularly if talking on the phone or reading, but should not be used for typing.

The seat pan should be horizontal or tilted gently forward. It is believed that a seat pan tilted slightly forward can help with lower back pain.

Micropostural changes are encouraged throughout the day to reduce static loading and encourage blood flow. Even a perfect chair is not appropriate for extended use.

A common cause of discomfort in the back and upper extremities is an improperly adjusted chair or a chair that is inappropriate for the task. It is very important to train employees on the proper adjustment of their chairs. One should be able to operate it and understand the options and adjustability features to perform their tasks efficiently. In short, the chair should be easy to use.

Because people come in different sizes and shapes, there is no one chair that is perfect for an entire user population. The best approach is to have vendor's loan chair samples and allow the employees to select their own chairs.

When you are considering a new chair, you should look at several different factors:

- Ergonomic factors (comfort and size)
- Adjustability (range and ease)
- Durability (cost, expected life, maintenance, etc.)
- Gliding force (ease of motion), tipping force (stability), and safety of adjustment (controls must not present hazards)
- Other features (armrests, casters, and footrests).

Figure 4.12 illustrates a good working station posture with an adjustable chair.

Ergonomics Guidelines for Office Chair Selection

Chair design contributes to the comfort and productivity of the workers. The chair can be a critical factor in preventing back fatigue as well as improving employee performance and efficiency. People who sit for long periods of time run a high risk of low-back injury, second only to those who lift heavy weights (Demographics, 2014). Management, professional, and office workers accounted for 23% of injuries and illness involving days away from work. The back was the primary body part affected and working position was the second highest source of injuries (National Safety Council, 2012). To reduce this risk, the user must be able to sit and maintain the spine in a neutral posture. A properly designed and adjusted chair is essential to maintaining a neutral posture.

Some manufacturers are eager to label furniture and accessories "ergonomically correct" or "ergonomically designed", much like food products are liberally labeled "all natural" or "new and improved". In reality, a chair that meets the body type of

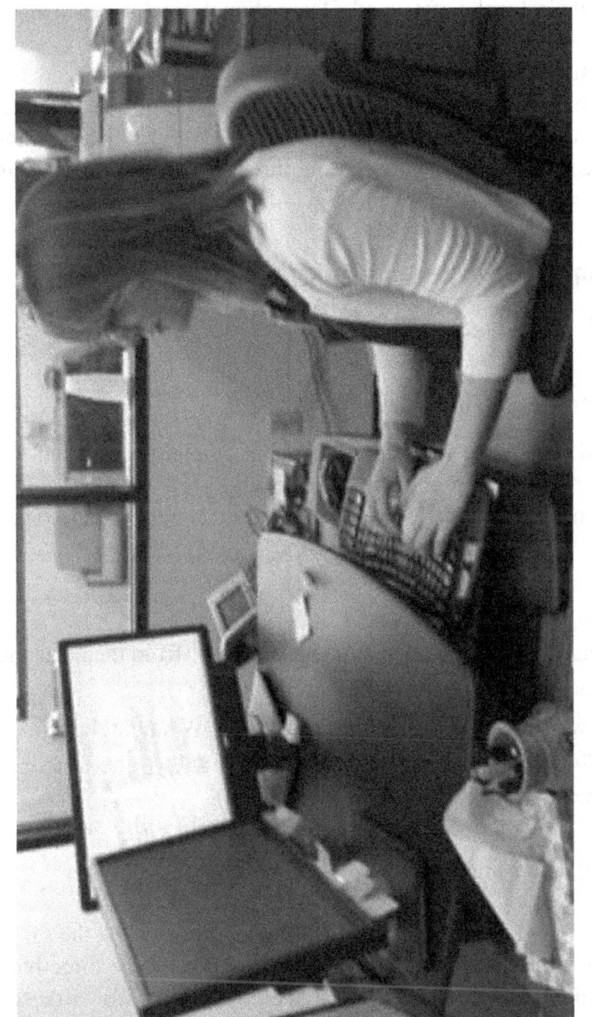

Figure 4.12 Good working posture with an adjustable chair (Permission from Jennifer Flynn)

one person might not fit the next. Therefore, what is "ergonomically correct" for one individual may cause injury to another.

Chair selection is best when based upon personal testing. People vary widely in their shapes and sizes, and manufacturers offer a range of sizes to meet these needs. The following chart contains key criteria to consider in chair selection. All adjustments should be made easily from the seated position.

Workers should use the chair in accordance with manufacturer's instructions and can contact their local Safety and Occupational Health office for additional information.

Note: Alternative seating such as exercise balls, ball chairs, and kneel chairs do not meet the minimum requirements below and are not considered acceptable office seating. The dimensions below are intended to fit 90% of the population, special accommodations may be required for petite or tall individuals (BIFMA, 1998).

Ergonomics Guidelines for Office Chair Selection

See table in Guide Appendix B

Keyboard and Mouse Height

The keyboard and the mouse should be located at a height equivalent to the user's seated elbow height to allow the user to float naturally over the keys. The keyboard height and position affects their posture. The keyboard should be placed in a horizontal position:

- Directly in front of the operator.
- Sloped toward the operator at an angle of up to 15° from the horizontal.
- Mobile so you can shift the keyboard on the desk.
- The home row on the keyboard should be at ELBOW height. It should be located at a height that provides the elbows to be at a 90–110° angle.
- Wrists and hands should be straight (not arched).
- Shoulders relaxed and lined up with the hips.
- Arms relaxed at the sides.

Since we started with the feet, the elbows should already be at the same height as the keyboard (when the back is straight). The mouse should be directly adjacent to the keyboard. Again, the mouse is at the seated elbow height. We will discuss how to measure the seated elbow height in the workstation section below.

The goal for the wrists is to keep them in a neutral posture as much as possible. Figure 4.13 illustrates a few deviations in the wrist and the problems associated with each one.

Wrist supports can help maintain the neutral position; however, they can be misused. Wrist rests are not generally recommended because they are used improperly. People use them as a parking place for their wrists instead of a resting place. In addition, keying while using wrist rests can increase pressure within the Carpal tunnel

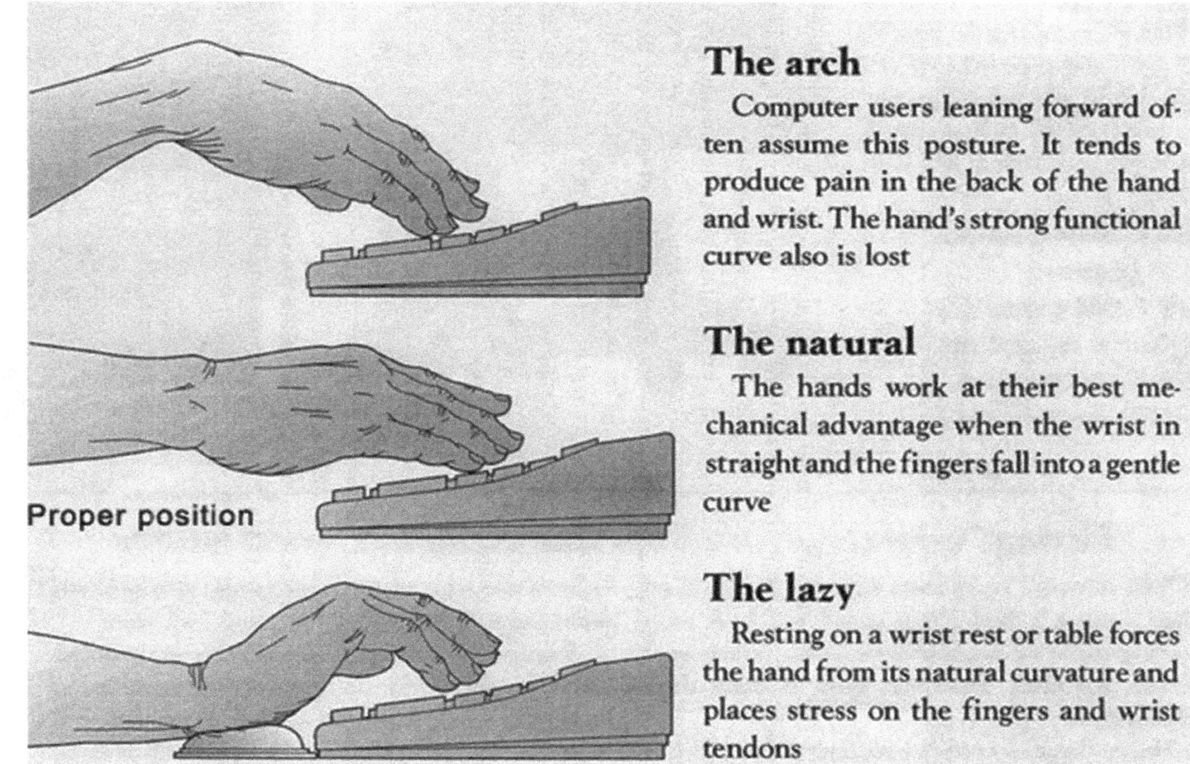

The arch
Computer users leaning forward often assume this posture. It tends to produce pain in the back of the hand and wrist. The hand's strong functional curve also is lost
The natural
The hands work at their best mechanical advantage when the wrist in straight and the fingers fall into a gentle curve
Proper position
The lazy
Resting on a wrist rest or table forces the hand from its natural curvature and places stress on the fingers and wrist tendons

Figure 4.13 Wrist deviations

Figure 4.14 Wrist rest

(up to 140%). If wrist supports are used, training in the proper use should be provided. The soft gel type is preferred over the hard plastic type. Figure 4.14 illustrates a wrist rest.

You can retrofit a fixed-height workstation with an adjustable keyboard tray. Every keyboard tray needs an arm (often purchased separately). This arm should have room for a mouse. A height-adjustable arm is easy to adjust and retract under the desk. Figure 4.15 shows a keyboard tray where the mouse can be slid from right to left. If possible, the keyboard tray should provide a negative tilt, swivel at the desk and the keyboard, and adjust to fit a sit to stand workstation. A keyboard tray with a ratchet mechanism and no knobs or levers can help people with disabilities or CTS. Stability is very important in a keyboard tray.

Mouse

A mouse that is too high can cause discomfort in the neck, shoulder, or arm. Locating the mouse on the desk surface when the person is using a keyboard tray can cause over-reaching and contact stress to the underside of the arm. If there is no room next to the keyboard, a mouse bridge can be placed over the number keys to provide space for operating the mouse. This option is only feasible if the number keys are not used

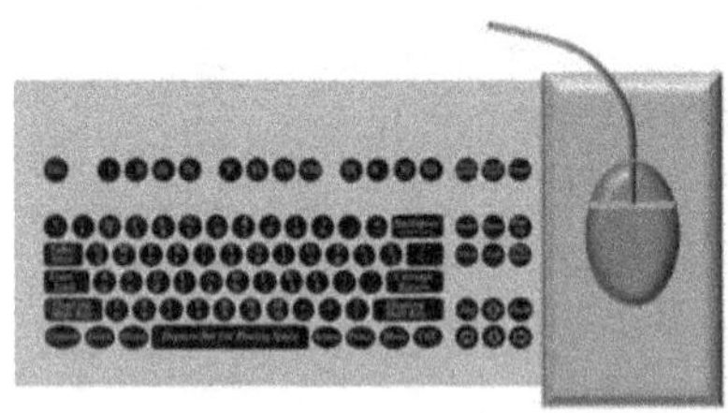

Figure 4.15 Mouse platform (Original artwork by Lee Ostrom)

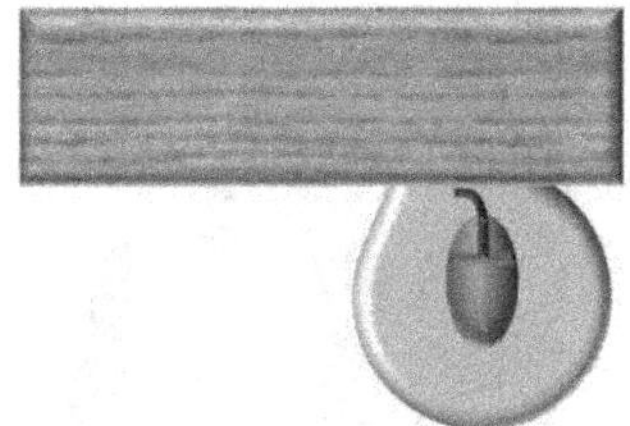

Figure 4.16 Another mouse platform (Original artwork by Lee Ostrom)

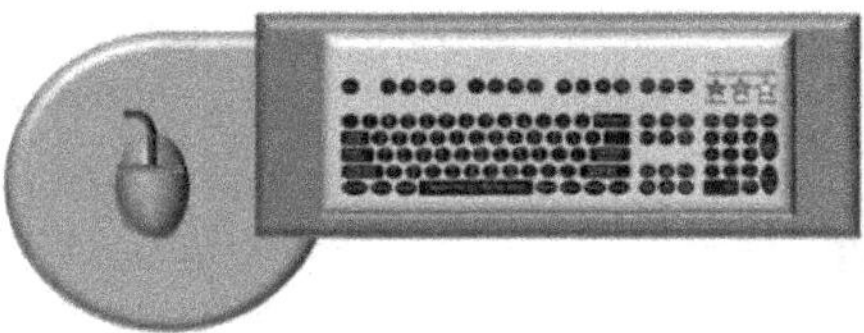

Figure 4.17 Mouse platform that can be attached to the right or left of the keyboard (Original artwork by Lee Ostrom)

frequently. A separate mouse tray can also be attached to the desk. Figures 4.15–4.17 illustrate mouse platforms.

The mouse should be sized for the hand, the user should feel comfortable, and the user should not have to fight with it. There are many sizes and shapes, both right and left handed, available today. You need to find the one that fits you. The mouse should be located at the same level as the keyboard and at a 90–110° angle, close to the body. You want to keep the wrists straight or slightly flexed. Figures 4.18–4.20 illustrate a few examples of input devices.

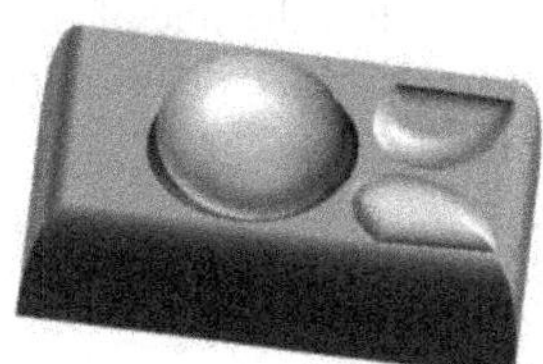

Figure 4.18 Input device – trackball (Original artwork by Lee Ostrom)

Figure 4.19 Input device – mouse (Original artwork by Lee Ostrom)

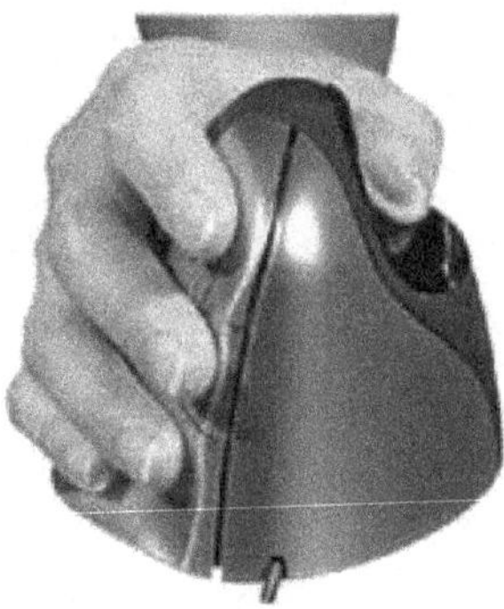

Figure 4.20 Input device – Evolvent™ VerticalMouse 4 (Permission from Evoluent)

MONITOR HEIGHT

The monitor should be located so the top of the viewing area is at or below eye level. Monitor height and distance is different if the user is wearing bifocals. This is discussed in more detail in the measurement section of this chapter.

The monitor should be directly in front of the worker, not to the side. If it is not directly in front, they tend to constantly tilt your head up and down, turning to the side, which may contribute to stiffness in the neck, back pain, and other strains.

The old rule of thumb is to place their monitor at approximately arm's length in front of you. This was before the advancement of the larger monitors. Some monitors are so large that if you followed this rule, it would be too close for viewing. The distance should be dependent on the size of their monitor, but they will know if it is too close or too far back due to either tilting forward to viewing the text or tilting backward because the text is too close. The monitor height should be equal to, or 20° below, the seated eye height in order to maintain a neutral head position. The mid-monitor should be about 4 in. below the line of sight. A monitor placed too high may lead to stiffness in the back or the neck, due to neck extension. Figures 4.21 and 4.22 illustrate examples of monitors.

Ergonomics for Dual Monitors

Many computer workstations have dual monitors as computer programs become even more complicated and LCD monitors become more affordable. Dual monitors are very useful for engineering drafting programs, movie and graphics editing, and emergency response centers. Below are some simple ergonomic setup guidelines for using dual monitors. Figures 4.23 and 4.24 illustrate dual monitor positioning.

Position the primary monitor directly in front of the computer user at approximately an arm's length away or at a comfortable viewing distance.

The primary monitor is the display where work is actually performed. Position the secondary monitor or monitors to the sides at about a 30° angle to the primary monitor. The secondary monitor is used to display computer libraries or items needed on an intermittent basis. Always follow proper ergonomic guidelines for setting up the overall computer workstation.

Figure 4.21 Flat screen monitor (Original artwork by Lee Ostrom)

Figure 4.22 Cathode ray tube (CRT) monitor (Original artwork by Lee Ostrom)

Measurements

If the workstation is adjustable, then after they have adjusted the chair height so the user's feet are resting comfortably, measure the person's seated elbow height. This measurement is taken with the person's hands resting in their lap and elbows at their sides. Measure from the floor to the bony point of the elbow. This is the seated elbow height, as illustrated in Figure 4.25.

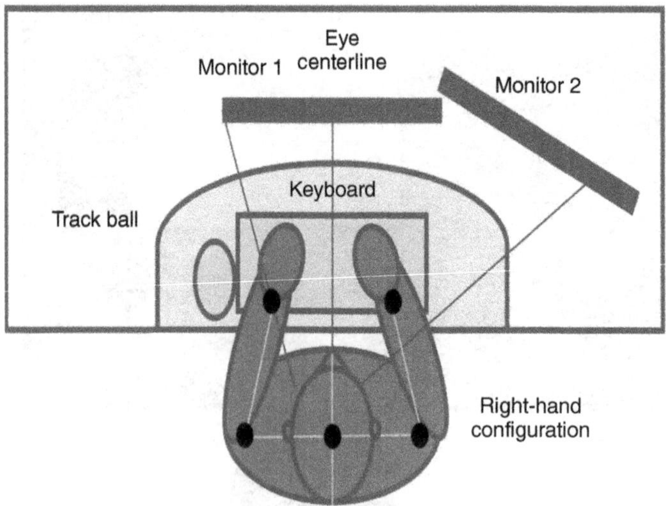

Figure 4.23 A and B: Dual monitor left- and right-hand workstation positioning

Next, we measure from the floor to the middle of the home row of the keyboard. This is the keyboard height. In order to maintain the neutral posture with straight wrists, the keyboard height should be equivalent to the user's seated elbow height. When a person's elbows are at their side in a 90° angle with their wrists and arms straight, the keyboard should be located just underneath the fingertips, and the person should float over the keys.

If the workstation is not adjustable, then the user's chair should be adjusted so that the keyboard height is equivalent to the user's seated elbow height. This may require a footrest if the chair is adjusted up. For taller users, the entire desk may need to be raised up on blocks. A fixed height workstation can often be retrofitted with an adjustable height keyboard and/or monitor arm.

With the user sitting up straight, shoulders relaxed, measure the height from the floor to the corner of the user's eye. The distance from the floor to the top row of characters on the screen is the monitor height. The monitor height should be equivalent to, or 20° below, the seated eye height. The eyes naturally fall down so we can easily read what is below eye height. Anything above eye height requires the user to extend their neck for viewing. Monitor height may need to be adjusted for bifocal wearers depending on which part of the glasses they use for viewing.

In the work setting, sustained or repeated forward head position and trunk flexion increase the risk of developing spinal disorders.

Figures 4.26 and 4.27 illustrate a couple of bad workstation setups.

OTHER ENVIRONMENTAL FACTORS

Lighting

The lighting will affect the monitor placement. The following lists a few guidelines to follow when dealing with lighting:

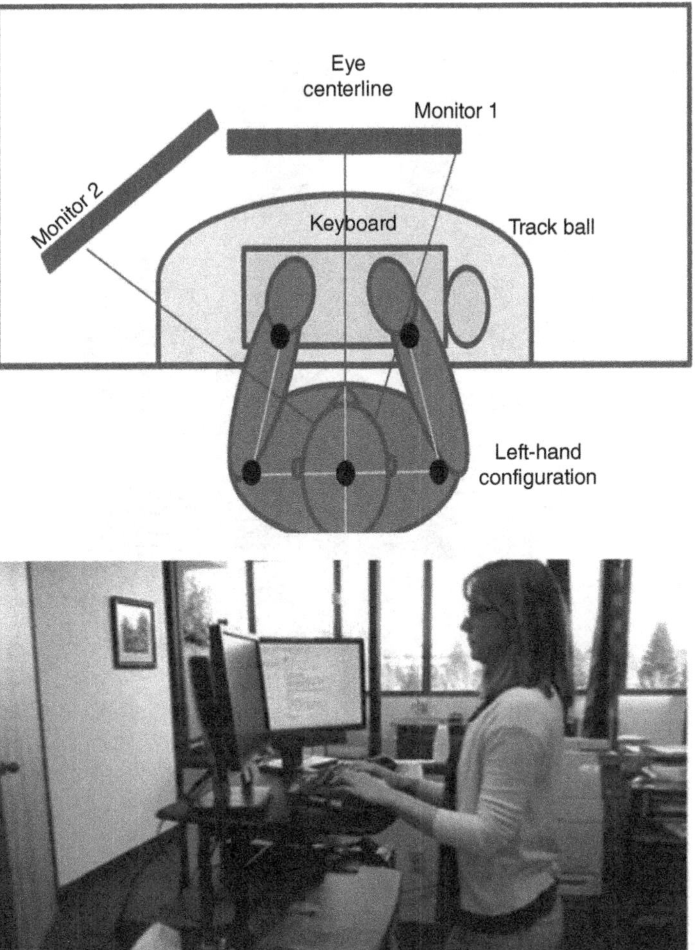

Figure 4.24 Good example of a dual-monitor workstation (Permission from Jennifer Flynn)

- Adequate lighting should always be provided dependent on the task.
- The ambient lighting (which is the general illumination of a room) should be between 300 and 500 lx.
- Task lighting should be provided for difficult-to-read hard copies.
- Care must be taken that the task lighting does not cause glare on the monitor screen (e.g., a lamp at your desk).
- Generally speaking, in an office the monitor is a light; therefore, the ambient lighting can be kept low and task lighting used for specific viewing needs.

Some Useful Definitions

Illuminance – The amount of light falling on a work surface or task from ambient or local light.

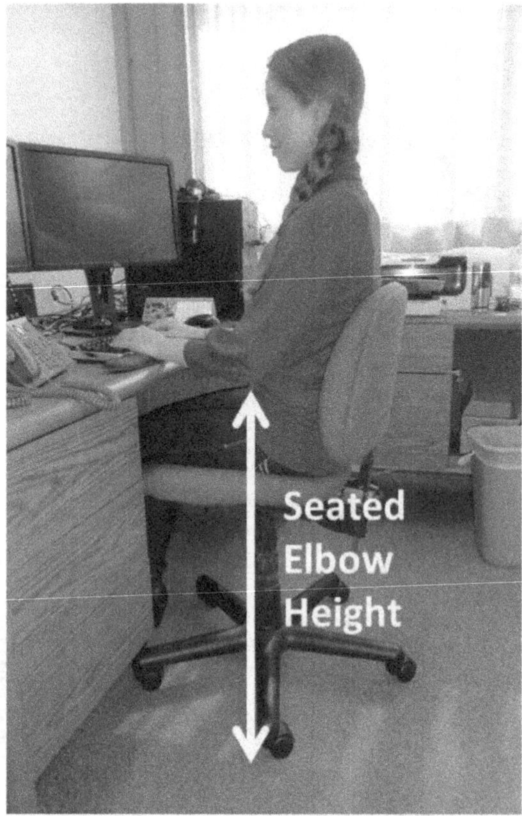

Figure 4.25 Seated elbow height

Luminance – The amount of light leaving a surface, may be emitted or reflected.

Candela – Power emitted by a light source in a particular direction, weighted by the luminosity function (a standardized model of the sensitivity of the human eye to different wavelengths). A common candle emits light with a luminous intensity of roughly 1 cd. The candela is sometimes still called by the old name *candle, such as foot-candle.* Foot-candle is the illuminance on a 1-ft surface of which there is a uniformly distributed flux of 1 lm (lumen). This can be thought of as the amount of light that actually falls on a given surface. The foot-candle is equal to 1 lm/ft^2. One foot-candle ≈ 10.764 lx. The name "foot-candle" conveys, "the illuminance cast on a surface by a 1-cd source 1 ft away" (OSHA).

Lumen – A measure of the power of light perceived by the human eye.

Lux – A measure of the intensity, as perceived by the human eye. 1 lx = 1 lm/m^2. Lux is a derived unit based on lumen, and lumen is a derived unit based on candela.

Contrast – The ratio of the luminance of the object and the luminance of the background.

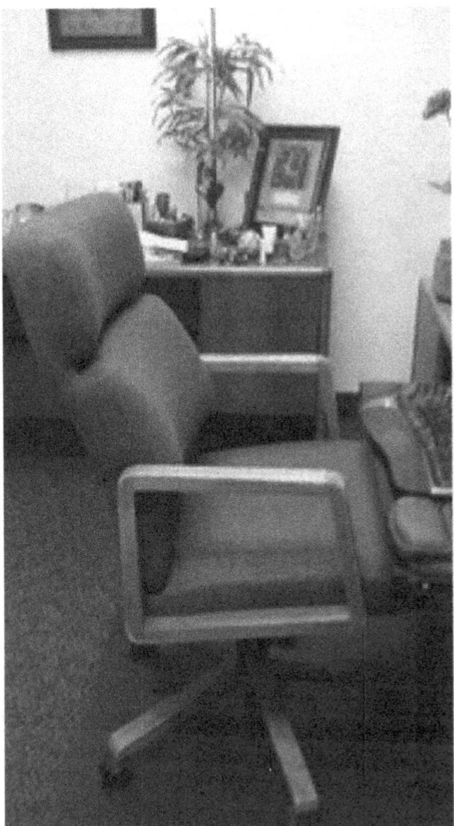

Figure 4.26 Poor chair – no adjustments and stationary armrests

Glare – The sensation produced by luminance within the visual field that is sufficiently greater than the luminance to which the eyes are adapted to cause annoyance, discomfort, or loss of visual performance and visibility.

Direct Glare – Caused by one or more bright sources of light that shine directly into the eyes.

Reflected Glare – Caused by light reflected from an object or objects that an observer is viewing.

Glare

Glare appears as mirror images or white spots on their computer screen. It is the result of bright steady light or reflection, which creates distractions from work. Glare causes the operator's iris to contract and the eyes to squint. The amount of light entering the eye is reduced, causing eyestrain (Grandjean & Kroemer, 1997). Reducing glare can improve the quality of the lighting.

In order to reduce or eliminate glare:

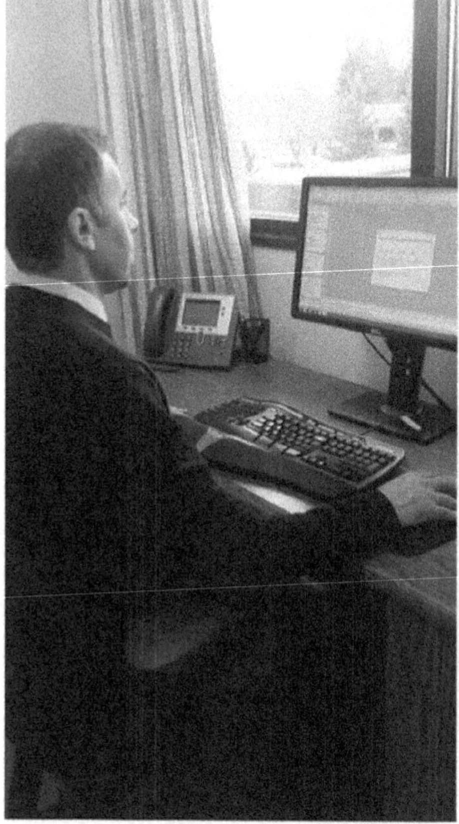

Figure 4.27 Poor workstation (Permission from Alex Vakanski)

- Position monitor at right angles from the windows.
- Focus task lights onto the documents directly, not on the screen or the operator's eyes.
- Install drapes or window coverings.
- Use a glare hood on top of the monitor, which prevents the light from above hitting the screen.
- Use a glare guard or glare screen.

The following evaluation form for task lighting is a useful tool in helping to select the proper task lighting. Use this form as a guide to help you choose the best form of task lighting for your office.

Cornell Task Lighting Evaluation Form (Hedge, 2004)

	Yes	No
1. Does the task light have an asymmetric reflector or lens?		
2. Is the task light dimmable?		
3. Is the task light flicker free?		
4. Does the task light source provide a good color rendering (CRI > 85)?		
5. Is the task light lamp/source long lived?		
6. Is the task light source cold to the touch (e.g., compact fluorescent, LED)?		
7. Is the task light source energy efficient (e.g., compact fluorescent, LED)?		
8. Is the task light easy to move and reposition?		
9. Does the task light stay in place after repositioning?		
10. Does the task light provide adequate glare-free light for your work needs?		
Total		

Noise

Noise is defined as "any disturbing sound" (Grandjean & Kroemer, 1997). The hardest noises to control in large office spaces are printers, copiers, and other workers' conversations. Many office areas today are cubicles, which do not block noise. The optimum noise level should be below 55 dB. There are several suggestions in reducing noise in the office:

- Turn down telephone ringers.
- Hold casual conversations and meetings in separate rooms.
- When on the phone try to direct voice away from coworkers.
- Sound-absorbing hoods can be provided for printers, which can deaden the noise emission by about 20 dB (Grandjean & Kroemer, 1997).

OFFICE ERGONOMICS: ADMINISTRATIVE CONTROLS

Unlike engineering controls that are physical changes to the office workspace, administrative controls are procedures and practices that limit exposure to stressors by control or manipulation of work schedule, or the manner in which work is performed.

Eleven tips for an ergonomic computer workstation are listed below (Hedge, 2014):

1. Use a chair with a dynamic chair back and sit back.
2. Top of monitor at eye level.
3. No glare on screen, relocate monitor or use shades to control glare.
4. Feet on floor or stable footrest.
5. Use a document holder, preferably in-line with the computer screen.
6. Wrists flat and straight in relation to forearms to use keyboard/mouse/input device.
7. Arms and elbows relaxed close to body.
8. Keyboard set equal to seated elbow height with feet flat on the floor.
9. Center monitor and keyboard in front of you.
10. Use a stable work surface and stable (no bounce) keyboard tray.
11. Take frequent short breaks (microbreaks).

What About Using Laptop Computers?

Laptops have become more affordable and many people have turned to using them in place of the traditional home or office computers. Because of the design of the laptops, it actually violates a basic ergonomic requirement for a computer, namely that the keyboard and screen are separated. When the personal computing desktop devices were integrated (the screen and keyboard) into a single unit, it resulted in widespread complaints of musculoskeletal discomfort. By the late 1970s, a number of ergonomics design guidelines were written, and all called for the separation of screen and keyboard. The reason was simple. With a fixed design, if the keyboard is in an optimal position for the user, the screen is not, and if the screen is optimal, the keyboard is not. Consequently, laptops are excluded from current ergonomic design requirements because none of the designs satisfy this basic need. This means that one need to pay special attention to how use their laptop because it can cause problems. They need to ask oneself what type of laptop user am I? Do I use it occasionally for short periods, or do I use it all of the time as my computer? Full-time users may experience more problems.

The posture when using a laptop is a trade-off between poor neck/head posture and poor hand/wrist posture. For the occasional user, because the neck/head position is determined by the actions of large muscles, you are better off sacrificing neck posture rather than wrist posture. There are a few things one can do to help the posture:

- Find a chair that is comfortable and that one can sit back in.
- Position laptop in their lap for the most neutral wrist posture that one can achieve.
- Angle the laptop screen so that one can see this with the least amount of neck deviation.
- If one use the laptop computer as your full-time device, then these are a few things can do to help alleviate the bad posture:
  - Position the laptop on their desk/work surface in front so that one can see the screen without bending your neck. This may require that you elevate the

laptop off the desk surface using a stable support surface, such as a computer monitor pedestal.

- Use a separate keyboard and mouse connected directly to the laptop or docking station.
- Use the keyboard on a negative-tilt keyboard tray to ensure wrist is in a neutral posture.
- Use the mouse on an adjustable position mouse platform (Hedge, 2004).

Many laptops offer large screens (15 in. plus) and can work as desktop replacements (giving the viewing area of a 17 in. monitor). However, think about where you will most use laptop to help you choose the best size. The larger the screen, the more difficult it will be to use this in mobile locations (e.g., airplane, car, train). On the other hand, the smaller the screen, the harder it is to see the text and this may cause eyestrain. One also need to consider the weight of the device and the accessories you need, such as the power supply, spare battery, external disk drive, zip drive, CD_RW. Many lightweight portables can become as heavy as regular laptops when you add the weight of all of the components together. A good rule of thumb is, if laptop and the components weigh 10 lb or more, then they should certainly consider using a carry-on bag with wheels. If they want a smaller bag and can comfortably carry laptop, consider a good shoulder bag. Figure 4.28 illustrates a good example of proper posture when using a laptop computer.

SELF-GUIDE FOR CONDUCTING AN OFFICE ERGONOMICS ASSESSMENT

General Information

When conducting an office computer workstation evaluation, it is important to start with a good introduction. One should always start by introducing and asking the

Figure 4.28 Proper posture for using a workstation (Permission from Raghunath Kanakali)

employees if they know what ergonomics is and why an evaluation is being conducted. This is a good place to use they succinct definition. A computer workstation evaluation checklist can be found in the guide Appendix B.

Ask the employees if they are having any problems or concerns that prompted them to request an evaluation.

Always ask the employees if they are experiencing any pain or discomfort. If you know where they are hurting, you can better assess what part of the workplace may be causing or aggravating their symptoms. Unless you are a physician, you should never try to diagnose an employee or cast doubt on a diagnosis they have already received.

One should always make sure the supervisor is aware of an evaluation before you enter the work area. Their evaluation can always be used in legal proceedings, please watch what and how you record.

One should never assert that this workstation caused a person's medical conditions. It may put them at risk of WMSDs or aggravate an MSD, but it is very difficult to substantiate a direct cause-and-effect relationship. If the employee gives personal information or asks you not to disclose a medical condition, do not record it. Always be courteous, establish a good rapport, listen well, and take good notes.

Step One: Workstation

For notes, one should record what type of workstation a person is using. A quick drawing, noting where items such as the monitor, keyboard, and phone are located can help when they are writing their recommendations. A fixed-height desk may require an adjustable-height keyboard, whereas an adjustable-height desk may just need proper adjustment. The types of questions one should ask are as follows:

- What type of workstation does the employee have?
- Is it a fixed-height or an adjustable-height workstation?
- Is it a corner desk, an L-shaped desk, or a straight desk?
- Where are the keyboard, mouse, and monitor located?
- Where is the lighting?
- Where are the most commonly used items? (Telephone, reference books, etc.)

Step Two: Adjust the Chair

If the workstation or the keyboard tray is adjustable, start at the bottom with the person's feet. Correct adjustment is one of the best methods of preventing employees from developing WMSDs.

- Adjust the **seat height**, so the user's feet rest comfortably on the floor.
- Thighs parallel to the floor.
- Hips at the same height, or slightly higher than the knees.
- If the **seat pan** depth is adjustable, it should be adjusted to allow for 2–3 fingers to fit between the edge of the chair and the backs of the knees.
- If the **seat pan** tilts, it should be horizontal or gently tilted forward.

- The seat **backrest** should be adjusted so that the lumbar area of the back is supported (retain natural curve).
- If the **armrests** can be adjusted up and down, they should support the arms at a 90–110° angle without hunching the shoulders or bending the back.
- If the **armrests** get in the way, remove them or put them all the way down.
- The chair should have a stable five caster base with appropriate casters for the floor.
- The chair should be large enough to allow the worker to make micropostural changes throughout the day.

Step Three: Work Surface

- Is there adequate clearance beneath the workstation for the user to get close enough to the task, and not contact obstructions?
- Are the monitor and keyboard in alignment (directly in front of) the user?

If the workstation is adjustable, after they have adjusted the chair height so the user's feet are resting comfortably, measure the person's **seated elbow height**. This measurement is taken with the person's hands resting in their lap and elbows at their sides. Measure from the floor to the bony point of the elbow. Figure 4.29 illustrates the seated elbow height.

Next, measure from the floor to the middle of the home row of the keyboard. This is the **keyboard height**. In order to maintain the neutral posture with straight wrists, the keyboard height should be equivalent to the user's seated elbow height. When a person's elbows are at their side in a 90° angle with their wrists and arms straight, the keyboard should be located just underneath the fingertips and the person's wrist should float over the keys. The mouse should be located at the same height as the keyboard and adjacent to the keyboard. Figure 4.30 illustrates the seated keyboard height.

If the workstation is not adjustable, then the user's chair should be adjusted so the keyboard height is equivalent to the user's seated elbow height. This may require

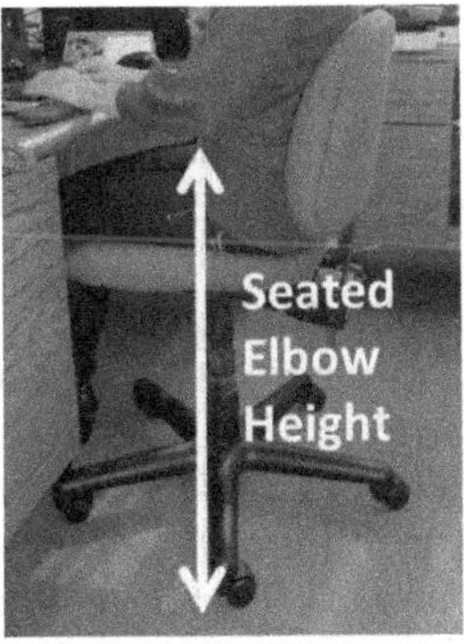

Figure 4.29 Seated elbow height

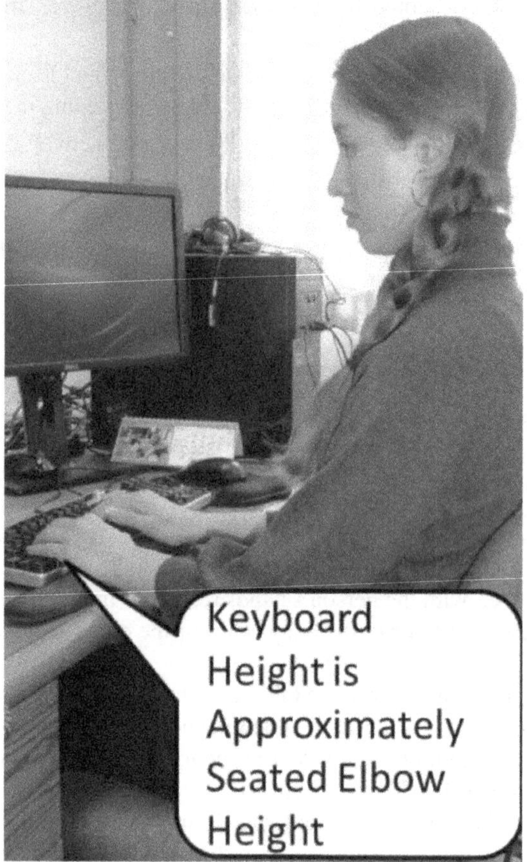
Keyboard
Height is
Approximately
Seated Elbow
Height

Figure 4.30 Elbow height = keyboard height

a footrest if the chair is adjusted up. For taller users, the entire desk may need to be raised up on blocks. A fixed-height workstation can often be retrofitted with an adjustable-height keyboard and/or monitor arm.

If the monitor is also on a fixed-height desk, then adjust the monitor after the chair height is adjusted so that **Monitor Height = Seated Eye Height** (or slightly below), if corrective vision is not used (You can remove the CPU or other items to decrease the height or add the CPU, monitor risers, monitor arm, or phone books, etc., to increase the height).

With the user sitting up straight, shoulders relaxed, measure the height from the floor to the corner of the user's eye. This is the user's **seated eye height**.

The distance from the floor to the top row of characters on the screen is the monitor height. The monitor height should be equivalent to or 20° below the seated eye height. The eyes naturally fall down so we can easily read what is below eye height. Anything above eye height requires the user to extend their neck for viewing. Monitor height may need to be adjusted for bifocal wearers depending on which part of the glasses they use for viewing. Figure 4.31 illustrates the monitor height.

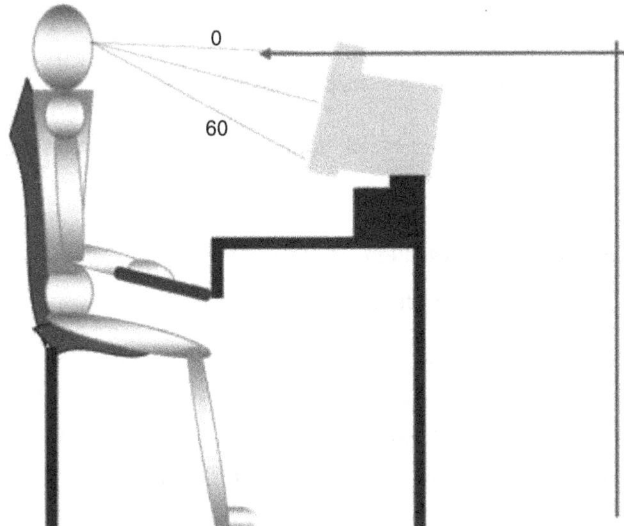

Figure 4.31 Monitor height = seated eye height

Step Four: Other Equipment

Telephone headsets are recommended for employees who are on the phone for extended periods or receive frequent calls. They are important for people who write or use the computer while talking on the phone. These people tend to cradle the phone with an awkward neck posture. Shoulder rests are less expensive but still cause an awkward neck posture and are not recommended for heavy phone use.

Document holders are useful for people that type from documents. Documents placed flat on the desk require the user to repeatedly glance from the horizontal surface to the monitor, which creates repetitive neck motions. The document should be placed at the same distance as the monitor as close as possible, unless the font is small.

Depending on the employee's tasks, certain items may decrease repetitions or improve posture:

- Ergonomic staple remover
- Automatic stapler or hole punch
- Box cutter
- Ergonomic pens
- "Y" connector for two input devices.

Step Five: Administrative or Behavioral Improvements

Some vision-related work habits to keep in mind:

- 20/20 Rule – refocus eyes for 20 s every 20 min.
- Blink/refocus eyes regularly.

- Keep the monitor screen clean (most dry/itchy eyes are a result of dust on the computer monitor).
- Regular vision care can help reduce eye strain.

It is very important to get the blood circulating and muscles moving throughout the day.

Step Six: Conclusion of Assessment

Make sure the employee knows the following:

- Where to go for additional information?
- How to get in touch with questions?
- Where to report pain or discomfort associated with his/her job?
- What to expect from the evaluation (report, follow-up, etc.)?

Thank the employee for their time.

What about Sit–Stand Workstations?

Cornell University conducted a study with participants using height-adjustable workstations. The results of this study suggested there were substantial decreases in the severity of upper body musculoskeletal disorder symptoms. One of the problems with standing workstations is when they raise the keyboard and mouse they also need to raise the height of the monitor or you start experiencing neck flexion discomfort. It seems there is also a greater wrist extension when standing, so people start to lean compromising their wrist postures. The overall conclusion is sit–stand workstations are cost–effective. It is best to sit while performing computer work using a properly set up workstation. The best thing is to stand and move for 2 min, every 20 min. Standing alone does not help, they need to move. The key according to Cornell is to build frequent movement variety into their normal workday (Hedge, 2004).

KEY POINTS

- One major problem with computer workstations is they keep an operator at that workstation in virtually the same posture all day long.
- The goal is to prevent musculoskeletal disorders by surveying the workplace.

REVIEW QUESTIONS

1. At what part of the body do you start your assessment?
2. How would you describe a neutral seated posture?
3. How do you measure the seated elbow height?

REFERENCES

BIFMA (1998). International Organization for Standardization 9241-5.

Case and Demographics. (2014). 2013 Nonfatal Occupational Injuries and Illnesses: Cases with Days Away from Work.

Grandjean, E., and Kroemer, K. (1997). Fitting the Task to the Human, *A Textbook of Occupational Ergonomics* 5th edn. CRC Press.

Hedge, A. (2004). *Effects of an Electric Height-Adjustable Worksurface on Self-Assessed Musculoskeletal Discomfort and Productivity in Computer Workers.* Cornell University Human Factors and Ergonomics Research Laboratory.

Hedge A. (December 06, 2014) *5 Tips for Using a Laptop Computer*, Cornell University.

The Vision Council (2015). Hindsight is 20/20/20: Protect Your Eyes from Digital Devices.

ADDITIONAL SOURCES

Andersson, G. B. (1981). "Epidemiological aspects of low back pain in industry" *Spine*, 6(1).

OSHA.

5

ADMINISTRATIVE CONTROLS AND STRETCH AND FLEX PROGRAM

LEARNING OBJECTIVE

The students will recognize the various administrative controls, their advantages, limitations, and several policies regarding the reduction of work place hazards.

INTRODUCTION

Administrative controls are procedures and practices that limit exposure by control or manipulation of work schedule or the manner in which work is performed. An example would be to rotate their workers so no one person is doing all of the heavy lifting tasks. Administrative controls reduce the exposure to ergonomic stressors and thus reduce the cumulative dose to any one worker.

If they are unable to alter the job or work place to reduce the physical stressors, administrative controls can be used to reduce the strain and stress on the work force. Administrative controls are most effective when used in combination with engineering controls.

Occupational Ergonomics: A Practical Approach, First Edition.
Theresa Stack, Lee T. Ostrom and Cheryl A. Wilhelmsen.

HIERARCHY OF HAZARD MITIGATION

Occupational Safety and Health Instruction (OSHA) lists the preferred priorities for corrective actions of ergonomic risk factors, which include the following:

- Ergonomic risk elimination
- Engineering controls
- Substitution of materials/tools/equipment
- Improved work practices
- Administrative controls.

The instructions also state that effective design or redesign of a task or workstation is the preferred method of preventing and controlling harmful stresses.

The methods of intervention (in order of priority) are listed in Table 5.1.

Solutions to Controlling Hazards

Interventions have included making changes in the work place by modifying existing equipment, purchasing new tools, or other devices to assist in the production process. Simple, low-cost solutions are often available to solve problems.

Making changes in work practices and policies is another solution that falls under administrative controls. Making these changes can reduce physical demands, eliminate unnecessary movements, lower injury rates and their associated workers' compensation costs, and reduce employee turnover, according to OSHA (OSHA, n.d.). In many cases, work efficiency and productivity will increase as well.

TABLE 5.1 Levels of Hazard Mitigation

Levels of Hazard Control
1. **Elimination** – A redesign or procedural change that eliminates exposure to an ergonomic risk hazard; for example, using a remotely operated soil compactor to eliminate vibration exposure
2. **Engineering controls** – A physical change to the work place; for example, lowering the unload height of a conveyor
3. **Substitution** – An approach that uses tools/material/equipment with lower risk; for example, replacing an impact wrench with a lower vibration model
4. **Administrative** – This approach is used when none of the above can be used or are impractical to implement. Administrative controls are procedures and practices that limit exposure by control or manipulation of work schedule or the manner in which work is performed. Administrative controls reduce the exposure to ergonomic stressors and thus reduce the cumulative dose to any one worker. If they are unable to alter the job or work place to reduce the physical stressors, administrative controls can be used to reduce the strain and stress on the work force. Administrative controls are most effective when used in combination with other control methods; for example, requiring two people to perform a lift

Administrative controls when dealing with hazardous chemicals may include the following:

- Microscaling the size of the experiment to reduce the amount of chemical usage
- Substituting in less hazardous chemicals such as using toluene instead of benzene
- Isolating or enclosing an experiment within a closed system such as a glove box
- Requiring all laboratory personnel have been provided with adequate training to perform their work safely
- Restricting access to areas in which certain hazardous chemicals are used
- Posting signs identifying specific hazards
- Requiring standard practices for chemical safety be observed
- Maintaining good housekeeping practices at all times in the laboratory.

How Can We Reduce Injury?

If we can curtail fatigue, we can reduce the probability of injury. Let us look at a typical injury progression triangle in Figure 5.1.

If we surveyed the work force, we would likely see a large percentage with fatigue at the end of the day, with fewer folks saying they were to the point of pain. If we can reduce the levels of those leaving work feeling fatigued, then we can reduce the discomfort level. If the discomfort level is reached, typically they rest or take anti-inflammatory medication. Figure 5.2 illustrates the cumulative effects of fatigue.

After the discomfort stage, we fall into the pain stage. If we curtail the percentage of those in the discomfort stage, then fewer folks would experience pure pain, which is the feeling that keeps us from performing an activity.

Finally, injury would not be as common as fatigue or discomfort by curtailing fatigue in stage one. As we see, we automatically reduce the probability of injury by breaking the cycle.

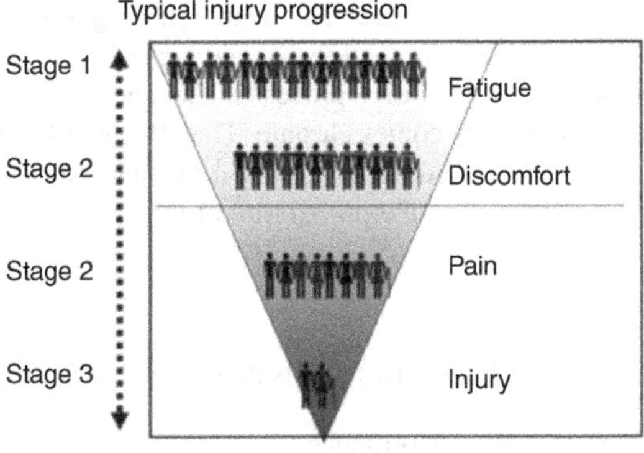

Figure 5.1 Injury progression triangle

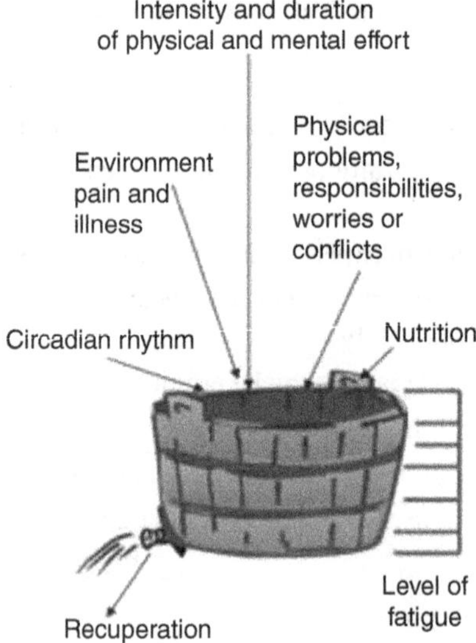

Figure 5.2 Cumulative effects of fatigue

Muscle Fatigue Physiology – Localized Muscle Fatigue

Muscle fatigue and discomfort is familiar to all of us. Physiologists use fatigue to define the result of energy consumption and loss of contractile function of the muscle. There are two types of fatigue: acute and chronic.

Acute versus Chronic Fatigue Acute fatigue has a recent onset and is temporary in duration. It is usually related to excessive physical activity, lack of exercise, insufficient rest or sleep, poor diet, dehydration, increase in activity, or other environmental factors. Acute fatigue can be a protective body function, alerting a person to rest. It is anticipated to end in the near future, with interventions such as rest or sleep, exercise, and a balanced diet. Chronic fatigue persists, and recovery is not quickly anticipated (Palliative Care, n.d.).

Localized muscle fatigue is a result of the accumulation of waste products such as lactate and results in an oxygen debt, which ultimately leads to fatigue. Figure 5.2 shows a good representation of fatigue. The degree of fatigue is an aggregate of all the different stressors of the day, both physical and mental.

To maintain health, the recuperative process must cancel out stress. When we do not rest or recuperate, fatigue becomes chronic. The effects of fatigue depend on the extent of fatigue, but in short, fatigue is usually a leading indicator to injury. If we can curtail fatigue, we can reduce the probability of injury.

Mental Fatigue

The problem with allowing fatigue to set in is that it does the following:

- Degrades physical work performance
- Causes workers to perform at a slower rate

- May hinder one's ability to perform certain exertions
- May cause one to suffer from tremor and other symptoms that interfere with precision
- Degrades mental capacity
- Increases error rate.

Stopping fatigue stops the progression of injury. If engineering controls are not feasible, then an administrative control is put into place. The best combination to reduce the dose is to use both engineering controls and administrative controls. Duration is a contributing risk factor that seems to be the constant factor in the development of WMSDs.

ADMINISTRATIVE CONTROLS

Job Rotation or Alternate Work Activity

There are many ways to reduce exposure in the work place. The specific goal, in the context of ergonomics, is to use different muscle groups, thus giving specific body parts a chance to rest and recover. Alternating sitting and standing is an easy method for increasing blood flow. For example, if an individual is bending and squatting using the leg and back muscles, consider rotating to an activity that uses the upper part of the body. This allows the employee to continue working but gives the lower body and back a rest.

If an engineering control cannot be put into place to alter the job or work place, then putting an administrative control in place can reduce strain by reducing exposure to a certain muscle/body group.

If possible, rotate the work so the heavy work is performed in the morning and the lighter tasks are performed in the later part of the day. Use the big muscle groups in the morning, such as unloading a truck, and the lighter tasks, such as completing paperwork in the afternoon.

A good example is a lifeboat certification and packing operation.

When life rafts arrive in the shop, they are deployed from their pods, the survival gear is removed and restocked, the compressed air bottle is checked and refilled, and the life rafts are checked and certified. To certify the rafts, they are deployed (inflated) and must hold a certain pressure for a certain period. Once they are tested, the rafts are deflated and survival gear bags are restocked and put into the rafts along with the compressed air bottle. The rafts are then folded and put back into the pods. This process is completed daily.

Figure 5.3 shows the posture of the workers as they work with the pod. As you can see, they work on the ground and struggle with putting the contents back into the pod. The workers are very fatigued and in pain at the end of the day.

Administrative controls were used by rotating the activities. Instead of performing all of the tasks daily on Mondays, Wednesdays, and Fridays, they packed the life rafts, which took 30–40 min for each pod. Tuesdays and Thursdays, they refilled the survival bags and completed the certifications. This gave them a break from the exhausting task of certifying the life rafts.

Figure 5.3 Repacking the life raft into containment pod, worker performs the task M, W, F; to rest on Tuesdays and Thursdays – this is an example of work rotation

Engineering controls were also put into place such as using a pod cart so the life rafts are simply transferred and never lifted or carried. A compressed gas cylinder transportation cart eliminated the need to carry the cylinders to the refilling area as shown in Figure 5.4. Using an elevated table for the certifications and repacking to reduce the awkward postures of being low to the floor is shown in Figure 5.5. Carpet was put on the surface to protect the knees. In this example, they used both

Figure 5.4 Life raft-compressed air cart eliminates carrying bottles

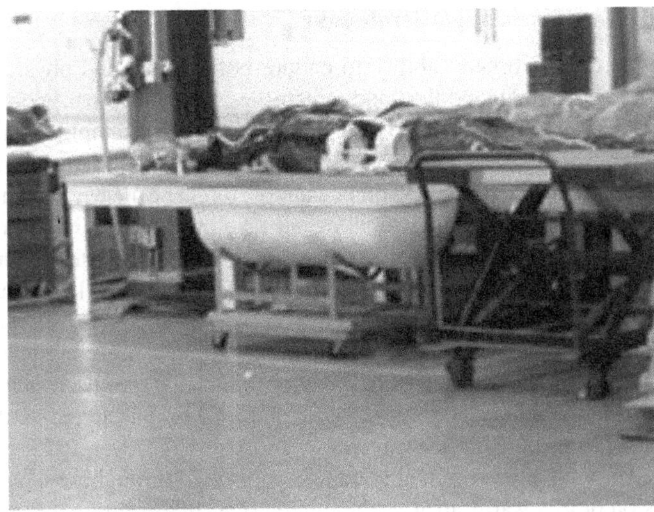

Figure 5.5 Life raft shop-elevated worktable and pod cart reduces bending to pack life rafts

engineering and administrative controls; and with new equipment, there have been no injuries reported to date.

Training/Worker Education

Provide information on proper body positioning (neutral posture), or proper lifting techniques to the workers. Educate them on how to identify WMSD risk factors, signs, and symptoms before an injury occurs. Training requires constant reinforcement to be effective in changing behaviors.

Stretch and Flex Programs

Research has shown the frequency of rest pauses determines their effectiveness; the higher the frequency, the more effective the rest becomes. Higher frequency does not necessarily mean a long time. Encouraging stretching increases blood flow and whisks away waste products. It increases flexibility, and they are less likely to injure a warm muscle. They can encourage people in various ways and by using different methods.

Many ergonomists believe that at least a 5-min rest is required for the maximum benefit. The break schedule is generally 50–55 min of work, and then a 5-min rest. Of course, this all depends on the type of task; if the task is very demanding, then more breaks are required.

Scientific research has shown that stretching can alter viscoelastic properties by decreasing stiffness and increasing tissue compliance of the musculotendinous unit leading to reduced risk of injury (Palliative Care, n.d.).

Stretching can lead to less fatigue and discomfort, which results in improved productivity. Work place stretching programs are most effective with sedentary tasks or as warm-ups to dynamic tasks. Stretching programs need to be part of an overall ergonomics program to be effective.

Stress Reduction/Stretching/Microbreaks

Stretching increases workers' ability to endure both mental and physical stresses. Stretching increases the blood flow and replenishes nutrients in our bodies.

Microbreaks can be implemented throughout the day. An example might be a community printer, which makes someone get out of his or her chair, which is a rest break from sitting, to access the printer. Another example might be the worker simply stands up when using the phone periodically. We all take brief pauses in work and these pauses can be energy pauses, as in doing something such as stretching, which aids in recovery by re-oxygenating the muscles. An individual is less likely to injure a warm muscle.

Microbreak stretching can reduce fatigue due to sustained positioning. It is difficult to maintain proper body mechanics in many positions due to muscle ache or strain. Sustained positions reduce the oxygen and blood flow to muscle tissue. Alternating positions or tasks throughout the day allows muscles to recover and perform more efficiently (Training, n.d.). Stretching can be performed throughout the workday to relieve tension or as a warm-up at the beginning of a shift or difficult task/lift.

Olympic athletes do not start an event without warming up first. It is a good idea to begin work day with some sort of warm-up program. Preshift warm-up programs can increase the temperature of muscles, increase healthy blood flow, and improve muscle coordination. Stretching increases a joint's ability to move through a greater range of motion, therefore, reducing the risk of musculoskeletal injury (Henry, 2013). In addition, balance programs can reduce the risk of falls in the work place.

A good example is the Ergo Joe Stretch n' Flex at Puget Sound Naval Shipyard (PSNS) 1996–2006.

PSNS was one of the first to implement a stretch and flex program. They started with the comparison of two dry docks, one with a stretch program and one without.

What is important to remember is that in 1996 PSNS' major mission was decommissioning vessels. This was done with cutting torches and a reciprocating saw (i.e., Sawzall). They cut the ships up into pieces and disposed of them. The manual material handling tasks were extraordinarily physically demanding, day in and out.

When they compared the dry dock over a few months period, they could see reduced injury rates in those areas with the stretch and flex program. The program was expanded Navy wide, and from 1996 to 2000 they had a 20% reduction in back injuries and a 50% participation rate. The workers were required to muster but not required to participate in the stretching program. Figure 5.6 illustrates an example used in a flex program.

Controls for Temperature Extremes

Administrative controls are a method of reducing exposure by changing the way work is performed or scheduled.

- Wear proper attire (especially the feet, hands, and head).
- Ensure proper hydration – NIOSH recommends 8 oz. of cool water every 20 min.
- Establish a work rest regimen based on task specific conditions.

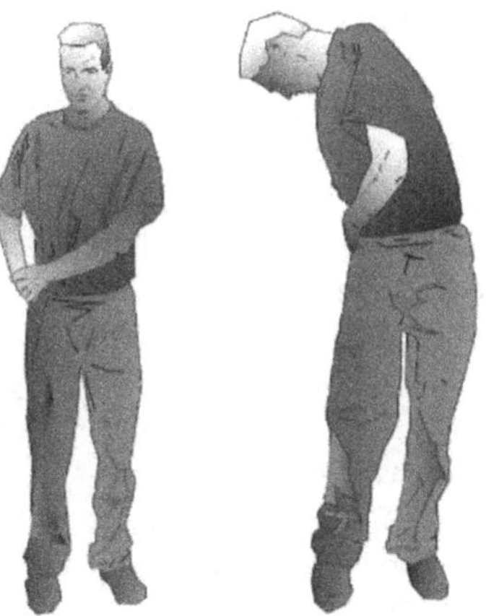

Lateral felxion stretch
1. Place right and left hands on right hip
2. Rotate upper body and head to right
3. Slowly bend from waist
4. Repeat on opposite side

Figure 5.6 Common stretches to remind workers the benefits of stretching

- Alternate
  - Light work in warm temperatures or at the end of the day
  - Heavy work in cooler temperatures and after a light warm-up
- Allow breaks outside of the extreme environment.
- Schedule work throughout the day.
- Avoid concentrations of strenuous work.
- Implement fitness for work standards to avoid placing high-risk individuals in potential heat stress situations.
- Acclimate new or return to work individuals (5 days in hot environments).

KEY POINTS

- Administrative controls are procedures and practices that limit exposure by control or manipulation of work schedule or manner in which work is performed.
- Administrative controls reduce the exposure to ergonomic stressors and thus reduce the cumulative dose to any one worker.
- Administrative controls are most effective when used in combination with engineering controls and is part of an ergonomics program.
- Administrative controls, such as rest breaks, reduce exposure, curtail fatigue, decrease recovery time, and the bottom line saves $$$.

REVIEW QUESTIONS

1. What are some of the problems with allowing fatigue to reoccur?

2. What are the benefits of stretching?

3. To maintain good health, what can cancel out our mental or physical stressors?

REFERENCES

Ergo Joe Stretch n' Flex at Puget Sound Naval Shipyard (PSNS). 1996–2006.

Henry, B. M. (2013). Occupational Injury Prevention Keeping Employees on the Job Through Safe and Effective Care. Retrieved February 2015, from advanceweb.com: http://occupational-therapy.advanceweb.com/Features/Articles/Occupational-Injury-Prevention.aspx.

OSHA, O. S. (n.d.). Retrieved February 2015, from Occupational Safety and Health Administration: http://www.osha.gov. /SLTC/ergonomics/controlhazards.html.

Palliative Care. (n.d.). Retrieved 2015, from StopPain.org: http://www.stoppain.org/palliative_care/content/fatigue/default.asp.

Training (n.d.). Retrieved February 2015, from NAVFAC: http://www.navfac.navy.mil/navfac_worldwide/pacific/fecs/far_east/contact_us/yokosuka/environmental_division/training.html.

6

ELEMENTS OF ERGONOMICS PROGRAMS

LEARNING OBJECTIVE

At the end of this chapter, students will define the elements of a good ergonomic program and apply the principles and guidelines in designing a good ergonomics program.

INTRODUCTION

The material at the beginning of this chapter follows the National Institute of Occupational Safety and Health (NIOSH) Suggested Elements of Ergonomics Programs (NIOSH, 1981).

Work-related musculoskeletal disorders (disorders of the muscles, nerves, tendons, ligaments, joints, or spinal discs) have increased dramatically in the past decade. These disorders are not typically the result of an acute event but reflect more gradual development, and the severity can range from mild to chronic and debilitating.

Ergonomics is defined as the science of fitting workplace conditions and job demands to the capabilities and limitations of the working population. Effective ergonomics promotes productivity, reduces injury risks, and increases worker comfort and satisfaction.

It is important to recognize and identify MSD problems. Some of the identifiers could be as follows:

- Injury records review (OSHA logs or workers compensation claims)
- Comparison to industry averages

Occupational Ergonomics: A Practical Approach, First Edition.
Theresa Stack, Lee T. Ostrom and Cheryl A. Wilhelmsen.

- Worker visits to clinic
- Jobs with repetitive, forceful exertions in awkward postures, frequent or heavy lifting, or vibrating equipment.

After identifying the problems, it is time to set the stage for action and integrate ergonomics into the company safety and health program. It is important to employ management commitment right from the beginning. A commitment of adequate resources including training the workforce, bringing in outside experts, and implementing improvements are vital to a successful program. Other vital aspects of a successful program are as follows:

- Treat ergonomics efforts as furthering the company's goals.
- Expect full cooperation of the total workforce.
- Assign lead roles to designated persons.
- Give ergonomics efforts priority with other cost reduction, productivity, or quality efforts.
- Set goals to address specific operations and prioritize the riskiest jobs.
- Release time or other compensatory arrangements for employees expected to handle assigned tasks for ergonomics efforts.
- Provide information to all involved, including injury data and productivity data.

Worker Involvement Benefits

There are great benefits for the worker involvement that cover areas such as enhancing the worker motivation and job satisfaction; a greater knowledge of the work, organization, and workers are frequently the best source of ideas to fix the problem jobs. Involving the workers enables them to feel ownership in the program.

So who should participate? A list of individuals from various organizations within the companies may include the following:

- Safety and hygiene personnel
- Healthcare providers
- Human resources personnel
- Engineering personnel
- Maintenance personnel
- Ergonomists or ergonomics specialists
- Worker and management representatives.

Another approach may be to form committees or teams such as a joint labor-management committee approach or a work group approach (team approach). The individual input approach provides employees with a communication facility and can respond to input received.

ERGONOMICS AWARENESS TRAINING

It is very important to provide training in ergonomics awareness, job analyses and control measures, and some problem solving. The overall objectives that could be included in the ergonomics awareness training should be as follows:

- Recognize risk factors for MSDs and understand methods for controlling them.
- Identify signs and symptoms of MSDs and be familiar with company healthcare procedures.
- Know the process the employer is using to address and control risk factors.
- Know the procedure for reporting risk factors.

The overall objectives that could be included in the job analyses and control measures training should be as follows:

- Demonstrate job analysis for identifying risk factors for MSDs.
- Select ways to implement and evaluate control measures.

The overall objectives that could be included in the problem solving training should do the following:

- Identify departments, areas, and jobs with risk factors through records, walk-through observations, and surveys.
- Identify tools and techniques for conducting job analyses.
- Develop skills in team building and problem solving.
- Recommend ways to control hazards.

So how do you know if the program is needed? How do you know if the employees are experiencing problems? Some of the key indicators of MSDs are found in the health and medical department, where the employee reports of physical stress are housed. Other indicators may be to review the OSHA logs and other existing records as well as calculating rates for comparisons such as the following:

- Plant medical records
- Insurance claims records
- Absentee records
- Job transfer applications.

Surveys are a great way to target symptoms where the respondents are asked to rate their level of discomfort for different areas of the body, type, onset, and duration of the symptoms reported. Periodic medical examinations and employee interviews are also great indicators of MSDs. When identifying the risk factors in a job, there are several specific factors to look for:

- Awkward postures (extremes of joint movement)
  - Twisting or bending while lifting or carrying
  - Wrist deviations
  - Overhead work (arms raised)
  - Extended reaching, etc.
- Forceful exertions (including lifting, pushing, and pulling)
- Forces increase with the following:
  - Weight or bulkiness of loads
  - Speed of movements
  - Use of awkward postures
  - Presence of vibration, etc.
- Repetitive motions
  - Frequent and similar motions every few seconds
  - Increased risk when repeated forceful exertions in awkward postures
- Duration of exposure
  - Amount of time a person is exposed to risk factors
  - The longer the period of continuous work, the longer the required recovery or rest time
- Contact stresses
  - Physical contact of body areas with hard or sharp objects
  - Desk edges, tool handles, and so on
  - Can inhibit nerve function and blood flow
- Vibration
  - Localized exposure to vibrating object, such as a power hand tools
  - Whole-body exposure to vibration when standing or sitting on vibrating equipment.

Other factors to be aware of are the conditions the workers are placed within such as the following:

- Cold temperatures
- Insufficient pauses or rest breaks for recovery
- Machine paced work
- Unfamiliar or unaccustomed work.

Screening the various tasks/jobs for risk factors could be performed through walk-through observational surveys to determine the obvious risk factors. It is fairly easy to see if the worker is experiencing pain or discomfort while performing their tasks. Other screening methods involve interviews with workers and supervisors and the use of checklists for scoring job features against a list of risk factors.

Job Analysis – Steps

When trying to analyze the job/task, one should first break the job into various elements or actions the job requires. Next, measure or quantify the risk factors and identify conditions contributing to risk factors. These should all be performed by individuals with considerable experience and training. The following is a list of the steps:

- Complete description of the job is obtained.
- Employees are interviewed.
- Job is divided into discrete tasks.
- Each task is then studied to determine specific risk factors.
- Risk factors may be further evaluated.

Job Analysis – Tasks

When looking at the tasks the workers need to perform in order to do their work, one could describe the tasks in terms of the following:

- Tools, equipment, and materials used to perform the job
- Workstation layout and physical environment
- Task demands and organizational climate.

Job Analysis – Detailed Data Collections

There are various sets of data that need to be collected to fully analyze the problems/issues within a company. The first one is collecting the data when observing the workers performing their tasks. The following illustrates the breakdown of the data collection information:

- Observe workers performing tasks to furnish time activity analysis
  - Job/task cycle data
  - Use videotape
  - Still photos of postures, workstation layouts, tools, and so on
- Workstation measurements
  - Work surface heights, reach distances, and so on.
  - Measure tool handle sizes, weighing tools and parts, measure parts.
  - Determine characteristics of work surfaces, such as slip resistance, hardness, and edges.
  - Measure exposures to cold, heat, whole-body vibration, and so on.

➢ Biomechanical calculations (muscle forces required to complete task or pressure on spinal discs based on load lifted, for example, NIOSH lifting guide)

➢ Special questionnaires, interviews, and subjective rating procedures.

After analyzing the data, the next step would be to develop controls to help reduce the problems/issues. The three types of controls are as follows:

- Engineering controls: Reduce or eliminate potentially hazardous conditions
- Administrative controls: Changes in work practices and management policies
- Personal equipment.

Engineering Controls

When we look at the engineering controls of actually designing the job, we should look at (1) workstation layout, (2) selection and use of the tools, and (3) work methods. Some useful strategies for job design are as follows:

- Change the way materials, parts, and products can be transported (e.g., use mechanical assist devices rather than manual handling).
- Change the process or product to reduce risk factors (e.g., maintain the fit of plastic molds to reduce the need for manual removal of flashing).
- Modify containers and parts presentation (e.g., height-adjustable material bins).
- Change workstation layout (e.g., use height-adjustable workbenches).
- Change the way parts, tools, and materials are to be manipulated (e.g., use fixtures to hold work pieces).
- Change tool designs (e.g., pistol handle grips for knives to reduce wrist deviations).
- Change assembly access and sequence (e.g., remove physical and visual obstructions).

Administrative Controls

- Reduce shift length or curtail overtime.
- Rotate workers through several jobs with different physical demands.
- Schedule more breaks for rest and recovery.
- Broaden or vary job content.
- Adjust the work pace.
- Train workers to recognize risk factors for MSDs.
- Instruct workers in work practices that can ease task demands.

Personal Equipment

Is the personal equipment effective for the individual to perform their job? Hard-hats, safety shoes, safety goggles, and so on are barriers against hazards while wrist supports, back belts, and vibration attenuation gloves are not barriers against risk factors for MSDs. There has not been enough evidence presented to verify the effectiveness for these types of personal equipment.

Controls

The following is just a small sample of gathering ideas for implementing controls within the workplace.

- Trade associations that may have information about good control practices
- Insurance companies that offer loss control services
- Consultants and vendors who deal in ergonomic specialty services and products
- Visits to other worksites known to have dealt with similar situations.

The next step is to implement the controls within the workplace. Again, there are many ways to implement the different controls and the list provided below is just a sampling.

- Trials or tests of selected solutions.
- Making modifications or revisions.
- Full-scale implementation.
- Follow up to evaluate control effectiveness.
- Designate the personnel responsible.
- Create a timetable.
- Consider the logistics necessary for implementation.

When they have implemented the controls, it is important to evaluate their effectiveness. Some of the best ways to do this is to:

- Use risk factor checklist or other job evaluation methods.
- Repeat the symptoms survey and compare with prior results (often in conjunction with checklist or other job analysis method).
- Evaluation should occur 1–2 weeks after implementation (short-term evaluation).
- Long-term evaluations
  - Reduction in incidence rate of MSDs
  - Reduction in severity rate of MSDs
  - Increase in productivity or quality of products or services
  - Reduction in job turnover or absenteeism.

Healthcare Management – Employer/Employee/Healthcare Management Responsibilities

Responsibilities the employer should own are as follows:

- Provide education and training regarding recognition of symptoms and signs of MSDs.

- Encourage early reporting of symptoms and prompt evaluation by care provider.
- Give care provider opportunities to become familiar with jobs and tasks.
- Modify jobs/tasks or accommodate limitations.
- Ensure privacy of medical information.
- Follow workplace safety and health rules.
- Follow work practice procedures related to their jobs.
- Report early any signs or symptoms of MSDs.

The employee responsibilities are as follows:

- Follow workplace safety and health rules.
- Follow work practice procedures related to their jobs.
- Report early any signs or symptoms of MSDs.

The healthcare management or care provider responsibilities are as follows:

- Acquire experience and training in evaluation and treatment of MSDs.
- Seek information and review materials regarding employee job activities.
- Ensure employee privacy and confidentiality.
- Evaluate symptomatic employees.
- Evaluate symptomatic employees, including ...
    - Medical histories and symptoms.
    - Descriptions of work activities.
    - Physical examinations.
    - Initial assessments or diagnoses.
    - Consider opinions as to whether occupational risk factors caused, contributed to, or exacerbated conditions.
- Follow-up examinations to document improvements.
- Become familiar with employee's job or tasks.
- Do periodic walk-throughs of the plant.
- Review job analysis reports or job descriptions.
- Review photographs or videotapes of jobs or tasks.

The best way to reduce issues and problems before they become serious is by early reporting. The employers should encourage early reporting and they should avoid establishing policies that discourage the reporting of symptoms. When the employee reports symptoms, the employer should provide prompt access to an evaluation by the company healthcare provider.

When the employee reports to the care providers, they should determine the worker's physical capabilities and work restrictions and recommend the employer assign the worker to tasks that are consistent with any restrictions. The employer could assign light-duty or temporary job transfers for the individual. Do not completely remove the worker from work unless recommended from the healthcare provider.

Proactive Ergonomics

It is always better to be proactive versus reactive. A good proactive ergonomics program would do the following:

- Emphasize ergonomics at the design stage of work processes.
- Design operations that ensure proper selection and use of tools, job methods, workstation layouts, and materials.
- Build a more prevention-oriented approach using knowledge gained from the ongoing ergonomics process.
- Design strategies should emphasize fitting job demands to the capabilities and limitations of workers.
- Design strategies should target causes of MSDs – engineering approaches are preferred over administrative approaches.

The next portion of this chapter addresses the return on investing in an ergonomics program.

ERGONOMIC RETURN ON INVESTMENT METHODS

- Review the costs, both direct and indirect, typically associated with ergonomic injuries/illnesses.
- Review techniques for quantifying return on investment to justify ergonomic interventions.
- Share interventions successfully implement at worksites.

The objectives are to learn the costs associated with ergonomic injuries and learn techniques for quantifying return on investment to justify ergonomic interventions. Ergonomics is more than the right thing to do – it pays off. They will need to be able to make a good defensive argument when asking for funding for ergonomic interventions or an ergonomics program. In this module, we are going to give the numbers one need to justify their request. We will discuss success stories and show how to calculate return on investment. This is called throwing beans at the bean counter. Everyone is fighting for a share of the same budget and they will need to arm yourself with good data and a well thought-out argument for why they deserve part of the pie.

Direct costs are those costs that you can directly measure, whereas the indirect costs often go unnoticed and uncalculated. The iceberg theory is that most of the problem is under water and hard to detect but that's the part that can sink ship.

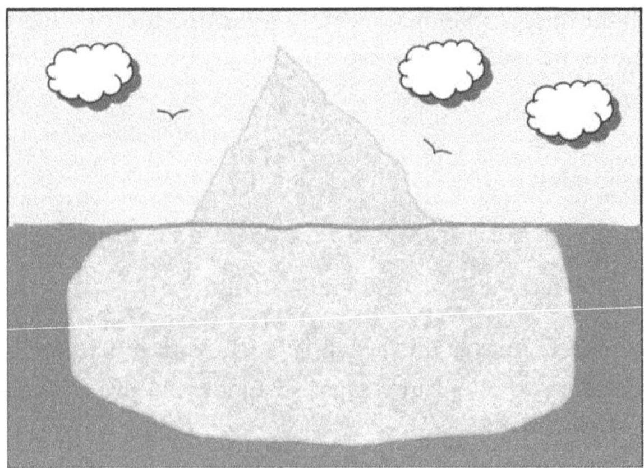

Figure 6.1 The rest of the iceberg what sinks the ship? (Original artwork by Lee Ostrom)

The Iceberg Theory

What sinks the ship? (see Figure 6.1)

Direct Costs = the tip of the iceberg vs **Indirect Costs** = the rest of the iceberg

Indirect costs:

1. Loss of good will lower morale
2. Inefficiency costs of restricted work
3. Hiring and training replacement
4. Overtime to other employees because of injury
5. Administrative costs
6. Costs arising from violation or injury investigation/follow-up.

Indirect costs include restricted or light duty, hiring and training costs, administrative costs, and so on.

When one have a disability claim, it takes time from everyone whose desk it crosses – from the administrative assistant to the company executive officer. There are also costs associated with advertising a new job, filling the new job (HR costs), and training a new employee (trainer costs) as well as lost productivity as the new employee becomes familiar with the job. An injury hurts the morale of everyone involved from co-workers to safety specialists to family members – because cumulative trauma injuries are avoidable. Injuries are a negative experience for everyone involved. Employees on long-term disability cost the company money every year, and they do not contribute to the productivity – therefore, it is money spent every year with no gains. Figure 6.2 illustrates the indirect costs in the iceberg theory.

For every **$1** expended on **Direct Costs,** an additional **$4** is expended on **Indirect Costs.** The following is taken from the Liberty Mutual Safety Index 1998 (Liberty Mutual Insurance Company).

Figure 6.2 The rest of the iceberg (indirect costs) sinks the ship (original artwork by Lee Ostrom)

Value of Direct Cost Saving per Musculoskeletal Disorder Averted

Value of lost production	$14,763
Medical costs	$3,080
Insurance administrative costs	$1,872
Indirect costs to employers	$2,832

OSHA estimates a 100% **Return on Investment** for preventive ergonomics over a 20-month payback period for most solutions.
Adapted from the OSHA Preliminary Economic Analysis and Initial Regulatory Flexibility Analysis for the Proposed Ergonomics Program Standard (OSHA, n.d.).

OSHA estimates that the total savings per prevented MSD is $22.5K.

One can calculate the average injury costs for your activity or use BLS data.

Potential injury is when there are no recorded injuries but potential still exists. One use this data to cost justify an intervention. For instance a $600 lift cart has the potential to prevent one $20K back strain.

These BLS numbers are from a report (Bureau of Labor Statistics, 2011)

Back strains	$20,618
Other strains	$14,600
CTS	$9,300

Bureau of Labor Statistics (BLS) Cost Factors

First-aid injury	$169 ea
Lost workday injury	$8,413 ea
Lost workday back injury	$100,000 ea

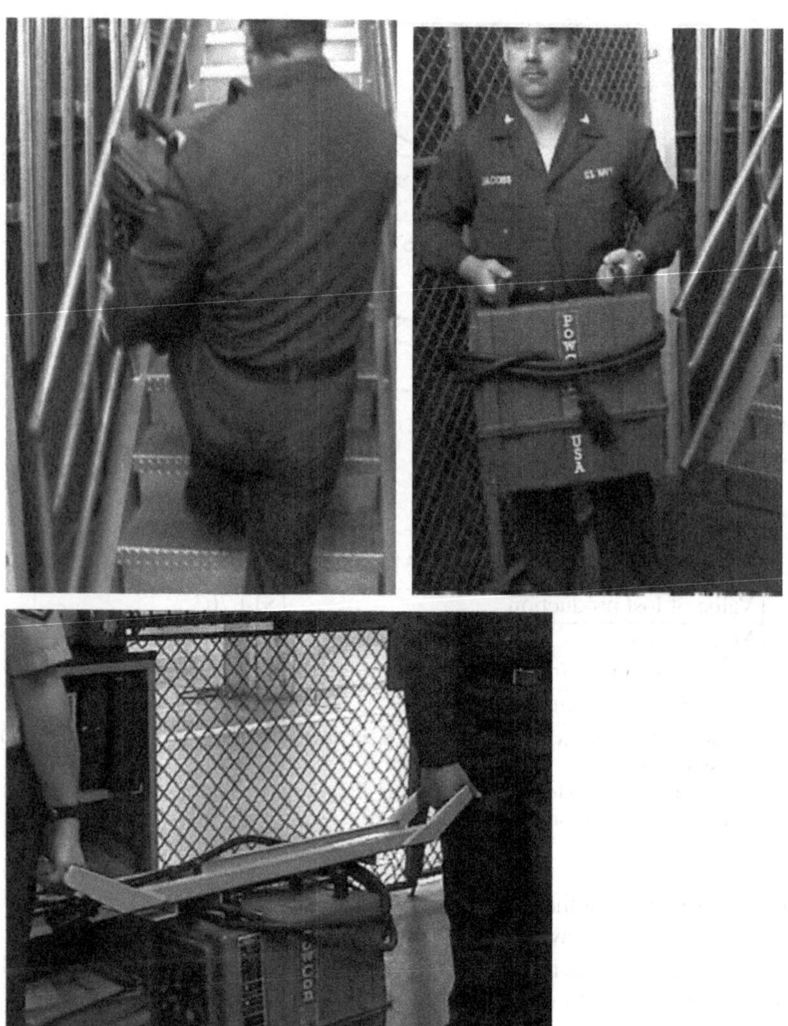

Figure 6.3 Solution: Carrying fixture designed by an employee and fabricated on-site 150 × 12 = 1800

A potential injury aversion to a back injury could be one where you provide equipment to facilitate lifting heavy or awkward objects. Figure 6.3 illustrates a method to help reduce the load on the worker.

One adverted back injury

$$20,618 \div 1800 = \mathbf{11.4 : 2\ \ pay\ backs}$$

ROI method – Collect data, analyze, and relay information to the highest levels. When reporting the information, be specific and show the numbers.

Managers weighing decisions expect specifics:

- Chargeback totals (5–10 years show cumulative costs)
- Most frequently occurring injury type (WMSDs)
- Most costly injury type (medical/compensation/productivity)
- Average cost per injury type per year
- Most efficient interventions

- Proof that practical, low-cost intervention is successful
- Cost per unit of ergonomic improvement = cost of intervention/number of workers helped.

One cannot predict specific injuries, but they **can** show increases in productivity, reduction in overall injury rates or severity, and costs. Do you know what you are spending on ergonomics? Go back 5–10 years in company records to find which injuries are ergonomic-related injuries.

WMSDs are usually the most costly injury type.

The most efficient solutions are usually low cost and can be used to address low hanging fruit. Inexpensive solutions, such as knee pads, are good protective equipment. Track expenditures in terms of cost per unit of ergonomic improvement = cost of intervention/# people helped. Some suggestions for low-cost solutions may be platforms and gloves.

Conduct your own mini study and give rivet gloves to riveters and compare injury rates to show improvement, $ spent, and people served versus injury rates. Budgets for safety glasses, respirators and steel toed shoes are not questioned, why not the leading cause of injuries? We know that the other preventative programs work, but no one ever asks how many injuries were prevented. Figure 6.4 illustrates a hazard at a facility dealing with the laundry.

ROI Method – Intervention versus Injury Costs – Example Project

Hazards: Throwing dirty linens and trash over the side of the building, and carrying supplies and furniture up three decks.

In this example, we will use injury rates to justify an ergonomic fix. This example is from a transitional housing facility, which is three floors tall with no elevator.

A bag of linen, particularly wet towels, can weigh between 40 and 50 lb. Reaching into a cart to lift heavy bags of linen places the employee at risk of developing WMSDs of the back and shoulders. Linen is carried downstairs, pulled downstairs in a cart, or thrown over the balcony. When linen is thrown over the balcony, it has to be lifted over a 43 in. high railing and tossed past a 53 in. deep ledge, as shown in Figure 6.4. Throwing a heavy load over shoulder height for a distance places the employee at risk of injury to the upper extremities and back. Objects falling from overhead place people and property below at risk of damage or injury. Pushing or pulling a cart on stairs is dangerous for the employees and guests in the stairwell. Carts in the stairwell can inhibit emergency egress. There are currently 30 employees in transitional housing with a maximum of 34 employees during the summer due to an increase of persons staying in the facility over the summer.

Figure 6.4 Throwing laundry over three flights of stairs

ROI Method

Intervention versus injury costs

Ergonomic Intervention:

$$\text{Outside 3-story lift} = \$60,500$$

Cost Justification:

Injury data FY97–FY01

14 recorded back injuries

Average cost of a Navy back injury $15,590

Injury data FY97–FY01

$$14 \times \$15,590 = \$218,260$$

Intervention cost/Average yearly back claim cost

60,500 ÷ 43,652 = 1.38

1.38 = 1 **year 138 day payback period**

The solution is a 3-story outdoor lift as shown in Figure 6.5. The cost is about 60K. This may seem like a lot, but if you look at the injury data from 1997 to 2001 in this facility, you will see 14 recorded back injuries.

ROI – 10-year cost savings

ROI method – Lifetime WMSD costs

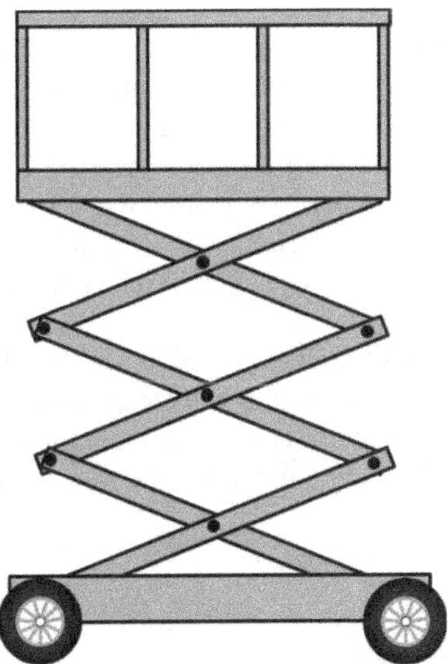

Figure 6.5 Lift (Original artwork by Lee Ostrom)

What happens when work practices are not altered after an injury?

The following are examples of injuries sustained at a facility (Six Sigma Costs and Savings, n.d.).

Example 1 In 1978, a supervisor suffered a back injury while helping move sheet metal. Metal sheets were stacked and braced. There was no lost time except the day of the injury, and just $1000 in medical costs. A recurrent injury in 1992 cost $18K. This injury stayed on the books for many years and totaled over half a million dollars. The employee is still bent over and miserable and in pain. A 1K injury did not warrant a fix at the time, but a simple fix could have saved 500K+.

Example 2 Another employee was injured doing the same thing in 1996. At that time an ergonomics program was in place, and a single incident triggered an evaluation of the job.

There is now a metal clamp to hold the sheets and a roller table to move them. Sheets of metal are now stored in wheeled metal storage that pull out like books. Once the sheets are pulled out, the hoist can be brought over and clamped down. The total cost of this fix was $21K. Figure 6.6 illustrates the metal storage and roller feed table system put into place within the facility.

Freestanding crane with magnetic clamp $12.2K

Roller feed table $1K

Metal sheet storage $8K

ROI Methods – Small Investment/Grand Payoff Ergo Joe Stretch n' Flex Program at Puget Sound Naval Facility 1996

As part of the program, 6000 posters were displayed throughout the facility that reminded workers to stretch. Awareness training and Ergo Mentor and Ergonomic Orientation Training were conducted. The Ergo mentor training reminds work groups to stretch and educates through better working techniques (body mechanics). The data showed a **22% reduction** in back injuries with a 90% voluntary participation. The program requires all employees to gather for 10 min at the beginning of each shift where stretching will be conducted, but the employees are not required to stretch, just be present. Figure 6.7 is an example of a lateral stretch exercise.

The next portion of this chapter will look at the benefits of Lean Engineering. A good ergonomics program allows the worker to function at their highest level of productivity, quality, and efficiency. Looking at Lean Engineering allows the process to function at its highest level of quality and efficiency as well.

- Both need a culture to be effective.
- Both are focused on the voice of the customer.
- Both strive to reduce waste.

Figure 6.6 Example of the sheet metal storage and roller feed table

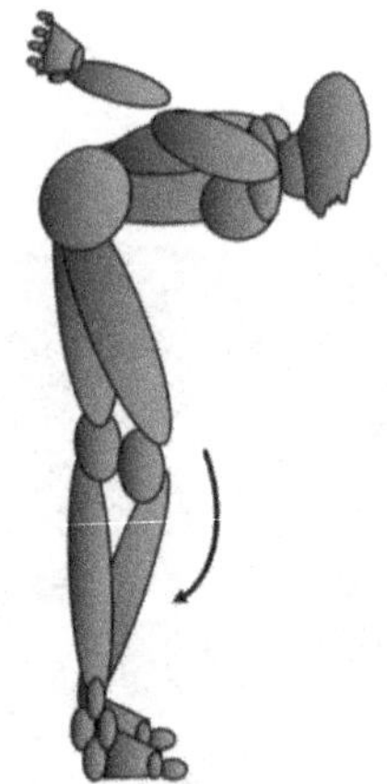

Figure 6.7 Lateral stretch exercise

ERGONOMICS PARALLELS LEAN 6 SIGMA

- OSH/ergonomist are important elements in the cross-functional/cross-departmental team.
  - Vital in the improve/recommend phase.
  - Injury/illness metrics can be included in the cross-functional business metrics to measure progress.
- Origins in industrial engineering.
  - LEAN concerned with manufacturing efficiency.
  - Ergonomics concerned with human efficiency.
  - LEAN 6 sigma strives to **increase** process **performance and reduce** output variation by removing **waste**.
- Ergonomics strives to reduce the risk of injury and illness by matching the task … to the workers' capabilities to **maximize** human **performance** (i.e., **reduces wasteful** human effort).

Ergonomics is most important at the level of work cell design and workstation design. By their very nature, well-designed cells relieve many of the risk factors associated with traditional workstations and functional layouts. For example, work cells often rotate workers through an entire process on each cycle. This reduces repetition and static postures. Workstations also have a direct influence on musculoskeletal disorders.

- Both need a culture to be effective and sustainable; both are engineering-based disciplines.
- Both strive to reduce waste in a sense.
- Benefits of both methodologies
  - Reduced direct and indirect costs
  - Increased productivity.

- Maximize performance.
- Reduce waste.
- Increase customer satisfaction.
- Ergonomics efforts also increase retention and employee morale.

- The mismatch between tasks ... and employees capabilities may be key variables that introduce or maintain wasteful processes, reduce customer satisfaction, and affects process variation.
- Poor quality or slow production can be linked to ergonomics.

The last portion of this chapter is an example of an occupational ergonomics safety program that could be developed using a regulatory standard.

MODEL SAFETY PROGRAM

DATE: ____________

SUBJECT: Occupational Ergonomics

REGULATORY STANDARD: OSHA – 29 CFR 1910.XXX (To Be Determined)

RESPONSIBILITY: The company ______________________ is ________ ________. He/she is solely responsible for all facets of this program and has full authority to make necessary decisions to ensure success of the program. The ________________ is the sole person authorized to amend these instructions and is authorized to halt any operation of the company where there is danger of serious personal injury.

Contents of the (YOUR COMPANY) Occupational Ergonomics Program

1. Written program
2. General requirements
3. Health surveillance
4. Ergonomic Assessment Committee
5. Program review and evaluation
6. Worksite analysis
7. Job hazard analysis
8. Hazard prevention and control
9. Periodic ergonomic surveys
10. Work practice controls
11. Administrative controls
12. Medical management
13. Training and education
14. Definitions.

(YOUR COMPANY) Occupational Ergonomics Program

1. **Written program.** (YOUR COMPANY) will review and evaluate this standard practice instruction on an annual basis, or when changes occur that prompt revision of this document, or when facility operational changes occur that require a revision of this document. Effective implementation requires a written program for job safety, health, and ergonomics that is endorsed and advocated by the highest level of management within this company and that outlines its goals and plans. This written program is designed to establish clear goals and objectives. It encompasses the total workplace, regardless of the number of workers employed or the number of work shifts and will be communicated to all personnel at every level.
2. **General requirements**. The goal of this program is to engineer problems out of the workplace wherever possible. Understanding that poor ergonomics in the workplace can result in cumulative trauma disorders (CTDs), and a host of other occupational injuries and conditions, (YOUR COMPANY) will establish ergonomic controls and operational procedures using this document as guidance.
3. **Health surveillance**.

3.1 Employee baseline. Prior to assignment, all new and transferred workers who are to be assigned to positions involving exposure of a particular body part to ergonomic stress will receive baseline health surveillance. The purpose of this baseline health surveillance is to establish a base against which changes in health status can be evaluated, not to preclude people from performing work.

3.1.1 Employee notification. Employees will be notified when they are placed in job descriptions where it is known or suspected that ergonomic hazards exist. These positions will be identified through the worksite analysis program discussed in Section 6 of this document and from the list of known high-risk jobs compiled by the company's healthcare provider.

NOTE: Due to variations in personal health at any given time, the use of medical screening tests or examinations have not been validated as predictive procedures for determining the risk of a particular worker developing a CTD.

3.2 Baseline health surveillance. The baseline health surveillance will include the following:

3.2.1 A medical and occupational history.

3.2.2 A physical examination will be performed by ________________ who is knowledgeable of the hazards associated with ergonomic stressors in general, and with the specific tasks to which the employee will be assigned. The examination will include the musculoskeletal and nervous systems as they relate to ergonomic stressors, palpation, range of motion (active, passive, and resisted), and other pertinent maneuvers of the upper extremities and back. Unless mitigating circumstances exist, laboratory tests, X-rays, and other diagnostic procedures will not be included as routine parts of the baseline assessment.

3.3 Conditioning period follow-up. Supervisors will ensure that new and transferred employees will be given the opportunity to perform anticipated normally assigned tasks during a 4–6 week break-in period to condition their muscle–tendon groups prior to working at full capacity. A follow-up assessment of these workers after the break-in period (or after 1 month, if the break-in period is longer than a month) will be conducted to determine the following:

3.3.1 If conditioning of the muscle–tendon groups has been successful.

3.3.2 Whether any reported changes in physical well-being such as soreness or stiffness is transient and consistent with normal adaptation to the job, or whether it indicates the onset of stressors associated with ergonomic hazards that cannot be changed.

3.3.3 If problems are identified, determine what appropriate action is required to change salient factors within the assigned task or the assigned employee that might be possible to better match the work to the worker.

3.4 Periodic Health Surveillance. Periodic health surveillance (every 2–3 years) will be conducted on all employees who are assigned to positions involving exposure of a particular body part to ergonomic stress. The content of this assessment will include the following:

3.4.1 A detailed update of the employee's medical and occupational history.

3.4.2 A physical examination by ____________________ who is knowledgeable of the hazards associated with ergonomic stressors. The examination will include a personal assessment by the employee of his/her continuing capability of performing assigned tasks, as well as examination of musculoskeletal and nervous systems as they relate to ergonomic stressors, palpation, range of motion (active, passive, and resisted), and other pertinent maneuvers of the upper extremities and back. Unless mitigating circumstances exist, laboratory tests, X-rays, and other diagnostic procedures will not be included as routine parts of the baseline assessment.

3.4.3 A detailed update of the employee's medical and occupational history. See 3.4.1 above – repetitive.

3.5 Documentation. Data gathered on employees as a result of health surveillance will be documented and filed in individual employee medical records.

4. **Ergonomic Assessment Committee**. In order for this to be an effective program, (YOUR COMPANY) will provide for and encourage employee involvement in decisions at all levels that affect worker safety and health, including the following:

4.1 Employee complaints, suggestions, or concerns will be brought to the attention of management. Feedback without fear of reprisal will be provided to all employees at every level.

4.2 The ______________________________ will maintain statistical data concerning reporting of signs and symptoms of ergonomic stressors by employees so that they can be evaluated and, if warranted, treated. This data will also direct managers to causal factors that may be evaluated for alteration, re-engineering or elimination. This data will be provided to the safety and health committee.

4.3 The safety and health committee will, as necessary, analyze statistical data concerning ergonomic stressors, and make recommendations to specifically trained individual monitors assigned to the affected area or to an Ergonomic Assessment Committee for corrective action.

4.4 An Ergonomic Assessment Committee will be established and/or specifically trained individual monitors will be assigned to conduct reviews of specifically identified jobs and tasks to analyze ergonomic stressors and recommend solutions. The ____________________ who is trained and locally certified as knowledgeable of the hazards associated with ergonomic stressors, will head the committee and direct activities. The committee will be composed of the following personnel.

Ergonomic Assessment Committee

Chairman ____________________________

Vice Chairman ____________________________

Safety Officer

Supervisory personnel

Union Committee personnel

Hourly “lead” personnel

Hourly “nonlead” personnel

Member(s)

- Trained in industrial ergonomic hazards (general)
- Trained in office and sedentary ergonomic hazards
- Trained in repetitive motion hazards
- Trained in force/lifting hazards
- Trained in posture hazards
- Trained in industrial engineering and process flow.

Representative from timekeeping and costs department(s)

Workforce Management Committee

Representative of medical management familiar with tasks performed and any injury or pain complaints by our employees.

5. **Program review and evaluation.** Senior company officers will review the ergonomics program regularly (semiannual reviews are recommended by OSHA) to evaluate implementation of the program, its success in meeting its goals and objectives, and to monitor costs and accomplishments. The results of the review will be in the form of a written progress report and program update. The report will be shared with all responsible parties and communicated to employees with special recognition of program participants who have contributed significantly to the success of ______________________ (YOUR COMPANY) by their participation and cooperation. New or revised goals arising from the review (identifying jobs, processes, and departments) will be provided to all employees. Information on deficiencies identified will include actions initiated. Evaluation techniques will include the following:

5.1 Analysis of trends in injury/illness rates with particular attention to ergonomic injuries and recovery statistics

5.2 Employee surveys

5.3 Before and after surveys/evaluations of job/worksite changes

5.4 Review of results of plant evaluations with recommended improvements

5.5 Up-to-date records or logs of Ergonomic Projects for job improvements tried or implemented including any pertinent injury/illness data for each. Cost data will be included wherever appropriate, both on production lost due to ergonomic challenges and costs related to eliminating ergonomic hazards.

***Decision point:** The following procedures should be adapted to each workplace.

6. **Worksite analysis**. Worksite analyses identify existing ergonomic hazards as well as conditions and/or operations that pose ergonomic challenges and areas where hazards may develop. This also includes close scrutiny and tracking of injury and illness records to identify patterns of trauma or strains that may indicate the development of CTDs. The objectives of worksite analyses are to recognize, identify, and correct ergonomic challenges. The following guidance provides a starting point for finding and eliminating those tools, techniques, and conditions, which may be the source of ergonomic problems. In addition to analyzing current workplace conditions, planned changes should be reviewed for potential ergonomic hazards before they are implemented. Changes to existing and new facilities, processes, materials, and equipment can be considered to ensure that changes made to enhance production will also reduce or eliminate risk factors. Worksite analysis is divided into four main parts: (1) Gathering information from all available sources. (2) Conducting employee baseline screening surveys to help determine which tasks need closer analysis. (3) Performing ergonomic job hazard analyses of those work stations with identified or potential risk factors. (4) After implementing control measures, conducting periodic surveys and follow-up to evaluate effectiveness of ergonomic improvement including costs and injury prevention as well as productivity improvement. Worksite analyses will be performed following the below listed procedures.

6.1 Information sources

6.1.1 Records analysis and tracking. Existing medical, safety, and insurance records, including OSHA-300 logs, will be analyzed for evidence of injuries or disorders associated with ergonomic stressors. Healthcare providers will be asked to participate in this process to ensure confidentiality of patient records and to identify correlation between ergonomic hazards in a specific task and the injury or condition reported. The purpose will be to develop the information necessary to identify ergonomic hazards in the workplace.

6.1.2 Incidence rates. Incidence rates for upper extremity disorders and/or back injuries should be calculated by counting the incidences of CTDs and reporting the incidences per 100 full-time workers per year per facility. In this calculation, the average employee works 2000 h/year. One hundred employees are reflected in the figure "200,000 work hours." Therefore,

$$\text{Incidence rate} = \frac{\text{Number of new cases/year} \times (200{,}000\,\text{work hours})\ \text{per facility}^*}{\text{Number of hours worked/facility/year}}$$

*The same method can be applied to departments, production lines, or job types within the facility by adjusting the number of hours actually worked by the employees within that subgroup or within that adjusted time period.

6.2 Screening surveys. Detailed baseline screening surveys will be conducted to identify jobs that put our employees at risk.

6.2.1 Checklist. The survey will be performed with an ergonomic checklist. This checklist will include components such as posture, materials handling, and upper extremity factors. The checklist will be tailored to the specific needs and conditions of a particular task or series of tasks within the workplace.

*Decision point examples of an ergonomic checklist are provided in the publication *Cumulative Trauma Disorders* by Putz-Anderson, p. 52. In addition, other examples of checklists will be given in OSHA's forthcoming *Ergonomics Program Management Guidelines for General Industry*. (A sample ergonomic checklist is attached for your review and clarification.)

6.2.2 Ergonomic risk factors. Identification of potential ergonomic hazards will be based on risk factors such as conditions of a job/task/process such as extreme heat/cold, height, confined area, bulky or uneven weights, lack of solid grip points, poorly designed work station, or work methods that contribute to the risk of developing problems associated with ergonomic stressors. Not all of these risk factors will be present in every job containing ergonomic challenges, nor is the existence of one of these factors necessarily sufficient to cause a problem associated with CTD's. Supervisors will ensure that known risk factors for specific employees, jobs or tasks are conveyed to the Ergonomic Assessment Committee for improvement or correction. Supervisors will use the following known risk factors to isolate and report suspected problem areas:

- Personal risk factors
  - Gender
  - Age
  - Anthropometry
  - Work method
  - Attitude
  - Training
  - Sight
  - Hearing
  - Smell
  - Physical strength
  - Weight

- Upper extremities risk factors
  - Repetitive and/or prolonged activities
  - Forceful exertions, usually with the hands
  - Pinch grips
  - Prolonged static postures
  - Awkward postures
    - Reaching above the shoulders
    - Reaching behind the back
    - Unusual twisting of wrists and other joints
  - Continued physical contact with work surfaces
  - Excessive vibration from power tools
  - Inappropriate or inadequate hand tools
- Back disorder risk factors
  - Bad body mechanics such as
    - Continued bending over at the waist
    - Continued lifting from below the knuckles
    - Continued lifting above the shoulders
    - Twisting at the waist
    - Twisting at the waist, while lifting.
    - Lifting or moving objects of excessive weight
    - Lifting or moving objects of asymmetric size
    - Prolonged sitting, with poor posture.
    - Lack of adjustable
      - Chairs
      - Footrests
      - Body supports
      - Work surfaces at work stations
    - Poor grips on handles
    - Slippery footing
    - Frequency of movement
    - Duration and pace
    - Stability of load
    - Coupling of load
      - Type of grip
    - Reach distances
    - Work height
- Environmental risk factors
  - Type of floor surface at work stations
  - Type of platforms
  - Work station temperature
    - Hot
      - Glove (reduces grip by up to 30%)
      - Fatigue
      - Sweat
      - Personal protective equipment (PPE)

        - Cold
            - Glove (reduces grip by up to 30%)
            - PPE
    - Lighting
        - Too bright
        - Too dim
    - Noise
        - Distractions
        - Associated fatigue
    - Vibration
        - Associated fatigue
        - Resonation through body
        - Resonation through tools
- Multiple risk factors. Jobs, operations, or work stations that have multiple risk factors have a higher probability of ergonomic risk. The combined effect of several risk factors is sometimes referred to as "multiple causation."

7. **Job hazard analysis**. This employer will identify through the use of information sources and screening surveys, jobs that place employees at risk. After a worksite analysis has been completed, a job hazard analysis for each job so identified will be conducted. Job hazard analyses will be routinely performed by qualified person(s) for jobs that put workers at risk. This analysis will help to verify lower risk factors at light duty or restricted activity work positions and to determine if risk factors for a work position have been reduced or eliminated to the extent feasible.

7.1 The following personnel or job positions are qualified to perform job hazard analysis surveys for this company.

Personnel Qualified to Perform Job Hazard Analysis

	Name	Title
(1)	__________	__________
(2)	__________	__________
(3)	__________	__________
(4)	__________	__________
(5)	__________	__________
(6)	__________	__________

7.2 Work station analysis. Work station analysis will be conducted to identify risk factors present in each job or workstation.

7.2.1 Upper extremities. For upper extremities, three measurements of repetitiveness will be reviewed; they are as follows:

7.2.1.1 Total hand manipulations per cycle.

7.2.1.2 Cycle time.

7.2.1.3 Total manipulations or cycles per work shift.

7.2.2 Force measurements. Force measurements will be noted as an estimated average effort, and a peak force. They will be recorded as "light," "moderate," and "heavy."

7.2.3 Tools. Tools will be checked for excessive vibration. (The NIOSH criteria document on hand/arm vibration should be consulted.)

7.2.4 The tools, personal protective equipment, and dimensions and adjustability of the work station will be noted for each job hazard analysis.

7.2.5 Postures. Hand, arm, and shoulder postures and movements will be assessed for levels of risk.

7.2.6 Lifting hazards. Work stations having tasks requiring manual materials handling will have the maximum weight-lifting values calculated. (The NIOSH *Work Practices Guide for Manual Lifting*, 1981, should be used for basic calculations. Note that this guide does not address lifting that involves twisting or turning motions.)

7.2.7 Videotape method. The use of videotape, where feasible, will be used as a method for analysis of the work process. Slow-motion videotape or equivalent visual records of workers performing their routine job tasks will used where practical to determine the demands of the task on the worker and how each worker actually performs each task. A task analysis log/form will be used to break down the job into components that can be individually analyzed.

8. Hazard prevention and control: This company understands that engineering solutions, where feasible, are the preferred method of control for ergonomic hazards. The focus of ABC's ergonomics program is to make the job fit the person, not to make the person fit the job. This is accomplished whenever possible by redesigning the work station, work methods, or tool(s) to reduce the demands of the job, including high force, repetitive motion, and awkward postures. This program will whenever possible research into currently available controls and technology. The following examples of engineering controls will be used as models for work station design and upgrade at (YOUR COMPANY).

8.1 Work station design. Work stations when initially constructed or when redesigned will be adjustable in order to accommodate the person who actually works at a given work station; it is not adequate to design for the "average" or typical worker. Work stations should be easily adjustable and either designed or selected to fit a specific task, so that they are comfortable for the workers using them. The work space should be large enough to allow for the full range of required movements, especially where handheld tools are used. Examples include the following:

8.1.1 Adjustable fixtures on work tables so that the position of the work can be easily manipulated.

8.1.2 Work stations and delivery bins that can accommodate the heights and reach limitations of various-sized workers.

8.1.3 Work platforms that move up and down for various operations.

8.1.4 Mechanical or powered assists to eliminate the use of extreme force.

8.1.5 Suspension of heavy tools.

8.1.6 Use of diverging conveyors off of main lines so that certain activities can be performed at slower rates.

8.1.7 Floor mats designed to reduce trauma to the legs and back.

8.2 Design of work methods. Traditional work method analysis considers static postures and repetition rates. This will be supplemented by addressing the force levels and the hand and arm postures involved. The tasks will be altered where possible to reduce these and the other stresses. The results of such analyses will be shared with employee healthcare providers; for example, to assist in compiling lists of "light-duty" and "high-risk" jobs. Examples of methods for the reduction of extreme and awkward postures include the following:

8.2.1 Enabling the worker to perform the task with two hands instead of one.

8.2.2 Conforming to the NIOSH *Work Practices Guide for Manual Lifting*.

8.3 Excessive force. Excessive force in any operation can result in both long-term problems for the worker and increased accident rates. Ways to reduce excessive force will be continually emphasized by first-line supervisors and employees. Examples of methods to reduce excessive force include the following:

8.3.1 Use of automation devices.

8.3.2 Use of mechanical devices to aid in removing scrap from work areas.

8.3.3 Substitution of power tools where manual tools are now in use.

8.3.4 Use of articulated arms and counter balances suspended by overhead racks to reduce the force needed to operate and control power tools.

8.4 Repetitive motion. All efforts to reduce repetitive motion will be pursued. Examples of methods to reduce highly repetitive movements include the following:

8.4.1 Increasing the number of workers performing a task.

8.4.2 Lessening repetition by combining jobs with very short cycle times, thereby increasing cycle time (sometimes referred to as "job enlargement.").

8.4.3 Using automation where appropriate.

8.4.4 Designing or altering jobs to allow self-pacing, when feasible.

8.4.5 Designing or altering jobs to allow sufficient rest pauses.

8.5 Tool design and handles. Supervisors will use ergonomic principles in the selection and or design of tools to minimize the risks of upper extremity CTDs and back injuries. In any tool design, a variety of sizes should be available. Examples of criteria for selecting tools include the following:

8.5.1 Matching the type of tool to the task.

8.5.2 Designing or selecting the tool handle so that extreme and awkward postures are minimized.

8.5.3 Using tool handles with textured grips in preference to those with ridges and grooves.

8.5.4 Designing tools to be used by either hand, or providing tools for left- and right-handed workers.

8.5.5 Using tools with triggers that depress easily and are activated by two or more fingers.

8.5.6 Using handles and grips that distribute the pressure over the fleshy part of the palm, so that the tool does not dig into the palm.

8.5.7 Designing/selecting tools for minimum weight; counterbalancing tools heavier than 1 or 2 lb.

8.5.8 Selecting pneumatic and power tools that exhibit minimal vibration and maintaining them in accordance with manufacturer's specifications, or with an adequate vibration monitoring program.
Note: Wrapping handles and grips with insulation material (other than wraps provided by the manufacturer for this purpose) is normally not recommended, as it may interfere with a proper grip and increases stress.

8.5.9 Selecting tools that minimize chronic muscle contraction or steady force.

8.5.10 Selecting tools that prevent extreme or awkward finger/hand/arm positions.

8.5.11 Selecting tools that minimize repetitive forceful motions.

8.5.12 Selecting tools that minimize tool vibration.

8.5.13 Selecting tools that minimize excessive gripping, pinching, pressing with the hand and fingers.

9. Periodic ergonomic surveys. Periodic surveys will be conducted to identify previously unnoticed risk factors or failures or deficiencies in work practice or engineering controls. Periodic surveys will be conducted on a(n) ____________________ basis. The periodic review process will include the following:

9.1 Feedback and follow-up. The company hazard notification system (safety program) will be used to provide employees with a system to notify management about conditions that appear to be potential ergonomic hazards. This process will allow this company to utilize their insight and experience to determine work practice and engineering controls.

9.1.1 Ergonomic questionnaires will be used to provide feedback on jobs, workstations, and tasks previously modified to incorporate ergonomic principles and upgrades.

9.1.2 Reports of ergonomic hazards or signs and symptoms of potential problem areas will be investigated by ergonomic screening surveys and appropriate ergonomic hazard analyses in order to identify risk factors and controls.

9.1.3 Trend analysis. Trends of injuries and illnesses related to actual or potential CTDs will be calculated, using multiple years of data where possible. Trends will be calculated for several departments, job titles, or work stations. These trends will also be used to determine which work positions are most hazardous and will be analyzed by the ergonomics assessment committee, and other responsible company personnel.

9.1.4 By using standardized job descriptions, incidence rates may be calculated for work positions in successive years to identify trends. Therefore,

(YOUR COMPANY) where possible will correlate previous and current job position titles. Using trend information can help to determine the priority of screening surveys and/or ergonomic hazard analyses.

9.1.5 Medical, safety, and insurance records, including the OSHA-200 log and information compiled through the medical management program will be used to provide evidence (or lack of) of ergonomic stressors. Company healthcare providers will be used to compile data from medical records to ensure confidentiality. This process should involve the identification and analysis of any apparent trends relating to particular departments, job titles, operations, work stations, or individual tasks.

10. Work practice controls. An effective program for hazard prevention and control also includes procedures for safe and proper work that are understood and followed by ABC managers, supervisors, and workers. Key elements of a good work practice program for ergonomics include proper work techniques, employee conditioning, regular monitoring, feedback, maintenance, adjustments and modifications, and enforcement.

10.1 Proper work techniques. Supervisor awareness and control of proper work techniques will improve safety. The following includes appropriate training and work practice controls for our employees:

10.1.1 Proper work techniques, including work methods that improve posture and reduce stress and strain on extremities.

10.1.2 Good tool care, including regular maintenance.

10.1.3 Correct lifting techniques (proper body mechanics).

10.1.4 Proper selection, use, and maintenance of all tools associated with the job.

10.1.5 Correct installation and use of ergonomically designed work stations and fixtures.

10.2 New employee conditioning period. Supervisors will ensure that new or transferred employees are allowed an appropriate conditioning period. New and returning employees will be gradually integrated into a full workload as appropriate for specific jobs and individuals. Employees will be assigned to an experienced trainer for job training and evaluation during the break-in period. Employees reassigned to new jobs should also have a break-in period. Important – Supervisors will closely monitor employees that fall into this category throughout their break-in period.

10.3 Monitoring. Regular monitoring at all levels of operation helps to ensure that employees continue to use proper work practices. This monitoring will include a periodic review of the techniques in use and their effectiveness, including a determination of whether the procedures in use are those specified; if not, then it should be determined why changes have occurred and whether corrective action is necessary.

10.4 Adjustments and modifications. Supervisors must continually be aware of changes in the dynamics of the workplace and to make appropriate operational changes. Such changes may include the following:

10.4.1 Line speeds

10.4.2 Staffing changes

10.4.3 Type, size, weight, or temperature of the product handled

10.4.4 Worker health and attitude changes.

10.5 Personal protective equipment (PPE). PPE used by employees of this company will be selected with ergonomic principles, and stressors in mind. Appropriate PPE will be provided in a variety of sizes, will accommodate the physical requirements of workers and the job, and will not contribute to extreme postures and excessive forces. Supervisors will consider the following factors when selecting PPE for personnel under their control:

10.5.1 Proper fit. For example, gloves that are too thick or that fit improperly can reduce blood circulation and sensory feedback, contribute to slippage, and require excessive or increased grip strength. The same is true when excessive layers of gloves are used (e.g., rubber over fabric, over metal mesh, over cotton). The gloves in use should facilitate the grasping of the tools needed for a particular job while protecting the worker from injury.

10.5.2 Protection against extreme cold is necessary to minimize stress on joints.

10.5.3 Protection against extreme heat is necessary to minimize slippage caused by perspiration. Also, to be considered, if slippage is occurring grip strength will be affected (reduced by up to 30%).

10.5.4 Other types of PPE that may be selected for use should be reviewed before purchase to ensure that there is no increase of ergonomic stressors.

11. Administrative controls. Company administrative controls will be used to reduce the duration, frequency, and severity of exposures to ergonomic stressors. Examples of administrative controls include the following:

11.1 Reducing the total number of repetitions per employee by such means as decreasing production rates and limiting overtime work.

11.2 Providing rest pauses to relieve fatigued muscle–tendon groups. The length of time needed depends on the task's overall effort and total cycle time.

11.3 Increasing the number of employees assigned to a task to alleviate severe conditions, especially in lifting heavy objects.

11.4 Using job rotation, used with caution and as a preventive measure, not as a response to symptoms. The principle of job rotation is to alleviate physical fatigue and stress of a particular set of muscles and tendons by rotating employees among other jobs that use different muscle–tendon groups. If rotation is utilized, the job analyses must be reviewed to ensure that the same muscle–tendon groups are not used when they are rotated.

11.5 Providing sufficient numbers of standby/relief personnel to compensate for foreseeable upset conditions on the line (e.g., loss of workers).

11.6 Job enlargement. Having employees perform broader functions that reduce the stress on specific muscle groups while performing individual tasks.

*Decision point OSHA recommendation – where healthcare providers are not employed full time, the part-time employment of appropriately trained healthcare providers is recommended.

12. **Medical management.** ______________________ will manage the program. Employees of each work shift should have access to healthcare providers or designated alternates in order to facilitate treatment, surveillance activities, and recording of information. The medical management program will as a minimum address the following issues:

12.1 Injury and illness recordkeeping

12.2 Early recognition and reporting

12.3 Systematic evaluation and referral

12.4 Conservative treatment

12.5 Conservative return to work

12.6 Systematic monitoring

12.7 Adequate staffing and facilities

12.8 Recordability criteria. Most conditions classified as CTDs will be recorded on the OSHA-200 form as an occupational illness under the "7f" column, which are "disorders associated with repeated trauma." These are disorders caused, aggravated, or precipitated by repeated motion, vibration, or pressure. In order to be recordable, the following criteria must be met:

12.8.1 The illnesses must be work related. This means that exposure at work either caused or contributed to the onset of symptoms or aggravated existing symptoms to the point that they meet OSHA recordability criteria. Simply stated, unless the illness was caused solely by a nonwork-related event or exposure off-premises, the case is presumed to be work related. Examples of work tasks or working conditions that are likely to elicit a work-related CTD are as follows:

12.8.1.1 Repetitive and/or prolonged physical activities

12.8.1.2 Forceful exertions, usually with the hands (including tools requiring pinching or gripping)

12.8.1.3 Awkward postures of the upper body, including reaching above the shoulders or behind the back, and angulation of the wrists to perform tasks

12.8.1.4 Localized contact areas between the work or work station and the worker's body; that is, contact with surfaces or edges

12.8.1.5 Excessive vibration from power tools

12.8.1.6 Cold temperatures.

12.8.2 A CTD must exist. There must be either physical findings or subjective symptoms and resulting action, namely:

12.8.2.1 There must be at least one physical finding (e.g., positive Tinel's, Phalen's, or Finkelstein's test; or swelling, redness, or deformity; or loss of motion).

12.8.2.2 There must be at least one subjective symptom (e.g., pain, numbness, tingling, aching, stiffness, or burning), and at least one of the following. (1) Medical treatment including self-administered treatment when made available to employees by this employer. (2) Lost workdays (includes restricted work activity). (3) Transfer/rotation to another job.

Note: **If the above criteria are met, then a CTD illness exists that must be recorded on the OSHA-300 form**. EXAMPLE. A production line employee reports to the health unit with complaints of pain and numbness in the hand and wrist. The employee is given aspirin and, after a follow-up visit with no change in symptoms, is reassigned to a restricted-duty job. Even though there are no positive physical signs, the case is recordable because work activity was restricted.

12.9 Occupational injuries. Injuries are caused by instantaneous events in the work environment. To keep recordkeeping determinations as simple and equitable as possible, back cases are classified as injuries, even though some back conditions may be triggered by an instantaneous event and others develop as a result of repeated trauma. Any occupational injury involving medical treatment, loss of consciousness, restriction of work or motion, or transfer to another job is to be recorded on the OSHA-300 form.

12.10 Evaluation, treatment, and follow-up of CTD's. If CTDs are recognized and treated appropriately early in their development, a more serious condition likely can be prevented; therefore, it is important to identify and treat these disorders as early as possible. The following systematic approach will be used to evaluate employees who report to the health unit.

12.10.1 Screening assessment. Upon the employee's presentation of symptoms, the healthcare provider's screening assessment will include, obtaining a history from the employee to identify the location, duration and onset of pain/discomfort, swelling, tingling and/or numbness, and associated aggravating factors. A brief noninvasive screening examination for the evaluation of CTDs will consist of inspection, palpation, range of motion testing, and various applicable maneuvers. (See Barbara Silverstein, *Evaluation of Upper Extremity and Low Back*, Selected Bibliography.)

12.10.1.1 Based on the severity of symptoms and physical signs, the ____________ or other healthcare provider will decide whether to initiate conservative treatment and/or to refer promptly to a physician for further evaluation.

12.10.1.2 If mild symptoms and no physical signs are present, conservative treatment is recommended. Examples include the following:

- Applying heat or cold. Ice is used to treat overuse strains and muscle/tendon disorders for relief of pain and swelling, thus allowing more mobility. Ice decreases the inflammation associated with CTDs even if no overt signs of inflammation (redness, warmth, or swelling) are present. The use of ice may be inappropriate for Raynaud's disease (vibration syndrome), rheumatoid arthritis, and diabetic conditions. Heat treatments should be used only for muscle strains where no physical signs of inflammation are present.

- Nonsteroidal anti-inflammatory agents. These agents may be helpful in reducing inflammation and pain. Examples of these types of agents include aspirin and ibuprofen.
- Special exercise. If active exercises are utilized for employees with CTDs, they should be administered under the supervision of the OHN or physical therapist. If these active exercises are performed improperly, they may aggravate the existing condition.
- Splints. A splint may be used to immobilize movement of the muscles, tendons, and nerves. Splints should not be used during working activities unless it has been determined by the OHN and ergonomist that no wrist deviation or bending is performed on the job. Splinting can result in a weakening of the muscle, loss of normal range of motion due to inactivity, or even greater stress on the area if activities are carried out while wearing the splint.

12.10.2 Follow-up assessment after 2 days.

12.10.2.1 If the condition has resolved, reinforce good work practices and encourage the employee to return to the health facility if there are problems.

12.10.2.2 If the condition has improved but is not resolved, continue the above treatment for approximately 2 days and reevaluate.

12.10.2.3 If the condition is unchanged or worse, check compliance with the prescribed treatment and perform a screening examination.

12.10.2.4 If the screening examination is positive, or if the condition is worse, the employee will be referred to the company physician, and reassigned to a light or restricted-duty position. If the screening examination is negative for physical signs, but the condition is unchanged, conservative treatment will be continued.

12.10.2.5 Job reassignments will be coordinated with the supervisor and must be chosen with knowledge of whether the new task will require the use of the injured tendons, or place pressure on the injured nerves. Inappropriate job reassignment can continue to injure the inflamed tendon or nerve, which can result in permanent symptoms or disability. The appropriate light-duty job can be selected from the low ergonomic risk jobs list maintained by the healthcare provider. These restricted or light-duty jobs are one of the most helpful treatments for CTDs. These jobs, if properly selected, allow the worker to perform while continuing to ensure recovery. Some CTDs require weeks (or months, in rare cases) of reduced activity to allow for complete recovery.

12.10.3 Follow-up assessment after 6 days.

12.10.3.1 After about 6 days, if the condition appears to have been resolved, supervisors will be notified, good work practices should be reinforced, and the employee should be encouraged to return to the health facility if problems resurface.

12.10.3.2 If the condition has improved but is not resolved, the above treatment will be continued for approximately 2 more days and reevaluated.

12.10.3.3 If the condition is unchanged or worse, compliance with prescribed treatment should be checked and a screening examination performed. If the screening examination is positive, the employee will be referred to the company physician.

12.10.4 Follow-up after 8 days.

12.10.4.1 If, after about 8 days, the condition has resolved, good work practices will be reinforced and the employee will be encouraged to return to the health facility if problems resurface.

12.10.4.2 If the condition has not resolved within approximately 8 days, the employee will be referred to the company physician automatically.

12.10.5 Other considerations

12.10.5.1 If an employee misses a scheduled reevaluation, the healthcare provider will contact the employee to assess the condition within approximately 5 days of the last presentation.

12.10.5.2 The referring physicians or healthcare providers will be furnished with a written description of the ergonomic characteristics of the job of the employee who is being referred.

12.10.6 Surgery. Recommendations for surgery will be referred for a second opinion. If surgery is performed, an appropriate amount of time off work is essential to allow healing to occur and prevent recurrence of symptoms. The number of days off work will depend on each worker's individual response and will agree with the recommendations of the treating physician.

12.10.7 Return to work. A physician evaluation of the employee after time away from work, to assess work capabilities, will be performed to ensure appropriate job placement. When an employee returns to work after time off, after an operation, or must rest an inflamed tendon, ligament, or nerve, there will be a reconditioning of the healing muscle–tendon groups. Supervisors and healthcare providers will give consideration to permanently reassigning the employee to an available job with the lowest risk of developing CTDs.

12.11 Other considerations

12.11.1 A case is considered to be complete once there is complete resolution of the signs and symptoms. After resolution of the problem, if the signs or symptoms recur, a new case is established and thus must be recorded on the OSHA 200 form as such. Furthermore, failure of the worker to return for care after 30 days indicates symptom resolution. Any visit to a healthcare provider for similar complaints after the 30-day interval implies reinjury or reexposure to a workplace hazard and would represent a new case.

12.11.2 It is essential that required data, including job identification, be consistently, fully, and accurately recorded on the OSHA-200 form. "Job identification" will include the appropriate job title for "Occupation" and the appropriate organizational unit for "Department" on the OSHA-200.

12.11.3 Periodic workplace walk-throughs. Healthcare providers will conduct periodic, systematic workplace walk-throughs on a monthly basis (OSHA recommended) to remain knowledgeable about operations and work practices, to identify potential light-duty jobs, and to maintain close contact with employees. Healthcare providers also should be involved in identifying risk factors for CTDs in the workplace as part of the ergonomic team. A record will be kept documenting the date of the walk-through, area(s) visited, risk factors recognized, and action initiated to correct identified problems. Follow-up will be initiated and documented to ensure corrective action is taken when indicated.

12.12 Trend analysis

12.12.1 Company healthcare providers should periodically (e.g., quarterly) review healthcare facility sign-in logs, OSHA-200 forms, and individual employee medical records to monitor trends for CTDs in our facilities. This ongoing analysis should be made in addition to a "symptoms survey" to monitor trends continuously and to substantiate the information obtained in the annual symptoms survey. The analysis should be done by department, job title, work area, and so on.

12.12.2 The information gathered from the annual symptoms survey will help to identify areas or jobs where potential CTD problems exist. This information may be shared with anyone in the plant, since employees' personal identifiers are not solicited. The analysis of medical records (e.g., sign-in logs and individual employee medical records) may reveal areas or jobs of concern, but it may also identify individual workers who require further follow-up. The information gathered while analyzing medical records will be of a confidential nature; thus, care must be exercised to protect the individual employee's privacy.

12.12.3 The information gained from the CTD trend analysis and symptoms survey will help determine the effectiveness of the various programs initiated to decrease ergonomic problems in our facilities.

12.13 Symptoms survey. A symptoms survey will be developed to provide a standardized measure of the extent of symptoms of work-related disorders for each area of the plant, to determine which jobs are exhibiting problems and to measure progress of the ergonomic program.

12.13.1 Design of survey. A survey of employees will be conducted to measure employee awareness of work-related disorders and to report the location, frequency, and duration of discomfort. Body diagrams will be used to facilitate the gathering of this information.

12.13.2 Surveys normally will not include employees' personal identifiers, this is to encourage employee participation in the survey.

12.13.3 Frequency. Surveys will be conducted on an annual basis. Conducting the survey annually should help detect any major change in the prevalence, incidence, and/or location of reported symptoms.

12.14 Low ergonomic risk jobs. This company will compile a list of light-duty jobs. Jobs will be analyzed to determine the physical procedures used in the performance of each job, including lifting requirements, postures, hand grips, and frequency of repetitive motion. For such jobs, the ergonomic risk will be described. This information will assist healthcare providers in recommending assignments to light- or restricted-duty jobs. The light-duty job should therefore not increase ergonomic stress on the same muscle–tendon groups. Supervisors should periodically review and update the lists.

Low Ergonomic Risk Job Listing

Department	Task/Job	Date Evaluated

12.15 High ergonomic risk jobs. This company will compile a list of high ergonomic risk jobs. Jobs will be analyzed to determine the physical procedures used in the performance of each job, including lifting requirements, postures, hand grips, and frequency of repetitive motion. For such jobs, the ergonomic risk will be described. This information will assist healthcare providers in determining jobs from which assignments to light- or restricted-duty jobs may be necessary. Supervisors should periodically review and update the lists.

High Ergonomic Risk Job Listing

Department	Task/Job	Date Evaluated

13. Training and education. The purpose of training and education is to ensure that our employees are sufficiently informed about the ergonomic hazards to which they may be exposed and thus are able to participate actively in their own protection.

13.1 Employees will be adequately trained about the (YOUR COMPANY)'s ergonomics program. Proper training will allow managers, supervisors, and employees to understand ergonomic and other hazards associated with a job or production process, hazard prevention and control, and their medical consequences. The training program should include the following individuals:

13.1.1 All affected employees

13.1.2 Engineers and maintenance personnel

13.1.3 Supervisors

13.1.4 Managers

13.1.5 Healthcare providers.

13.2 Program design. The program will be designed and implemented by ______________________________. Appropriate special training should be provided for personnel responsible for administering the program.

13.3 Learning level. The program will be presented in language and at a level of understanding appropriate for the individuals being trained. It will provide an overview of the potential risk of illnesses and injuries, causes and early symptoms, means of prevention, and treatment.

13.4 Evaluation. The program will also include a means for adequately evaluating its effectiveness. This will be achieved by using combinations of the following:

13.4.1 Employee interviews

13.4.2 Testing methods

13.4.3 Observation of work practices.

13.5 Training for affected employees will consist of both general and specific job training:

13.5.1 General training. Employees who are potentially exposed to ergonomic hazards will be given formal instruction on the hazards associated with their jobs and with their equipment. This will include information on the varieties of hazards associated with the job, what risk factors cause or contribute to them, how to recognize and report symptoms, and how to prevent these disorders. This instruction will be repeated for each employee as necessary. This training will be conducted on an annual basis. (OSHA's experience indicates that, at a minimum, annual retraining is advisable).

13.5.2 Job-specific training. New employees and reassigned workers will receive an initial orientation and hands-on training prior to being placed in a full-production job. Training lines may be used for this purpose. Each new hire will receive a demonstration of the proper use of and procedures for all tools and equipment. The initial training program will include the following:

13.5.2.1 Care, use, and handling techniques pertaining to tools

13.5.2.2 Use of special tools and devices associated with individual work stations

13.5.2.3 Use of appropriate guards and safety equipment, including personal protective equipment

13.5.2.4 Use of proper lifting techniques and devices

13.6 Training for supervisors. Supervisors are responsible for ensuring that employees follow safe work practices and receive appropriate training to enable them to do this. Supervisors, therefore, will undergo training comparable to that of the employees, and such additional training as will enable them to recognize early signs and symptoms of ergonomic stressors, to recognize hazardous work practices, to correct such practices, and to reinforce the ABC ergonomic program, especially through the ergonomic training of employees as may be needed.

13.7 Training for managers. Managers will be made aware of their safety and health responsibilities and will receive sufficient training pertaining to ergonomic issues at each work station, and in the production process as a whole, so that they can effectively carry out their responsibilities.

13.8 Training for engineers and maintenance personnel. Plant engineers and maintenance personnel will be trained in the prevention and correction of ergonomic hazards through job and work station design and proper maintenance, both in general and as applied to the specific conditions of the facility.

13.9 Employee training and education. Company healthcare providers will participate in the training and education of all employees. This training will be reinforced during workplace walk-throughs and the individual health surveillance appointments. All new employees will be given such education during orientation. This demonstration of concern and the distribution of information should facilitate the early recognition of CTDs prior to the development of more severe and disabling conditions and increase the likelihood of compliance with prevention and treatment.

13.9.1 Early report of symptoms. Employees will be encouraged by healthcare providers and supervisors to report early signs and symptoms of CTDs to the in-plant health facility or ______________________ (predesignated authority). This will allow for timely and appropriate evaluation and treatment. Supervisors and managers at all levels will be careful to avoid any potential disincentives for employee reporting, such as limits on the number of times an employee may visit the health unit.

14. Definitions. A wide variety of terms are currently used by employers, occupational safety and health professionals, and others in describing ergonomic programs. The following definitions are provided to clarify the terms used by OSHA in their ergonomic program management guidelines:

– **Cumulative trauma disorders (CTDs)** is the term used in these guidelines for health disorders arising from repeated biomechanical stress due to ergonomic hazards. Other terms that have been used for such disorders include "repetitive motion injury," "occupational overuse syndrome," and "repetitive strain injury." CTDs are a class of musculoskeletal disorders involving damage to the tendons, tendon sheaths, synovial lubrication of the tendon sheaths, and the related bones, muscles, and nerves of the hands, wrists, elbows, shoulders, neck, and back. The more frequently occurring occupationally induced disorders in this class include carpal tunnel syndrome, epicondylitis (tennis elbow), tendonitis, tenosynovitis,

synovitis, stenosing tenosynovitis of the finger, DeQuervain's disease, and low back pain.

– **Ergonomic hazards** refer to workplace conditions that pose a biomechanical stress to the worker. Such hazardous workplace conditions include, but are not limited to, faulty work station layout, improper work methods, improper tools, excessive tool vibration, and job design problems that include aspects of work flow, line speed, posture and force required, work/rest regimens, and repetition rate. They are also referred to as "stressors."

– **Ergonomic risk factors** are conditions of a job, process, or operation that contribute to the risk of developing CTDs.

– **Ergonomics Team/Committee** refers to those responsible for identifying and correcting ergonomic hazards in the workplace, including ergonomic professionals or other qualified persons, healthcare providers, engineers and other support personnel, plant safety and health personnel, managers, supervisors, and employees.

– **Ergonomist or ergonomics professional** means a person who possesses a recognized degree or professional credentials in ergonomics or a closely allied field (such as human factors engineering) and who has demonstrated, through knowledge and experience, the ability to identify and recommend effective means of correction for ergonomic hazards in the workplace.

– **Healthcare provider** is a physician who specializes in occupational medicine, or a registered nurse specializing in occupational health, or other health personnel (such as emergency medical technicians) working under the supervision of a physician or registered nurse. Healthcare providers will have the training outlined in Appendix B, "Medical Management Program."

– **Qualified person** means one who has thorough training and experience sufficient to identify ergonomic hazards in the workplace and recommend an effective means of correction. An example would be a plant engineer fully trained in ergonomics.

– **Systems approach** to safety and health management means a comprehensive program by the employer that addresses workplace processes, operations, and conditions as interdependent systems in order to identify and to eliminate or reduce all types of hazards to employees. Thus, complex ergonomic problems may require a combination of solutions.

REVIEW QUESTIONS

1. List five MSD problem identifiers.
2. What are the benefits of worker involvement in an ergonomics program?
3. Who should participate in a good ergonomics program?
4. What are some of the risk factors to look for in a job/task?

5. What is a job analysis? List some of the steps involved in performing a job analysis?

6. What are some of the return on investment methods?

REFERENCES

Bureau of Labor Statistics (2011). USDL-11-16-12. News Release.

Liberty Mutual Insurance Company. Manual Materials Handling.

NIOSH (1981). Work Practices Guide for Manual Lifting. Cincinnati, OH.

OSHA, O. S. (n.d.). Retrieved February 2015, from Occupational Safety and Health Administration: http://www.osha.gov./SLTC/ergonomics/controlhazards.html.

Six Sigma Costs and Savings. (n.d.). Retrieved 2015, from Six Sigma: http://www.isixsigma.com/implementation/financial-analysis/six-sigma-costs-and-savings/.

7

BIOMECHANICS

LEARNING OBJECTIVES

At the end of this module, students will be able to describe the basic elements of biomechanics as they relate to the human body.

INTRODUCTION

Biomechanics is the study of forces on the human body (Marras, 2006). Every task from the gait of the physically handicapped, the lifting of a load by a factory worker, to the performance of an athlete can be described in terms of the specific movements and loading, both muscular and structural (Winter, 1990). The specific movements and loading on the musculoskeletal system change from case to case.

BACKGROUND

Why Is Biomechanics Important to Ergonomics?

Biomechanics is a useful tool in situations where lifting, pushing, or pulling is performed with or without a load. In certain body postures, the body's own weight creates postural stress. The goal of occupational biomechanics is to quantitatively describe the musculoskeletal loading that occurs during work, in order to derive the degree of risk associated with the task.

Occupational Ergonomics: A Practical Approach, First Edition.
Theresa Stack, Lee T. Ostrom and Cheryl A. Wilhelmsen.

The Body Described as Levers

A major assumption of occupational biomechanics is that the body behaves according to the laws of Newtonian mechanics. Movement in the body is produced by a system of levers. The levers work together to produce coordinated action, some by actual movement (dynamic) and others by stabilization (static).

BIOMECHANICS OF THE HUMAN BODY

Parts of a Lever System

The following explanation of levers, as applied to the human body, is a simplified version to help all students understand the correlation between body posture and forces. Figures 7.1 and 7.2 explain the basic parts of a lever system.

Free body and force diagrams used.

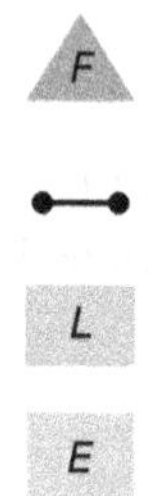

Figure 7.1 Parts of a free body diagram

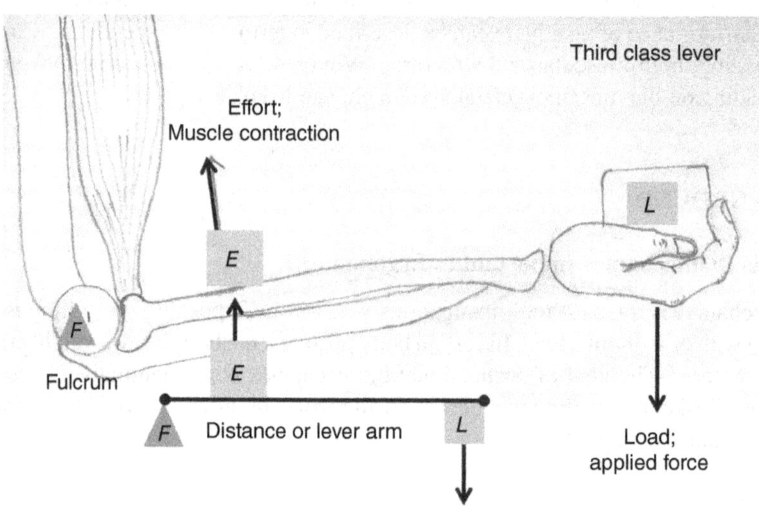

Figure 7.2 Parts of a lever system as displayed on the human body

- **Fulcrum**: pivot about which a **Lever** turns. In the human body, this is typically the joint center.
- **Lever:** rigid bar that turns about an axis of rotation or **Fulcrum**. In the human body, this is typically a combination of body parts, a limb for example (what anthropometry calls the bodies link lengths).
- **Lever arm:** distance from the fulcrum to either the load or the effort.
- **Load***:* *applied force*. In the human body, this can be the weight of the object in the hand, as well as the limb weight.
- **Effort***:* *resistance force*. In the human body, this is the force generated by the musculoskeletal system that is applied to cause movement against load and/or stabilize a joint.

Classes of Levers in the Human Body

The class of lever is determined by the relative position of the load or applied force, fulcrum, and effort force, as shown in Figure 7.3.

- In a first-class lever, the load and effort are at opposing ends of the lever and the fulcrum is located in the center, for example, a child's see saw. When the effort is applied, the load moves in the opposite direction.
  - The skull is a human example
- In a second-class lever, the fulcrum and effort are at opposing ends of the lever and the load is located in the center, for example, a wheelbarrow. When the effort is applied, the load moves in the same direction.
  - The foot is a human example

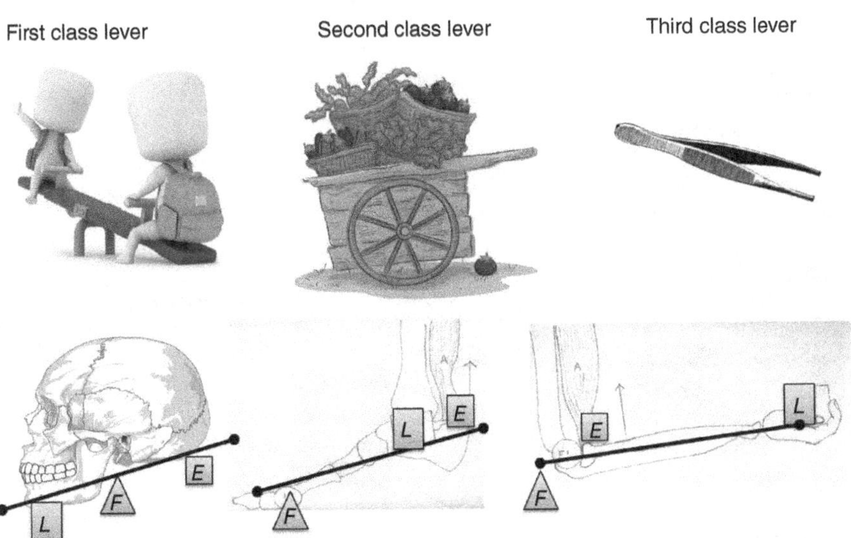

Figure 7.3 This figure displays the relationship between simple levers and the human body during static loading

- In a third-class lever, the fulcrum and load are at opposing ends of the lever and the effort is located in the center, for example, a staple remover. When the effort is applied, the load moves in the same direction.
    - The bent arm is a human example.

BIOMECHANICS MADE SIMPLE

Biomechanical loads or stresses on the body are not defined purely by the magnitude of the weight or applied load. The position of the weight relative to the fulcrum (or point rotation of the joint) and direction of force defines the muscular effort required by the body. For example, as shown in Figure 7.4, holding a 40 lb dumbbell does not produce 40 lb of force on the elbow, nor does the bicep respond with 40 lb of force. It does, however, create a tendency for the system to rotate, and those rotational forces are called moments.

Let's Take a Moment

The measure of a force's tendency to cause a body to rotate about a specific point, that is, *rotational force* is a moment. A moment is a vector quantity having both direction and magnitude. A moment is defined as the product of the applied force and the perpendicular distance (lever arm) through which the force is applied. The perpendicular distance between the fulcrum and the effort is the muscle effort lever arm; the perpendicular distance between the fulcrum and the load is the load lever arm.

Moments are commonly expressed in Newton-meters (N m). A Newton is a unit of force that takes mass into consideration. One Newton is equal to 0.225 lb. Moments can also be expressed as inch-pounds or foot-pounds.

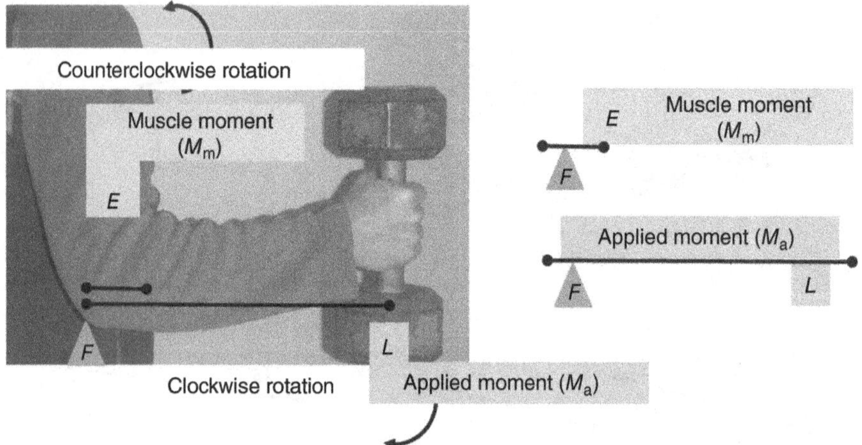

Figure 7.4 Holding a 40-lb weight results in 480 in.-lb of clockwise rotational force, the moment, at the elbow. The body must respond counterclockwise effort caused by the muscles contractile force

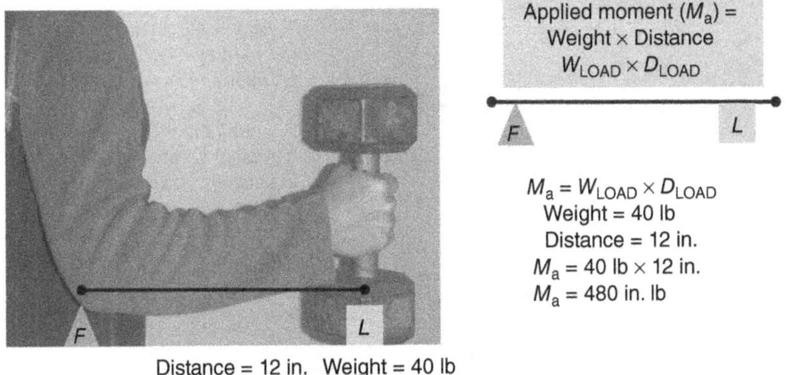

Figure 7.5 Moment (or force) = weight × distance

While this method is simplistic, the general relation of forces acting and the body's reaction to them is sound. For the purposes of this chapter, a moment or rotational force is the product of weight and distance.

$$\text{Moment} = \text{Weight} \times \text{Distance}$$

(*D*) *Distance* (or moment arm or lever arm) is the measurement from the point of rotation perpendicular to the direction of the applied weight or load. In Figure 7.5, the distance from the elbow to the object being held in the hand is applied force lever arm. Distance is expressed in inches (in.).

(*W*) *Weight* is the force generated by the gravitational attraction of the earth on the mass of an object. Weight is expressed in pound-force (lb) as opposed to pound-mass (lbm). In the example from Figure 7.5, a third-class lever can be seen in the arm with the point of rotation or fulcrum at the elbow.

Holding a weight in the hand creates a moment around the elbow, tending to make it extend. Muscles spanning the elbow create the opposite moment by contracting, so that the elbow is able to support the weight; the greater the weight in the hands, the larger the moment at the elbow.

$$M = W \times D$$

Holding a 40-lb object in the hand produces a 480 in. lb (40 lb × 12 in.) of rotational force at the elbow (Figure 7.5). The rotational force created by the effort must be equal in magnitude to a 480 in. lb to stop rotation, but applied in the opposite or counterclockwise direction (Figure 7.6).

This 480 in. lb applied moment from the load results in the muscle responding with 960 lb of effort (Figure 7.7).

As inferred in this example, the longer the distance between the object and the point of rotation, the greater the rotational force or moment. When lifting an object, the point of rotation in the torso is the L5/S1 spinal unit. The lever arm or distance is measured from the L5/S1 vertebra to the object in the hands, reference Figure 7.8.

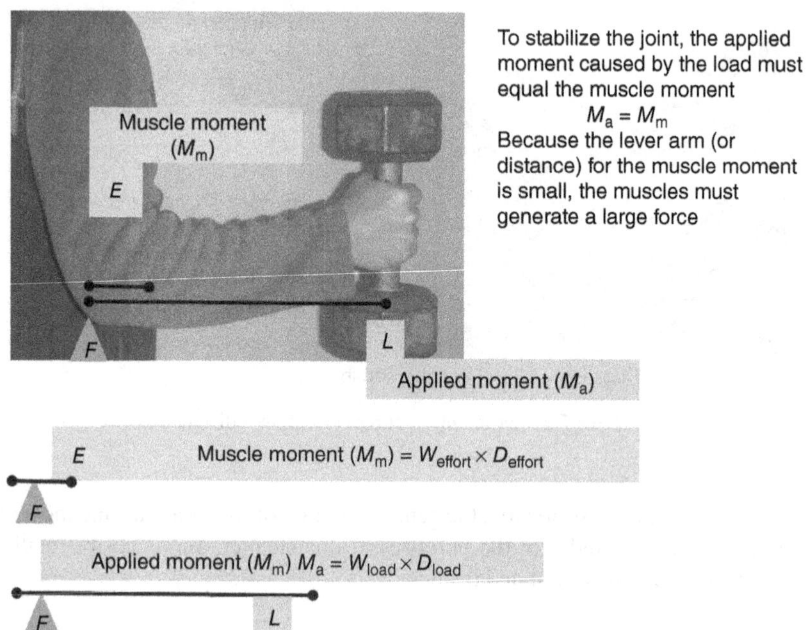
Muscle moment
(M_m)
E
F
L
Applied moment (M_a)
To stabilize the joint, the applied moment caused by the load must equal the muscle moment
$M_a = M_m$
Because the lever arm (or distance) for the muscle moment is small, the muscles must generate a large force
E
Muscle moment (M_m) = $W_{effort} \times D_{effort}$
F
Applied moment (M_m) $M_a = W_{load} \times D_{load}$
F
L

Figure 7.6 To stabilize a joint the resultant or muscle force (Mm) must equal the applied force (Ma)

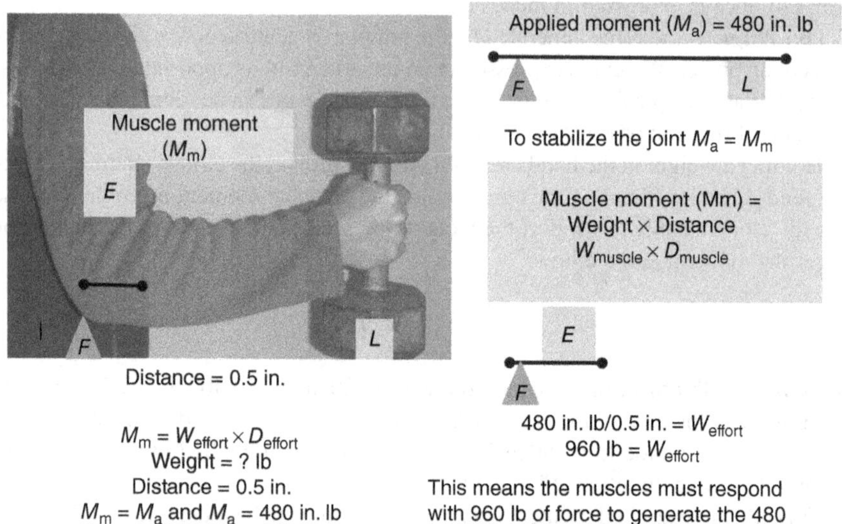
Muscle moment
(M_m)
E
F
L
Distance = 0.5 in.
Applied moment (M_a) = 480 in. lb
F
L
To stabilize the joint $M_a = M_m$
Muscle moment (Mm) =
Weight × Distance
$W_{muscle} \times D_{muscle}$
E
F
$M_m = W_{effort} \times D_{effort}$
Weight = ? lb
Distance = 0.5 in.
$M_m = M_a$ and M_a = 480 in. lb
480 in. lb = W_{effort} × 0.5 in.
480 in. lb/0.5 in. = W_{effort}
960 lb = W_{effort}
This means the muscles must respond with 960 lb of force to generate the 480 in. lb moment that is needed to stabilize the joint. the load, not the muscle has the biomechanical advantage

Figure 7.7 Resultant or muscle force is significantly higher than applied force

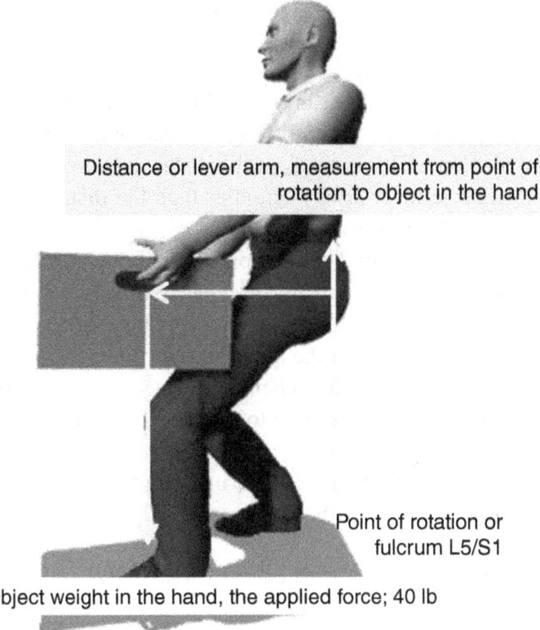

Figure 7.8 The spine as a third-class lever system (Adapted with permission The Ergonomics Image Gallery)

If rotational force or **moment = weight × distance,** how much rotational force or moment is generated on the L5/S1 spinal unit when the 40 lb weight is lifted?

It Depends on the Distance!

- Holding 40 lb, 20 in. from the L5/S1 results in 800 in. lb of rotational force.
- Holding 40 lb, 15 in. from the L5/S1 results in 600 in. lb of rotational force.
- Holding 40 lb, 10 in. from the L5/S1 results in 400 in. lb of rotational force.

Clearly, a practical application to understanding the forces on the body is demonstrated. Reducing the distance from the load to the point of rotation reduces the applied rotational forces on the spine. In other words, when lifting, the load should be held as close to the body as possible.

The muscles in our body need to generate the same amount of rotational force, in the opposite direction, to keep the body from rotating. The lever arm in the spine, the distance from the fulcrum to the muscle attachment, is small. Therefore, the muscle must generate a force greater than that which is held in the hands.

Third-Class Lever The musculoskeletal system can be represented by a lever system. Three types are present in the human body. Most of the joint rotations in our body behave as a third-class lever. A third-class lever has the benefit of good range of motion, speed, and power. However, large muscle forces are required to move even a small amount of weight.

In a third-class lever, the point of rotation or fulcrum is located at one end of the system. The applied load acts on the other end of the system and the force (muscle force) acts between the two (Figure 7.5).

A third-class lever system puts the body at a biomechanical disadvantage because the muscles have to generate considerably more rotational force than the rotational force generated by the load. This is because the distance from the point of rotation to the muscle action (muscle lever arm) is smaller than the distance from the point of rotation to load (applied lever arm).

The same lever system is found in the spine; the point of rotation or fulcrum is the L5/S1 spinal unit. The distance from the muscle attachment point to the L5/S1 is approximately 1 in. (see Figure 7.9). Given our greatest effort, we can never bring the load any closer to our spine than the depth of our torso.

To keep the weight of the object from bending the torso forward, the muscles need to generate a moment equal in magnitude to the moment of the applied weight (i.e., the load).

Figure 7.9 illustrates that holding 40 lb, 20 in. from the L5/S1 results in 800 in. lb moment. To calculate the forces on the spine:

$$\text{Muscle moment } (M_{\text{m}}) = \text{applied moment } (M_{\text{a}})$$

$$M_{\text{m}} = M_{\text{a}}$$

$$M_{\text{a}} = 800\,\text{in. lb (which is } 40\,\text{lb} \times 20\,\text{in.)}$$

$$M_{\text{m}} - 800\,\text{in. lb}$$

$$M_{\text{m}} = \text{distance} \times \text{force } (W_{\text{muscle}})$$

$$800\,\text{in. lb} = 1\,\text{in.} \times W_{\text{muscle}}$$

$$800\,\text{lb} = W_{\text{muscle}} \text{ or muscle force}$$

Figure 7.9 illustrates that the muscles generate 800 lb. of force to statically hold a 40-lb weight and that the muscles of the spine are at a biomechanical disadvantage.

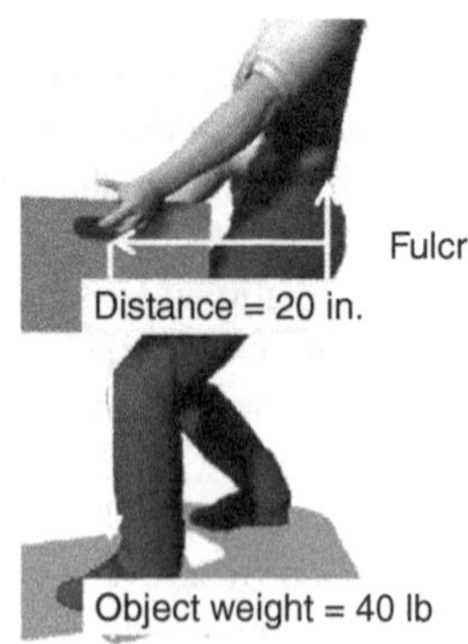

Applied moment (M_a) = Weight × Distance

M_a = 40 lb × 20 in.

M_a = 800 in. lb

The 40 lb weight is causing 800 in. lb rotational force

Muscle moment (M_m) = Applied moment (M_a)

$M_m = M_a$

M_m = 800 in. lb

Distance from the L5/S1 to the muscle insertion is 1 in.

$M_m = D_{muscle} \times W_{muscle}$

800 in. lb = 1 in. × W_{muscle}

800 lb = W_{muscle}

The 40 lb weight causes 800 ib of muscle force or effort

Figure 7.9 The spine as a third-class lever system displaying that the further the load is from body the higher the forces on the spine (The Ergonomics Image Gallery)

To increase the biomechanical advantage of the spine, keep the weight located close to the body, as this decreases the lever arm (D) of the applied force.

When bending to pick up a load, the weight of the torso adds considerably to the applied force especially if lifting in an awkward posture. The torso contains approximately 60% of your body weight. A 200-lb person carries 120 lb in their torso.

The moment or rotational force is calculated the same way (i.e., $M = W \times D$). The distance in this example is the distance from the L5/S1 spinal unit to the center of gravity of the torso. Note these calculations are simplified to illustrate the ratio of applied force to muscle force.

The moment of the body (M_{body}) = 120 lb × 6 in. or M_{body} = 720 in. lb. This value is added to the applied moment (Ma) or 800 in. lb + 720 in. lb = 1520 in. lb. To solve for forces acting on the spine when bending forward to lift:

$$\text{Muscle moment } (M_{\text{m}}) = \text{Total applied moment } (M_{\text{total}})$$

$$M_{\text{m}} = M_{\text{total}}$$

where

$$M_{\text{total}} = 800 \text{ in.-lb (Ma)} + 720 \text{ in.-lb } (M_{\text{body}})$$

$$M_{\text{total}} = 1520 \text{ in.-lb}$$

$$\text{Mm} = M_{\text{total}}$$

$$M_{\text{effort}} = \text{distance} \times \text{force } (W_{\text{muscle}})$$

$$1520 \text{ in.-lb} = 1 \text{ in.} \times W_{\text{muscle}}$$

$$1520 \text{ lb} = \text{Muscle Force}$$

Keeping the spine in a neutral posture reduces the muscle contraction forces and thus the forces on the spinal unit. In this example, the muscles generate 1520 lb of force to statically hold a 40-lb weight because of the added force caused by the weight of the torso when leaning forward. Clearly, the muscles are at a biomechanical disadvantage.

To increase the biomechanical advantage of the spine, keep the weight located close to the body, as this decreases the lever arm of the applied force (Ma), and keep the ears over the shoulders and shoulders over the hips to limit the contribution of the torso to the applied forces. Figure 7.10 shows an extraordinarily difficult posture to maintain for lifting.

Second-Class Lever A second-class lever has the fulcrum on one end, the muscle force or effort on the other, and the load between the two. A wheel barrel or nutcracker is an example of a second-class lever (see Figure 7.3). In the body, the foot is a second-class lever; the ball of the foot acts as the fulcrum or point of rotation. The load is applied through the tibia (or lower leg bone) and the muscle force is applied through the gastronomies or calf muscle.

The muscle has the biomechanical advantage in this lever system due to the length of the lever arm between the point of rotation and the muscle force. Relatively little muscle force is necessary to move a heavy weight (refer to Figure 7.11).

Figure 7.10 Awkward lifting posture results in high spinal loading

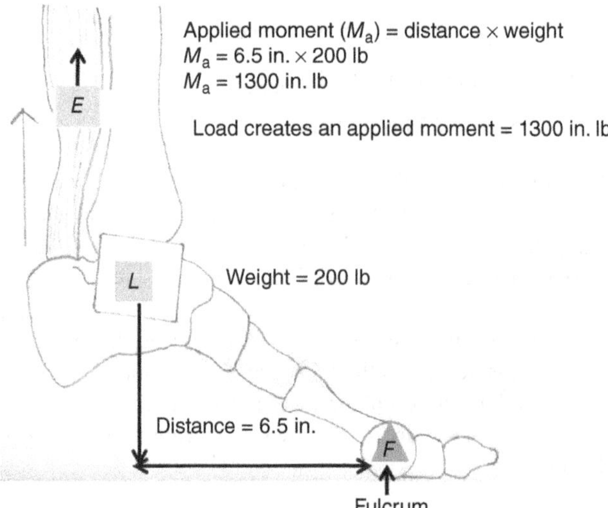

Figure 7.11 In the body, your foot has the mechanical advantage; this is an example of a second-class lever

How much muscle force does it take to move a 200-lb person? The distance from the ball of the foot (fulcrum) to the tibia is assumed to be 6.5 in. (for a 6-ft man of average build who weighs 200 lb).

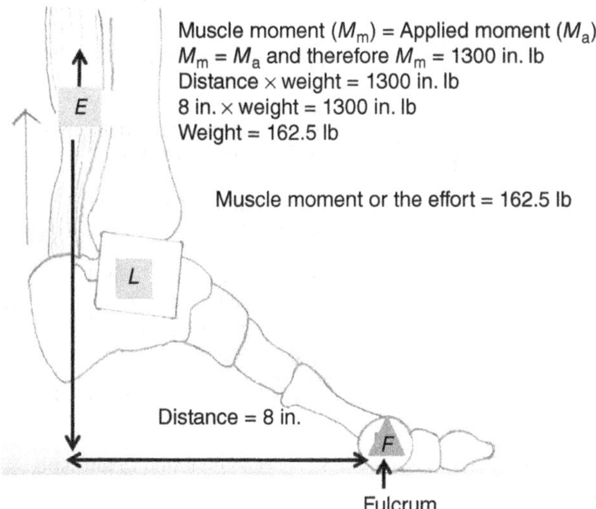

Figure 7.12 Second-class lever in the foot give us a biomechanical advantage, our muscles contract with far less force than is required to move the body

To calculate the applied moment:

To move the body, the calf muscle needs to generate a moment equal in magnitude to the moment of the applied weight but in the opposite direction; the body weight is pulling the foot downward, while the calf is lifting the foot up.

The muscles in a second-class lever system have the mechanical advantage. In this example, the muscles generate 162.5 lb of force to move a 200-lb person (see Figure 7.12).

First-Class Lever First-class levers are those that have a fulcrum in the middle of the system, an applied load on one end, and an opposing (muscle) force on the opposite end of the system. Sea saws, crowbars, and scissors are examples of a first-class lever (see Figures 7.3 and 7.13). This type of lever results in balanced movement, good force, and range of motion. Our skull is an example of a first-class lever in the body.

The applied force is the center of gravity of the skull located in the proximity of the chin, the fulcrum is the joint between the skull and the spine, and the muscle force is actuated by the muscles on the back of the head or neck that tilt the head backward.

How much muscle force does it take to hold my head in a neutral posture? Using the same 200 lb male, the distance from the center of the skull (fulcrum) to the chin (assumed center of gravity of the head) is 4 in. In addition, the average adult head weighs between 15 and 21 lb, for this example, we will use 18 lb.

To calculate the moment created by the applied weight (the skull) reference Figure 7.13:

$$\text{Applied moment}\,(M_a) = \text{distance} \times \text{weight}$$

$$M_a = 4\,\text{in.} \times 18\,\text{lb}$$

$$M_a = 72\,\text{in.\,lb}$$

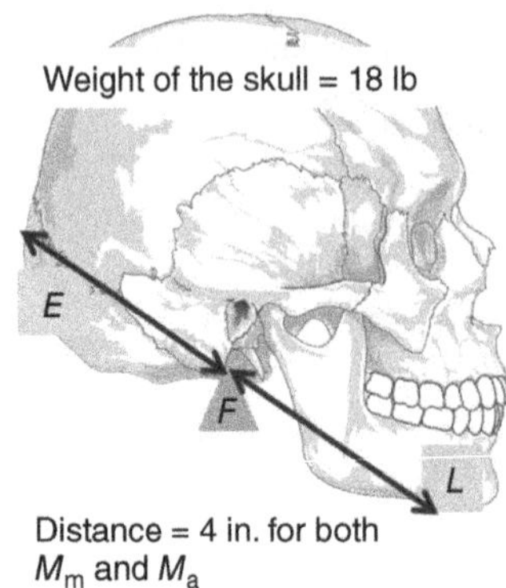

Applied moment (M_a) = distance × weight
M_a = 4 in. × 18 lb
M_a = 72 in. lb
The skull, if centered on the neck, results
In a 72 in. lb rotational force

Muscle moment (M_m) = Applied moment (M_a)
$M_m = M_a$, where M_a is 72 in. lb, therefore
M_m = 72 in. lb
Distance × weight = 72 in. lb
4 in. × weight = 72 in. lb
Weight or muscle force = 18 lb

It takes 18 lb of force to hold up the 18 lb skull
if the skull is centered on the spine

Figure 7.13 A first-class lever can be found in the skull

To stabilize the head in a neutral posture, neck muscle needs to generate a moment equal in magnitude to the moment of the applied weight.

$$\text{Muscle moment } (M_{\text{m}}) = \text{Applied moment } (M_{\text{a}})$$
$$M_{\text{m}} = M_{\text{a}}$$
$$M_{\text{a}} = 72\,\text{in.lb}$$
$$72\,\text{in.lb} = \text{distance} \times \text{weight}$$
$$72\,\text{in.lb} = 4\,\text{in.} \times \text{weight}$$
$$18\,\text{lb} = \text{muscle force}$$

In a first-class lever system, neither the muscles nor the load have the mechanical advantage as long as the head is held in a neutral posture. From this example, Figure 7.14, it is easy to see that the further the head leans forward, the greater the contribution of the applied moment and thus the greater the muscle force required to maintain the posture.

Back Injury Prevention Rules of Thumb

- If the load is not close, the pressure is gross.
- If the back is bent, one will not prevent.
- If muscles are slack, you will hurt your back.

SUMMARY

The goal of occupational biomechanics is to describe the degree of postural stresses on the body. A basic understanding of the degree of postural stress a person experiences is beneficial to the ergonomic practitioner for recognizing where workplace, tool, or task enhancements will best benefit the worker and the employer.

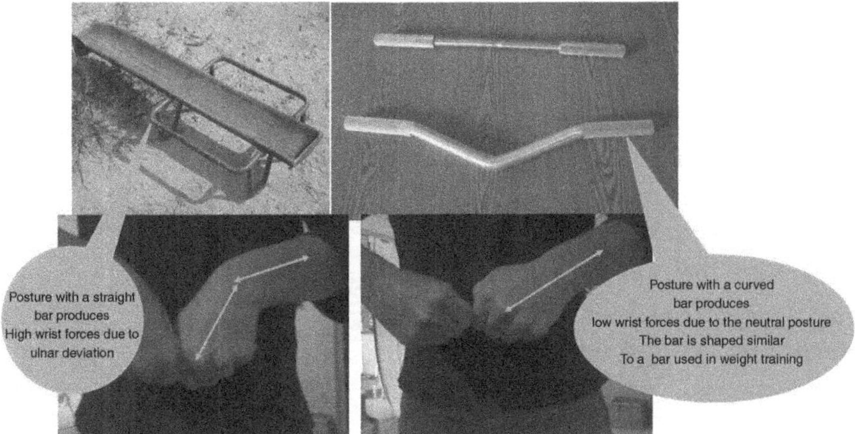

Figure 7.14 The loading tray has a straight handle that results in a deviated write posture, change the angle of the handle would result in a neutral posture and less stress on the wrists

CASE STUDY

Recommend searching www.youtube.com for "firing of the M198 Howitzer" to gain a better understanding of the highly repetitive and physical nature of this drill coupled with the heavy weight and high number of rounds fired. More detailed information can also be found in the Case Study Chapter 16.

The United States Marine Corps Artillery Instructional Battery (AIB) stationed at The Basic School Marine Corps Base Quantico, VA, was experiencing a high rate of injuries and our ergonomics team (Dr Lee Ostrom, Dr Cheryl Wilhelmsen and Theresa Stack) was asked to perform an analysis of the "Call for Fire" operation to help determine why these injuries were occurring. We designed and carried out an ergonomics study to determine causes of these injuries. The study was carried out using data collection methods that would collect anthropometric, psychophysical, biomechanical, and human error data.

The "Call for Fire" exercise takes place over 3 days. During the first day, the AIB sets up the M198 Howitzers (see Figure 7.15). The ammunition is also delivered during day 1. Approximately 1100 rounds of 155 mm ammunition, fuses, and sufficient powder are brought to the site and staged behind each gun. Days 2 and 3 are when the actual firing of the guns occurs; each gun fires between 350 and 400 rounds of ammunition. The rounds weigh between 95 and 105 lb each depending on whether they contain high explosive or white phosphorus.

The Marines move the rounds multiple times through the course of the 2-day shoots. Biomechanical modeling was used to determine if moving the rounds produces an increased risk of injury for the combination of force, posture, frequency, and duration.

Three-Dimensional Static Strength Prediction Program 6.0.2 (Michigan, 1999) was used to model the task as well as to compare the round weight, posture, frequency, and duration to the weight handling limits in MIL-STD 1472F (Defense, 1999).

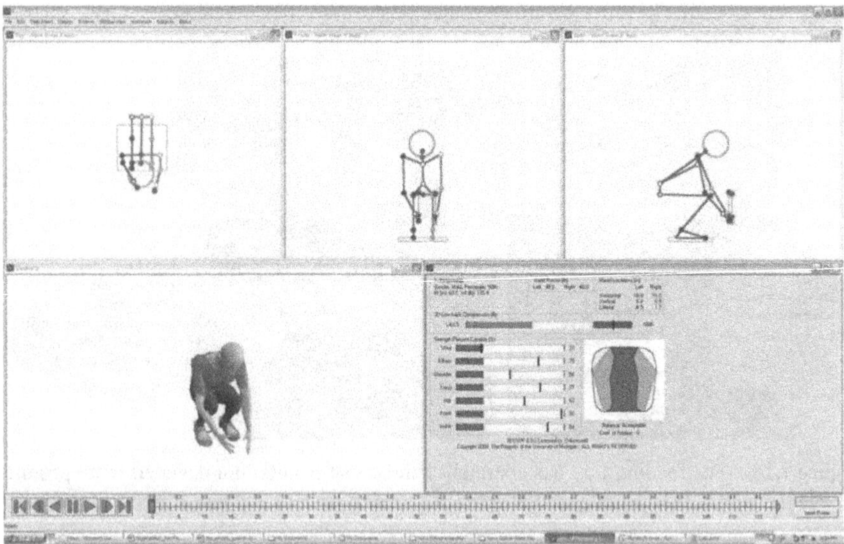

Figure 7.15 Biomechanical modeling of step 5 – moving the round from the pallet to the loading tray

The procedure for moving the round is as follows and can also be found on the Internet with the search term:

1. The fire instructions are radioed to the recorder.
2. The recorder announces the fire order.
3. The powder person adjusts the powder.
4. The type of round is verified.
5. A round is removed from the ready board and laid into the loading tray (Figure 7.11).
6. The two loaders pick up the round.
7. The rammer places the ram at the back of the round.
8. The breach is opened.
9. The loaders bring the round up to the breach, Figure 7.10.
10. The rammer pushes the round in the breach.
11. The loader on the right side of the loading tray releases his grip.
12. The other loader steps back and to the left of the gun.
13. The powder man brings up the powder and hands it to the A-gunner.
14. The A-gunner verifies the amount of powder and places it in the breach.
15. The A-gunner closes the breach and places a primer in the priming hole.
16. The lanyard is attached and then pulled.
17. The gun fires.
18. The A-gunner opens the breach and swabs the breach and breach plug.

TABLE 7.1 Loading Tray and Example of Changing the Angle on the Tray to Reduce the Force on the Wrist

Body Segment	Compressive Forces	Result
L5/S1	1762 lb ±124 lb	Spinal compression fall above the action limit of 770 lb and is therefore considered above hazard threshold for that body segment
	Population Capable of Performing the Task (%)	
Wrist	23	Above hazard threshold
Elbow	75	Marginal
Shoulder	50	Above hazard threshold
Torso	77	Below hazard threshold
Hip	62	Marginal
Knee	97	Below hazard threshold
Ankle	84	Below hazard threshold

TABLE 7.2 Move One Round from Staging to Pallet and Move One Round from Pallet to Loading Tray

Round Weight (lb)	Recommended Weight Limit (MIL-STD) (lb)	Task Factors	Result
97	64	Recommended weight limit for a one-man carry is 82 lb under optimal conditions This value is discounted (reduced) by 22% due to the frequency of the exposure Assume 200 lifts in a 9 h/day (2.7 lifts/min)	Round weight exceeds recommended carrying weight, the task is therefore considered above hazard threshold

Multiple steps were analyzed for the full study; as shown in Table 7.1, it was found that steps 5 and 6 produced the highest forces on the spine as well as the wrist.

To verify the results and provide an alternative evaluation, the Department of Defense Design Criteria Stand for Human Engineering (Defense, 1999; Section 5.9.11.3) was used (see Tables 7.2 and 7.3).

Based on the biomechanical modeling and comparison to the MIL-STD, moving the round from the pallet to loading tray produces an increased risk of injury to the wrists. Redesigning the carry tool may reduce these factors by improving not only the coupling of the hands to the tool but also the degree of hand/wrist deviation. Decreasing the frequency of moving the rounds from the staging area to the pallet,

TABLE 7.3 Moving the Round and the Loading Tray to the Gun (Start of Task)

Round Weight (lb)	Recommended Weight Limit (MIL-STD) (lb)	Task Factors	Result
97	136	Recommended weight limit for a two-man lift (below 36 in.) is 174 lb under optimal conditions This value is discounted (reduced) by 22% due to the frequency of the exposure. Assume 200 lifts in a 9 h/day (2.7 lifts/min)	Round weight does not exceed recommended lifting weight limit; the task is therefore considered below hazard threshold

for example, by allowing the LTs to assist, would decrease the risk of injury to the low back and torso by allowing time for the soft tissues to rest and recover.

Moving the round from the pallet to the loading tray produces the greatest spinal compression due to the squatted posture and twisting motion. Elevating the loading tray improves the task characteristics. The only feasible way to reduce the risk of injury is to allow frequent task rotation.

Loading the gun with the loading tray produces a moderate risk of spinal injury. The risk factors can be greatly reduced by allowing for frequent task rotation. For example, decreasing the frequency of the exposure (using the MIL-STD) reduces the risk level from unsafe to safe. In other words, when loading the gun, it is the frequency of the exposure that is causing the risk.

The Marines received a full report and implemented solutions to reduce the physical demand on the service members.

KEY POINTS

- A person holding an object closer to the body results in lower rotational forces on the spine and thus lower muscle forces. The following is qualitative as there are many factors that affect the spinal loading.
  - 40 lb × 20 in. = 800 in. lb
  - 40 lb × 15 in. = 600 in. lb
  - 40 lb × 10 in. = 400 in. lb
- Movement in the body can be described by a system of levers.
- The levers work together to produce coordinated action, some by actual movement (dynamic), and others by stabilization (static).
- Biomechanical loads or stresses on the body are not purely defined by the magnitude of the weight.
- The longer the lever arm (the distance from the fulcrum to the applied weight), the greater the force.

REVIEW QUESTIONS

1. What are the three lever systems used to describe the human body?
2. Which lever system most represents the human spine?
3. If the load is positioned at arm's length from the body, what happens to the forces on the spine?
4. What are the ways a person can reduce the forces on their spine without changing the weight of the load?
5. Which lever system is considered a biomechanical advantage to the human and why?

EXERCISE

1. Reference appendix.

REFERENCES

Department of Defense. (1999). MIL-STD-1472F, Department of Defense Design Criteria Standard: Human Engineering. United States Government Printing Office.

Marras, W. (2006). *Fundamentals and Assessment Tools for Occupational Ergonomics* 2nd edn. CRC Press.

University of Michigan. (1999). 3D Static Strength Prediction Program. http://www.umich.edu/~ioe/3DSSPP/background.html.

Winter, D. (1990). *Biomechanics and Motor Control of Human Movement,* 2nd edn. Wiley-Interscience.

ADDITIONAL SOURCES

Chaffin, D. (2006). *Occupational Biomechanics,* 4th edn. Wiley-Interscience.

McGinnis, P. (2013). *Biomechanics of Sport and Exercise,* 3rd edn. Human Kinetics.

Peterson, D. R. (2014). *Biomechanics: Principles and Practices.* CRC Press.

8

PSYCHOPHYSICS

LEARNING OBJECTIVE

This chapter will provide a discussion of psychophysics.

The learning objectives for this chapter are as follows:

- Understanding the origins of psychophysics
- How to conduct a psychophysical study
- How to use a psychophysical scale.

INTRODUCTION

On a Sunday, Bob participated in a marathon and then celebrated well into evening for finishing the race in the top 100. It is now Monday morning and he is supposed to perform 100% at his job as a loading dock worker. How might he perform? At 100%, 70%, or might he be exhausted? Many things play into how he might perform. Does he have physical issues because of the marathon or the partying afterward or does he feel motivated because he did so well in the marathon and had an enjoyable evening? Performance at work is not always based on the condition of the body, but also on how the mind controls the body. Psychophysics is the study of how the mind and body work together to perform a task.

Psychophysics is most commonly defined as the quantitative branch of the study of perception. Psychophysics examines the relations between observed stimuli and responses and the reasons for those relations. This definition is a very narrow

Occupational Ergonomics: A Practical Approach, First Edition.
Theresa Stack, Lee T. Ostrom and Cheryl A. Wilhelmsen.

view of the influence this branch of psychology has had on much of psychology in general. Psychophysics has been based on the assumption that the human perceptual system is a measuring instrument yielding results such as experiences, judgments, and responses that may be systematically analyzed. Psychophysics has a relatively long history of over 140 years, and its experimental methods, data analyses, and models of underlying perceptual and cognitive processes have reached a high level of refinement. Because of this, many of the techniques originally developed in psychophysics have been used to unravel problems in learning, memory, attitude measurement, social psychology, and, most importantly, ergonomics. Scaling and measurement theory have adapted psychophysical methods and models to analyze decision making in contexts entirely divorced from perception.

Psychophysics is an old branch of psychology that is concerned with the relationship between sensations and their physical stimuli (Snook, 1978). Modern psychophysical theory states that the strength of a sensation (S) is directly related to the intensity of its physical stimulus (I) by means of a power function: "$S = KI$". The constant (K) is a function of the particular units of measurement that are used. When plotted on log-log coordinates, a power function is represented as a straight line. Many exponents have been determined for various types of stimuli (NIOSH, 1981). Some of these are electric shock, 3.5; taste, 1.3; and loudness, 0.6. Stevens (1976) suggested that for lifting the exponent was 1.45. In a recent study, Baxter, Stalhammer, and Troup (Baxter, 1986) studied the Stevens' power function for the heaviness of boxes being lifted and lowered. This was determined for 91 firemen by using a ratio rating method. For 90/91 subjects, the equation was satisfactorily fitted to a power relationship despite considerable variation between individuals. The mean exponent for perceived heaviness was 1.35 and fractionation point 0.06 kg (fractionation being the k or constant). The authors concluded that doubling a weight being lifted resulted in a 2.6 increase in perceived weight.

Psychophysics has been applied to many areas. Psychophysics was used by Houghton and Yagloglou (1923) in the development of the effective temperature scale and by Stevens (1960) in the development of the scales of brightness. Foreman et al. (1984) applied acceptability scaling to vertical isometric force applications at knee and waist levels. Subjects were asked to select a comfortable level of force that they judged could be held for 2 min. This proved repeatable; however, subjects did express some difficulty in imagining the effort for that long of duration. Borg (1962, 1970) used psychophysics in the development of the ratings of perceived exertion (RPE scale). The current RPE scale, which was modified by Borg in 1970, is a scale consisting of 15 grades from 6 to 20 (see Table 8.1). The number 7 represents a very, very light workload; the number 9 a very light workload; the number 11 a fairly light workload; the number 13 a somewhat heavy workload; the number 15 a heavy workload; the number 17 a very heavy workload; and the number 19 a very, very heavy workload. Elkblom and Goldberg (1971) found in many situations the subject's heart rate mirrors the physical strain experienced subjectively. That is, the rating of perceived exertion a subject gave a task multiplied by 10 approximated the subject's heart rate observed for that task.

Davies and Sargent (1979), however, found that heart rate per se has little influence on RPE and is not an important factor underlying the perception of effort. The RPE scale has been used by numerous researchers in many areas of physical

TABLE 8.1 Borg's Rating of Perceived Exertion Scale

6	Very, very light
8	
9	Very light
10	
11	Fairly light
12	
13	Somewhat heavy
14	
15	Heavy
16	
17	Very heavy
18	
19	Very, very heavy
20	Maximal exertion

Source: Adapted from Borg (1970).

exertion and discomfort. Davies and White (1982) used the RPE scale to rate how subjects perceived their exertion while performing exercise running uphill, walking, and performing the box stepping test. The researchers found that the subjects' energy consumption and their rating of perceived exertion were well correlated. Nordin et al. (1984) studied the effect of the weight of paint on the workload of painters painting ceilings and found that all but one subject perceived their workload to be less with lighter paint. There were no significant differences in heart rate or total oxygen consumption between the paints, but oxygen consumption per area of ceiling painted was less. Although not specifically stated in the article, the data presented indicated that the rating of perceived exertion multiplied by 10 did not approximate heart rate. Balogun et al. (1986) used the RPE scale to determine the perceptual responses while carrying external loads on the head and by a yoke. The researchers defined two RPE scales. The first was the RPE-L or local rating of perceived exertion. This was used to determine the subjects' perception of exertion by the neck muscles while carrying loads on their head. The second scale was the RPE-0 scale or overall perception of perceived exertion. This was used to determine the subjects' overall perception of perceived exertion while performing a maximal exercise on a treadmill. In summary, the RPE has been proven to give good indications of subjects' perceived exertion while performing a wide variety of physical activities. Some of the first studies that utilized psychophysics in lifting were conducted by the U.S. Air Force (Emanuel et al., 1956; Switzer, 1962). The psychophysical approach to studying lifting varies from other approaches in that the subject has control of one of the task variables. This variable is usually the weight the subject is asked to lift. Before the experiment begins, the subject is instructed to lift as much as he or she can without overexertion or excessive fatigue. The weight they select to lift is referred to as the maximum acceptable weight of lift. Asfour (1980) instructed his subjects to adjust the load to the maximum amount they could lift without strain or discomfort, without being tired, weakened, overheated, or out of breath. Legg and Myles (1981) state that with good subject cooperation and firm experimental control, the psychophysical method can identify loads that subjects can lift repetitively for an 8-h workday without metabolic, cardiovascular, or subjective evidence of fatigue. Psychophysical lifting capacity studies have been conducted by Snook and Irvine (1967), Snook et al. (1970), McConville and Hertzberg (1966), Poulson (1970), McDaniel (1972), Dryden (1973), Knipfer (1974), Ayoub et al. (1978), Karwowski (1982) Bakken (1983), Mital (1983, 1986), Jiang (1984), Hafez (1984), Fernandez (1986), and others. Varied aspects of lifting have been studied, and many psychophysical lifting capacity prediction models have been developed. Ayoub et al. (1978) designed a study to develop predictive models for different height levels as a function of operator and task variables. The results from this study were used by Mital and Ayoub (1980) to develop lifting capacity models, which could accommodate varied paces of work. Karwowski (1982) conducted a study in which biomechanical and physiological stresses were measured simultaneously. He used the "fuzzy set" theory to develop a model that incorporated the interactions of those stresses in a moderate environment. Bakken (1983) studied the interaction of lifting range and frequency of lift. Hafez (1984) conducted a study to determine the effects of high heat levels, varied lifting frequencies, and their interaction on the ability to perform psychophysical work. Mital and Aghazadeh (1987) conducted a study to determine the acceptable weight of lift above reach height. The psychophysical approach is not above criticism, however. Garg & Ayoub (1980) observed that there was considerable disagreement among investigators in regard to the maximum acceptable weight of lift. Mital (1983) conducted an experiment to verify the psychophysical methodology used to determine the lifting capacity. The subjects estimated the load they could lift after a 26-min load adjustment period. The subjects then lifted for 8- and 12-h periods. On an average during the 8-h lifting session, males could only lift 65% of the weight they estimated during the load adjustment period. Females could only lift 84% of the weight they had estimated. During the 12-h session, male subjects only lifted 70% of the weight they had estimated and female subjects lifted only 77%. For the 12-h session, the male subjects selected a weight during the weight adjustment period that was considerably less than the weight selected for the 8-h session. Fernandez (1986) found the reduction for the 8-h period was an average of 86% of the weight the male subjects estimated they could lift. This study showed the difference to be much less than that Mital found. Ayoub and Selan (1983) stated that the psychophysical criterion is an appropriate single design criterion to use in the determination of lifting capacity. Inputs for psychophysical lifting prediction models include static and dynamic strength data, lifting frequency, and range of lift. The isoinertial incremental 6-ft lift strength test (6-ft lift) has been used as an input for many psychophysical lifting capacity models (Ayoub et al., 1986a,b, 1987a,b, 1988; Fernandez, 1986). The advantage of using a single dynamic strength test like the 6-ft lift strength test is its ease of use; it takes less than 2 min to conduct. Also, the isoinertial incremental 6-ft lift is a dynamic strength test and is an appropriate input for psychophysical lifting capacity models, since lifting is a dynamic activity. Finally, the isoinertial incremental 6-ft lift is a safe strength test.

Definitions

- The study of the relationship between physical activities and the human perception as to their difficulty
- In the context of ergonomics, it is how we perceive work.

Principles

- Psychophysics tends to show that people work at a place natural to them.
  - Approximately the same as a walking pace of 3 mph or
  - One liter of oxygen per minute (33% of the VO_2 max).
- Given control of at least one work variable (time, frequency, etc.), the worker will adjust that variable in order to perform the task at a "natural" level of output, while not becoming injured.

Psychophysical Scales

- Measurement of how perceived intensities vary with physical or physiological intensities
- Estimation of effort, exertion, and fatigue during physical work.

Benefits:

- Provides valuable information about a subject during any study
- Does not take much time to administer
- Can be used within a battery of tests.

Limitations:

- Subjects cannot be expected to give absolutely valid or reliable ratings.

DISCUSSION

As discussed in the introduction, psychophysical scales have been around a long time. The Borg Scale of Perceived Exertion is one of the most widely used scales. Around the world in health clubs on the walls beside treadmills, stationary bikes, and step machines, one often sees a scale going from 6 to 20. This is called an RPE Scale, which stands for "Rate of Perceived Exertion". It is a psychophysiological scale, meaning it calls on the mind and body to rate one's perception of effort. Understanding the meaning and use of this chart will benefit the average fitness enthusiast.

The RPE scale measures feelings of effort, strain, discomfort, and/or fatigue experienced during both aerobic and resistance training. One's perception of physical exertion is a subjective assessment that incorporates information from the internal and external environment of the body. The greater the frequencies of these signals, the more intense the perceptions of physical exertion. In addition, response from muscles and joints helps to scale and calibrate central motor outflow commands. The resulting integration of feedforward–feedback pathways provides fine-tuning of the exertional responses.

Perceived exertion reflects the interaction between the mind and body. That is, this psychological parameter has been linked to many physiological events that occur during physical exercise. These physiological events can be divided into respiratory/metabolic (such as ventilation and oxygen uptake) and peripheral (such as cellular metabolism and energy substrate utilization). Previous studies have demonstrated that an increase in ventilation, an increase in oxygen uptake, an increase in metabolic acidosis, or a decrease in muscle carbohydrate stores are associated with more intense perceptions of exertion. The scale is valid in that it generally evidences a linear relation with both heart rate and oxygen uptake during aerobic exercise.

THE BORG RPE SCALE AND THE BORG CR10 SCALE

Borg RPE Scale:

- Most commonly used scale (preferable over CR10 for perceived exertion)
- 15-point scale (between 6 and 20)
  - 9 corresponds to "very light" exercise. For a healthy person, it is like walking slowly at his or her own pace for few minutes.
  - 13 on the scale is "somewhat hard" exercise, but it still feels OK to continue.
  - 17 "very hard" is very strenuous. A healthy person can still go on, but he or she really has to push him- or herself. It feels very heavy, and the person is very tired.
  - 19 on the scale is an extremely strenuous exercise level. For most people, this is the most strenuous exercise they have ever experienced.
- Research found that there is a correlation between an athlete's RPE and their heart rate, lactate levels, $\%VO_2$ max, and breathing rate.
  - A general rule of thumb is to multiply the RPE by 10 to estimate the heart rate. For example, if a person's RPE is 12, then $12 \times 10 = 120$, so the heart rate should be approximately 120 beats/min.

Borg CR10 Scale:

- "Category-Ratio" Scale
- Anchored at the number 10, representing extreme intensities
- 9-point scale (between 1 and 10)
  - 1 corresponds to "very light" exercise. For a healthy person, it is like walking slowly at his or her own pace for few minutes.
  - 3 on the scale is "somewhat hard" exercise, but it still feels OK to continue.
  - 5 on the scale is associated with fatigue, but you don't have difficulties.
  - 7 "very hard" is very strenuous. A healthy person can still go on, but he or she really has to push him- or herself. It feels very heavy, and the person is very tired.
  - 10 on the scale is an extremely strenuous exercise level. For most people, this is the most strenuous exercise they have ever experienced.
  - 11 or higher is an "absolute maximum," which is more than ever experienced.
- Can also be used to rate pain (preferable over RPE).
- Smaller number scale can be limiting.
- Scale does not have a linear relationship with exercise intensity.

BORG SCALES WORKSHEETS

The following worksheets are provided as a framework for using the Borg Scales. The format of the worksheets includes both the instructions to be presented to the subject and the scale itself with descriptions.

Worksheet 1: Borg's RPE Scale

Instructions While doing physical activity, we want you to rate your perception of exertion, that is, how heavy and strenuous the exercise feels to you, combining all sensations and feelings of physical stress, effort, and fatigue. Do not concern yourself with any one factor such as leg pain or shortness of breath but try to focus on your total feeling of exertion.

Look at the rating scale below while they are engaging in an activity; it ranges from 6 to 20, where 6 means "no exertion at all" and 20 means "maximal exertion." Choose the number from below that best describes their level of exertion. This will give them a good idea of the intensity level of their activity, and can use this information to speed up or slow down movements to reach their desired range.

Try to appraise their feeling of exertion as honestly as possible, without thinking about what the actual physical load is. Try not to over- or underestimate the rating. Their own feeling of effort and exertion is important, not how it compares to other people. Look at the scales and the expressions and then give a number (Table 8.2).

TABLE 8.2 Borg Scale Worksheet

6	No exertion at all
7	Extremely light
8	
9	Very light (walking slowly at your own pace)
10	
11	Light
12	
13	Somewhat hard (exercise but feels OK to continue)
14	
15	Hard (heavy)
16	
17	Very hard (very strenuous, but you are fatigued)
18	
19	Extremely hard (extremely strenuous and cannot continue for long)
20	Maximal exertion

Source: Adapted from Borg's Perceived Exertion and Pain Scales.

TABLE 8.3 10-Point Borg Scale

0	Nothing at all
0.3	
0.5	Extremely weak (just noticeable)
1	Very weak (walking slowly at your own pace)
1.5	
2	Weak (light)
2.5	
3	Moderate (exercise, but feels OK to continue)
4	
5	Strong (heavy)
6	
7	Very strong (very strenuous, but you are fatigued)
8	
9	
10	Extremely strong (as hard as ever experienced)
11	
•	Absolute maximum (highest possible)

Source: Adapted from Borg's Perceived Exertion and Pain Scales.

Worksheet 2: Borg's CR10 Scale

Instructions 10, "Extremely strong" is the main anchor. It is the strongest perception one have ever experienced. It may be possible to experience something stronger; therefore, "Absolute maximum" is further down the scale without a fixed number (•). If one perceive intensity stronger than 10, they may use a higher number (Table 8.3).

Start with a "verbal expression" and then choose a number. If the perception is "Very weak," say 1; if "Moderate," say 3, etc. They can use fractions of points as well. It is important to use values that they perceive, not what they believe they should report. Be honest and try not to over- or underestimate the intensities.

Scaling perceived exertion: Rate their perceived exertion, in other words, how heavy or strenuous the physical activity feels to them. This depends mainly on the strain and fatigue in their muscles, feelings of breathlessness or aches in the chest. They must only attend to their subjective feelings and not to the physiological cues or what the physical activity actually is.

Scaling pain: Think of your worst experiences of pain. If they use 10 as the strongest exertion they have ever experienced, then think about the three worst pain experiences they have ever had.

10	10 "Extremely strong": Main point of reference as it is anchored in their previously experienced worst pain
•	The feeling that is somewhat stronger than the worst pain experienced (10). If much stronger, like 1.5 times, then 15.

CASE STUDY – ERGONOMIC ASSESSMENT OF THE FIRING OF THE M198 HOWITZER

Abstract

The United States Marine Corps Artillery Instructional Battery stationed at The Basic School Marine Corps Base Quantico, VA, was experiencing a high rate of injuries and our ergonomics team was asked to perform an analysis of the operation to help determine why these injuries were occurring. We designed and carried out an ergonomics study to determine causes of these injuries. The study was carried out using data collection methods that would collect anthropometric, psychophysical, biomechanical, and human error data. This paper focuses on the anthropometric and psychophysical aspects of the study. The 15-point and 11-point Borg Scales were used as the data collection instruments. At the end of the first day of the study, the Marines were experiencing a high level of fatigue and pain based on their perceptions. Recommendations were made based on the findings from the study.

Highlights

- Members of the Artillery Instructional Battery (AIB) were experiencing many injuries.
- The command requested an ergonomic assessment to determine the causes of the injuries.
- The Borg Scales were used to aid fatigue, exertion, and the ergonomic analysis of the tasks associated with firing of the M198 Howitzer.
- The study showed that the gun crews were experiencing a high level of fatigue followed by pain.
- Recommendations were made to reduce the apparent causes of the fatigue and pain.

Acknowledgments

We wish to thank Cathy Rothwell of Navy Ergonomics with the Naval Facilities Command for her support of this work.

We also wish to thank Anna Smith of The Basic School Safety Office at Marine Corps Base Quantico, VA, for her great help during the data collection phase of the project.

Introduction

The first assignment a United States Marine Corps officer has after obtaining his commission is at The Basic School (TBS) at Camp Barrett, Marine Corps Base Quantico, VA.

TBS further prepares young officers for their military careers by providing them training in a range of topics. The course has 1585 h of training: spending 932.6 h (60%) in the classroom and 652.5 h (40%) in the field. Classroom events include platform instruction, tactical decision games (TDGs), sand table exercises (STEXs), and small group discussions. There are various field events, beginning with fire team and squad level, moving all the way up to platoon-reinforced events. The field events consist of realistic blank-fire training, SESAMS (similar to paintball), and live fire range (Fleet Marine Force, n.d.).

One of the exercises is a Call for Fire from field artillery. This exercise is conducted approximately every 45 days. In this exercise, the new Second Lieutenants (2nd LTs) determine what type of artillery support they need for a scenario and call in the coordinates into a fire support center. The fire support center determines the fire solution and sends this information to the AIB. The AIB then provides the fire support via one of three 155-mm M198 Howitzers. The AIB is a company that consists of 34 very professional, dedicated, and hardworking Marines.

The AIB was experiencing a high number of injuries and our ergonomics team was asked to perform an analysis of the operation to help determine why these injuries were occurring and provide solutions with regard to military-specific constraints.

Description of Activity

The Call for Fire exercise takes place over 3 days. During the first day, the AIB sets up four M198 Howitzers. Three will be active during the exercise and the fourth is considered a "ghost gun" and is kept in reserve. The ammunition is also delivered during day one. Approximately 1100 rounds of 155-mm ammunition, fuses, and sufficient powder are brought to the site. Days two and three are when the actual firing of the guns occurs. Each gun will fire between 350 and 400 rounds of ammunition. The rounds weigh between 95 and 105 lb each depending on whether they contain high explosive or white phosphorus. We did not survey the Marines on day one but much of the setup requires manual materials handling.

The normal gun crew consists of the following members:

- Chief of the gun
- Gunner
- A-gunner
- Recorder
- Powder man
- Ammo team chief
- Two ammo handlers
- Rammer.

Figures 8.1 and 8.2 show aspects of the M198 Howitzer loading process.

The crews of the guns during the shoot consisted of 6–7 members due to Marines on the injured list. The limited human resources made it difficult to rotate the crew as would be done when a full crew complement were available.

The fire process included the following:

1. The fire instructions are radioed to the recorder.
2. The recorder announces the fire order.
3. The powder man adjusts the powder.
4. The type of round to be fired is verified.

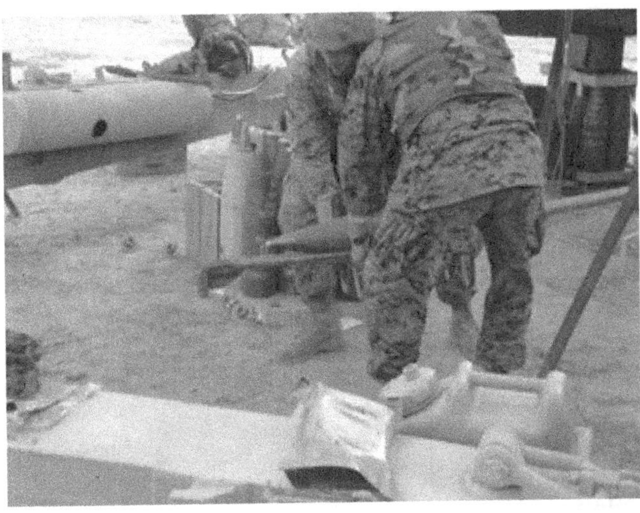

Figure 8.1 Marines pick up round in loading tray

Figure 8.2 Marines ram a round into the breach of the M198 Howitzer

5. A round is removed from the ready board and laid into the loading tray.
6. The two loaders pick up the round.
7. The rammer places the ram at the back of the round.
8. The breach is opened.
9. The loaders bring the round up to the breach.
10. The rammer pushes the round in the breach.

11. The loader on the right side of the loading tray releases his grip.
12. The other loader steps back and to the left of the gun.
13. The powder man brings up the powder and hands it to the A-gunner.
14. The A-gunner verifies the amount of powder and places it in the breach.
15. The A-gunner closes the breach and places a primer in the priming hole.
16. The lanyard is attached and then pulled.
17. The gun fires.
18. The A-gunner opens the breach and swabs the breach and breach plug.

Variations occurred in this process based on the health of the crew and the number of crew members available. These variations will be discussed in the course of the article.

Ergonomic Study

The ergonomic study was designed to collect the following data:

1. Anthropometric data
2. Psychophysical measures of exertion, fatigue, and pain using the Borg Scale
3. Qualitative data on activities
4. Biomechanical data
5. Human error data
6. Activity performance data.

Only the anthropometric data and psychophysical aspects of the study will be discussed in this paper due to the great amount of data collected.

A quasi-experimental design was used to conduct the study. As with any active data collection process, participants tend to alter their activities. Therefore, the ergonomics team did their best to not interfere with the activities. Naturalistic observation techniques were used, as well as noninvasive inquiry at times when the Marines were not actively engaged in loading and firing the Howitzers.

The setup was conducted on Monday, August 10, 2009. The AIB setup consisted of four 155-mm M198 Howitzers. The number one gun was a ghost gun or the spare. Guns two through four were to be used on the next 2 days. The AIB also received 1100 rounds of ammunition and corresponding powder and fuses. The setup began at approximately 0600 and finished at approximately 1600. During the afternoon, the ergonomics team collected the anthropometric data and conducted the training of the Marines. Table 8.4 contains the anthropometric data collected.

On Tuesday, August 11, 2009, the crews arrived at approximately 0600. The Call for Fire exercise began at 0820. The ergonomics team collected data on the activities the Marines had done the day before and whether the Marines ate that morning. The day was going to be hot, so heat stress precautions were going to be in place. A wet bulb globe temperature (WBGT) was positioned near the firing range to monitor the heat index level. The WBGT was only periodically monitored; however, heat

TABLE 8.4 Marines Demographic Data

Marine	Stature	Weight	Shoulder to Mid-hand Length	Shoulder Width	Age	Months in Marines	Months at This Assignment	Number of 2-day Shoots
1	73	200	32.0	19.0	23	59	14	24
2	68	195	30.3	17.5	22	56	9	6
3	69	185	28.0	18.5	31	88	36	48
4	73	235	32.0	19.0	25	73	14	26
5	71	200	28.6	18.5	24	68	4	4
6	70	190	29.6	19.0	26	63	12	10
7	72	215	31.0	19.0	26	73	16	25
8	67	168	28.6	17.5	25	90	19	30
9	72	195	28.5	19.5	23	62	10	20
10	67	166	26.5	19.0	24	60	10	10
11	68	180	29.0	20.0	24	62	14	30
12	64	132	25.5	18.0	27	96	4	16
13	73	165	32.3	18.5	23	61	16	32
14	66	168	26.3	19.0	26	67	17	34
15	72	188	29.9	19.0	22	62	15	30
16	67	180	27.8	19.0	26	89	36	48
17	81	245	34.5	21.5	27	67	24	48
18	68	170	29.6	19.0	23	60	12	24
19	69	170	29.3	19.0	25	68	16	25
20	71	202	30.0	19.0	23	63	15	30
21	70	210	28.4	20.0	24	68	18	36
23	71	130	30.0	17.0	25	68	18	30
25	69	170	29.4	18.5	25	61	16	25
28	70	189			25	84	24	24
31	65	193	27.0	19.0	26	84	12	24
Mean	69.8	185.6	29.3	18.9	24.8	70.1	16.4	26.4
Median	70	188	29.3	19	25	67	15	25
Mode	67	170	32	19	25	68	16	30
Std	3.4	26.3	2.1	0.9	1.9	11.5	7.7	11.7
Min	64	130	25.5	17	23	56	4	4
Max	81	245	34.5	21.6	31	96	36	48

index did reach black flag conditions (>90 °F or >32.2 °C) in the afternoon (Weather Information, n.d.).

The ergonomics team observed the entire shoot and collected data as activities permitted. The AIB rested for approximately 1200–1300 for lunch break and during an airstrike from approximately 1330–1415. Data collection ended at 1800.

On Wednesday, August 12, 2009, the shoot began very similar to how it had on Tuesday. The AIB arrived at approximately 0600 and the Call for Fire exercise began at approximately 0730. At approximately 1100, a ceremony was conducted and missions ceased during this time frame. Also, in the afternoon, the 2nd LTs were allowed to help move the rounds that gave the AIB crew adequate rest breaks and data collection ceased at this time. Data collection ended at 1800.

Anthropometric and Demographic Data

Psychophysical Measures of Exertion, Fatigue, and Pain – Borg's Scale There are a variety of methods for determining exercise intensity levels. Common methods include the target heart rate range (Activity Measurement, n.d.) and the Borg RPE (Borg, 1998). We needed a method of data collection that was noninvasive, and the RPE is a noninvasive technique.

The 15-point Borg Scale (6–20) was used to collect indications of exertion.
The 11-point Borg Scale was used to collect indications of pain and fatigue.

The following directions were given to the AIB crew prior to the exercise:

> At intermittent times during the shoot, the ergonomists will ask you to gauge your feeling of exertion. This feeling should reflect how heavy and strenuous the activity feels, combining all sensations and feelings of physical stress, effort, and fatigue. Do not concern yourself with any one factor such as leg pain or shortness of breath but try to focus on your total feeling of exertion.
>
> Look at the rating scale below while engaging in an activity; it ranges from 6 to 20, where 6 means "no exertion at all" and 20 means "maximal exertion." Choose the number from below that best describes the level of exertion you feel at this time. This is a good indication of the intensity level of the activity.
>
> Try to appraise the feeling of exertion as honestly as possible, without thinking about what the actual physical load is.

Analogous instructions were given concerning the pain and fatigue data for use with the 11-point scales. The key point to using these scales is the calibration of the participants in the study. Therefore, prior to the data collection efforts, every Marine who participated in the study was briefed as to what the scales and the level of the scales meant.

Exertion, fatigue, and pain data were collected at random intervals. However, data was not collected from every Marine at every interval. The reason was because of the activities the Marines were involved in at the data collection times.

Fire missions during the data collection varied from one round to seven, with most being 3, 4, or 6 rounds.

Results

As discussed above, indicators of exertion, fatigue, and pain were collected at random intervals and at times that impacted the Marines the least. However, they were documented as close to a 60-min interval as possible. Table 8.5 shows the means of these values.

Figure 8.3 shows the change in the mean value for exertion, Figure 8.4 shows the change in the mean value for fatigue, and Figure 8.5 shows the change in the mean value of pain over the course of day 1. The indicators trend with time and activity, increasing as the day goes on, leveling off during rest and then increases again with resumed activity. Figure 8.6 compares pain and fatigue as they trend.

TABLE 8.5 Means of Exertion, Fatigue, and Pain Values for Day 1

Value	Time of Day								
	0800	0900	1000	1100	1200	1300	1530	1700	1800
Exertion	6.14	6.14	8.4	14	9.4	11.88	13.8	14.25	15.53
Fatigue	3.14	3.2	2.0	6.0	3.7	5.85	6.9	7.34	8.37
Pain	1.0	0.71	1.4	1.0	2.83	2.88	3.95	5.42	6.5

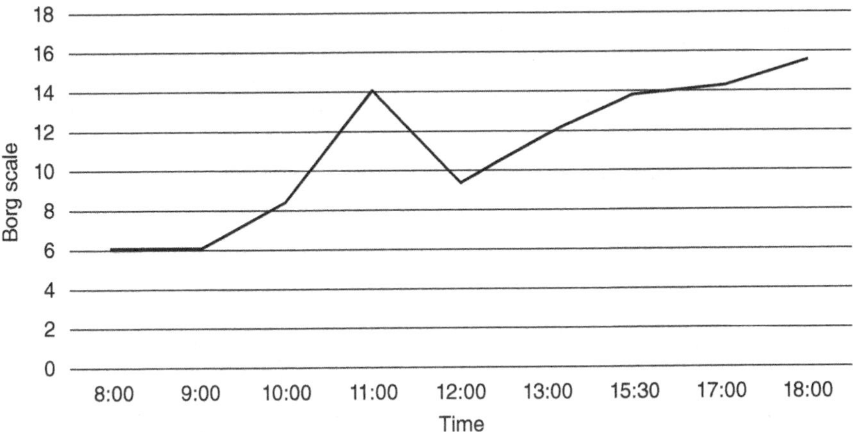

Figure 8.3 Change in mean exertion levels – day 1

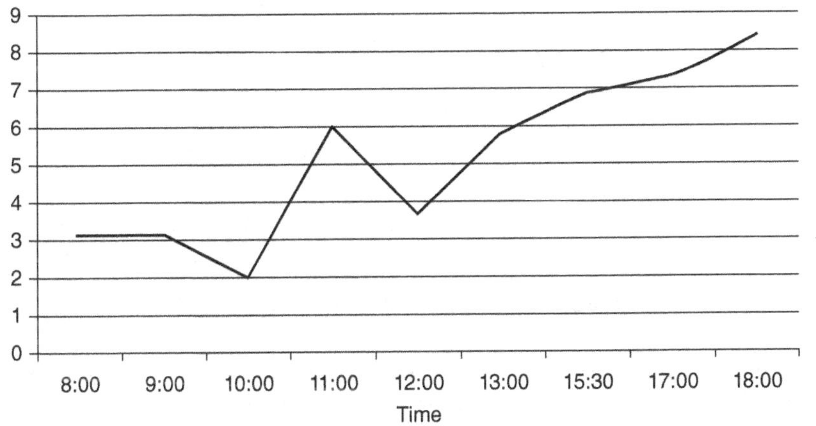

Figure 8.4 Changes in mean fatigue levels – day 1

It is apparent in these graphs that as the day progresses, the perceptions of exertion, fatigue, and pain increase. However, the level for pain stayed relatively low for a considerable amount of the day, but at approximately 1300 h it increased rapidly. Fatigue, on the other hand, increased in the morning, but after a break, around noon due to a simulated air support exercise in which the AIB did not participate, the fatigue

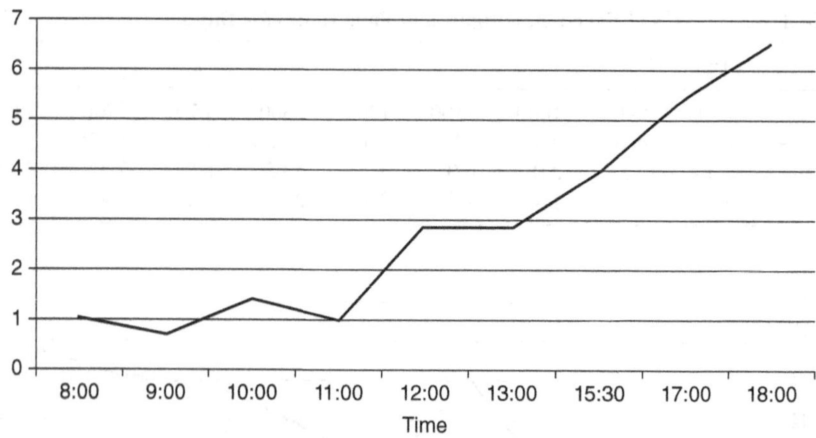

Figure 8.5 Changes in mean pain levels – day 1

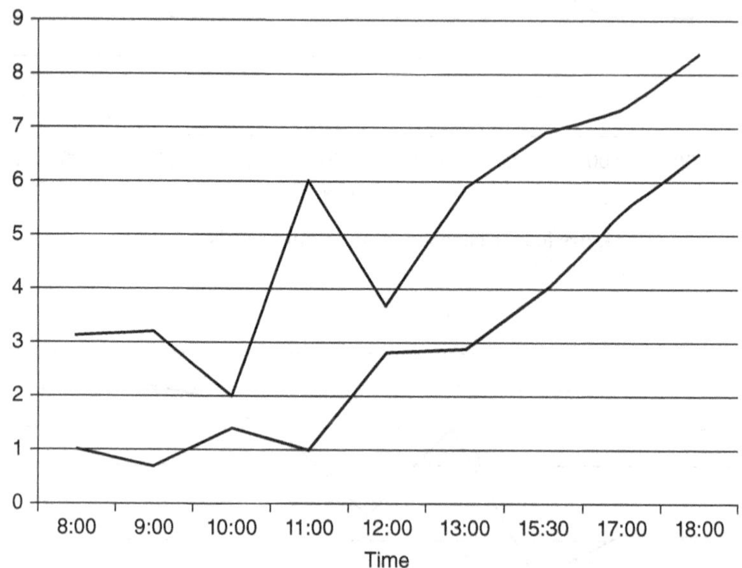

Figure 8.6 Comparisons of fatigue and pain levels

level dropped. It then increased rapidly after the firing began again. It is also apparent that pain lags slightly behind fatigue as the day progresses.

The question was asked whether these increases were statistically significant. In other words, the hypothesis is that there is no statistical difference between the perceptions of exertion, fatigue, and pain as the day progressed. The *t*-test for unequal variances was used to test this. Tables 8.6–8.8 show the detailed results of some of these tests. Further results are summarized in Tables 8.6–8.8. However, it is important to note that in this study statistical significance may or may not provide more information than is displayed in the proceeding graphs. This is because data were not gathered at uniform times from the Marines, and it appeared to us that the Marines

TABLE 8.6 Results of *t*-Test for Exertion at 0900 and 1300

Factor	0900	1300
Mean	6.142857	11.88889
Variance	10.47619	12.86111
Observations	7	9
df	14	
t-Statistic	3.35938	
$P(T \leq t)$ one-tail	0.002338	
t-Critical one-tail	1.76131	

TABLE 8.7 Results of *t*-Test for Fatigue at 0900 and 1300

Factor	0900	1300
Mean	3.222222	5.85
Variance	3.944444	3.669444
Observations	9	10
df	17	
t-Statistic	2.92841	
$P(T \leq t)$ one-tail	0.00469	
t-Critical one-tail	1.739607	

TABLE 8.8 Results of *t*-Test for Pain at 1300 and 1500

Factor	1300	1500
Mean	2.875	3.95
Variance	4.125	2.136111
Observations	8	10
df	12	
t-Statistic	1.25885	
$P(T \leq t)$ one-tail	0.116011	
t-Critical one-tail	1.782288	

got better at reporting values as time went along. Of course, the reason for the lack of statistical significance is due to the variability of the data.

Tables 8.9–8.11 show statistical comparisons of the data at 0900 on day 1 and 0900 on day 2. Only pain is statistically significantly different from day 1 to day 2. Exertion and fatigue levels are not statistically significantly different. This is an indication that the Marines recovered somewhat but not fully. Lasting pain can be a precursor to injury development (see Tables 8.12–8.14).

Table 8.15 shows statistical comparisons of the data at 1200 for pain for day 1 and day 2. Exertion and fatigue were not tested because of lack of data. There was no statistically significant difference between the 2 days at 1200.

TABLE 8.9 Statistical Significance of Pair-Wise Comparison of Exertion

Comparisons of Exertion Indicators at Time Intervals				
Factor	0900–1300	1300–1530	1530–1700	1700–1800
Mean	6.1 versus 11.9	11.9 versus 13.8	13.8 versus 14.3	14.3 versus 15.5
t-Statistic	−3.35938	1.35813	0.31641	0.67007
t-Critical one-tail	1.7613	1.76131	1.734064	1.717144
Statistical significance	Yes	No	No	No

TABLE 8.10 Statistical Significance of Pair-Wise Comparison of Fatigue

Comparisons of Fatigue Indicators at Time Intervals				
Factor	0900–1300	1300–1700	1500–1700	1700–1800
Mean	3.2 versus 5.9	5.9 versus 6.9	6.9 versus 7.3	7.3 versus 8.6
t-Statistic	2.92841	1.61751	0.7124	1.86881
t-Critical one-tail	1.739607	1.782288	1.745884	1.729133
Statistical significance	Yes	No	No	Yes

TABLE 8.11 Statistical Significance of Pair-Wise Comparison of Pain

Comparisons of Pain Indicators at Time Intervals				
Factor	0900–1300	1300–1500	1500–1700	1700–1800
Mean	0.7 versus 2.9	2.9 versus 4.0	4.0 versus 5.4	5.4 versus 6.5
t-Statistic	2.51158	1.25885	2.0668	1.31278
t-Critical one-tail	1.782288	1.782288	1.720743	1.705618
Statistical significance	Yes	No	Yes	No

TABLE 8.12 Results of t-Test for Exertion Day 1 to Day 2 at 0900

Factors	Exertion 0900 Day 2	Exertion 0900 Day 1
Mean	8.25	6.142857
Variance	0.25	10.47619
Observations	4	7
df	6	
t-Statistic	1.687552	
t-Critical one-tail	1.94318	

The data analysis shows that the Marines' perceptions of exertion, fatigue, and pain increase statistically significantly over time. The Marines are performing the same level of exertion on the second day at 0900 as the first day at 0900 and have the same level of fatigue. However, the perception of pain is increased. Once the Marines have additional help in the afternoon of the second day by the 2nd LTs, their perceptions of fatigue and pain levels drop significantly. In fact, data collection in the afternoon of day 2 ceased because the Marines stated they were no longer fatigued.

TABLE 8.13 Results of *t*-Test for Fatigue Day 1 to Day 2 at 0900

Factors	Fatigue 0900 Day 2	Fatigue 0900 Day1
Mean	4.083333	3.222222
Variance	2.041667	3.944444
Observations	6	9
df	13	
t-Statistic	0.975924	
t-Critical one-tail	1.770933	

TABLE 8.14 Results of *t*-Test for Pain at 0900 Day 1 to Day 2

Factors	Pain 0900 Day 2	Pain 0900 Day 1
Mean	2.25	0.714286
Variance	0.25	1.571429
Observations	4	7
df	8	
t-Statistic	2.866667	
t-Critical one-tail	1.859548	

TABLE 8.15 Results of *t*-Test for Pain at 1200 Day 1 to Day 2

Factors	Pain Day 2 1200	Pain Day 1 1200
Mean	2.8	2.833333
Variance	2.26	6.151515
Observations	6	12
df	15	
t-Statistic	−0.03535	
t-Critical one-tail	1.75305	

Regression analyses were performed between anthropometric data and the psychophysical indicators of pain and fatigue. Table 8.16 shows the results of the analysis for the relationship between stature and pain. The result of this analysis shows there is little to no relationship between these two variables because the R^2 value is only 0.23. Table 8.14 shows the results for the relationship between stature and fatigue and Table 8.17 shows the results for the relationship between weight and fatigue. These tables and Table 8.18, show there is little to no relationship between anthropometric variables and perceptions of pain. Similar comparisons were conducted between age and fatigue and pain and these comparisons also show no statistical relationship (see Table 8.19).

KEY POINTS

The findings from psychophysical aspects of the study were as follows:

1. The Marines' perceptions of exertion, fatigue, and pain increase statistically significantly over the course of the day. Once a break is provided for the Marines, their perceptions of exertion, fatigue, and pain decrease.

TABLE 8.16 Regression Analysis of Stature and Pain at 1800 Day 1

Regression Statistics	
Multiple *R*	0.479674
R^2	0.230087
Adjusted R^2	0.160095
Standard error	3.561562
Observations	13

TABLE 8.17 Regression Analysis of Stature and Fatigue at 1800 Day 1

Regression Statistics	
Multiple *R*	0.189022186
R^2	0.035729387
Adjusted R^2	−0.044626498
Standard error	4.204556406
Observations	14

TABLE 8.18 Analysis of Weight and Fatigue at 1800 Day 1

Regression Statistics	
Multiple *R*	0.118799
R^2	0.014113
Adjusted R^2	−0.06804
Standard error	30.96039
Observations	14

TABLE 8.19 Analysis of Weight and Pain at 1800 Day 1

Regression Statistics	
Multiple *R*	0.258599
R^2	0.066874
Adjusted R^2	−0.01796
Standard error	26.96009
Observations	13

2. There is no relationship between anthropometric measurements or demographics and exertion, fatigue, and pain. This indicates that the increases in these factors are due to the task and not due to anthropometrics or demographics of the Marines.
3. Upon being given additional help by the 2nd LTs on day 2 the Marines' perceptions of exertion, fatigue, and pain reduced to the point the Marines felt they were no longer overexerted, fatigued or in pain.

The recommendations based on these findings were as follows:

1. More Marines need to be added to the AIB Company to support the shoots. The optimal number would be 9–10.
2. The 2nd LTs should be allowed to move the rounds, thus significantly reducing the number of lifts the Marines perform.
3. Providing additional aid, either with more Marines or allowing the 2nd LTs to help, reduces the exposure of loading the gun to within acceptable levels according to the MIL-STD 1472F.

The US Marine Corps has implemented several changes to this process since the study was conducted. Most notably, they increased the number of Marines in the AIB Company.

REVIEW QUESTIONS

1. What are the inherent differences between the 15-point and the 10-point Borg Perceived Exertion Scales?

2. What is the advantage of using the Borg Scale over collecting physiological data?

3. Is the Borg Scale appropriate for assessing all tasks? Why or why not?

4. Develop a study to assess some activity, such as a lifting task. What are the advantages and disadvantages over using the NIOSH lifting equation?

REFERENCES

Activity Measurement. (n.d.). Retrieved August 30, 2011, from CDC: http://www.cdc.gov/physicalactivity/everyone/measuring/index.html.

Asfour, S. (1980). *Energy Cost Prediction Models for Manual Lifting and Lowering Task.* Lubbock, TX: Texas Tech University.

Ayoub, M. M., Bethea, N. J., Deivanayagam, S., Asfour, S. S., Bakken, G. M., Liles, D., Mital, A., & Sherif, M. (1978). Determination and modeling of lifting capacity. National Institute of Occupational Safety and Health, Final Report, Grant No. 5R01-OH-00545-02, Cincinnati, OH.

Ayoub, M. M., & Selan, J. L. (1983). *A Mini-Guide for Lifting*. Texas Tech University, Institute for Ergonomics.

Ayoub, M. M., Smith, J. L., Selan, J. L., & Fernandez, J. E. (1986a). *Materials Handling in Unusual Positions – Phase IV*. University of Dayton Research Institute.

Ayoub, M. M., Smith, J. L., Selan, J. L., Chen, H. C., Fernandez, J. E., Lee, Y. H., et al. (1986b). *Manual Materials Handling in Unusual Positions – Phase II*. University of Dayton Research Institute.

Ayoub, M. M., Jiang, B. C., Smith, J. L., Selan, J. L., & McDaniel, J. E. (1987a). Establishing Physical Criterion for Assigning Personnel to U.S. Air Force XE "Force" Jobs. *American Industrial Hygiene Association Journal*, 48(5), 464–470.

Ayoub, M. M., Smith, J. L., Selan, J. L., Chen, H. C., Lee, Y. H., & Kim, H. K. (1987b). *Manual Materials Handling in Unusual Positions – Phase III*. University of Dayton Research Institute.

Ayoub, M. M., Smith, J. L., Chen, M. E., Danz, M. E., Kim, H. K., Lee, Y. H., et al. (1988). *Manual Materials Handling in Unusual Positions – Phase IV*. University of Dayton Research Institute.

Bakken, G. (1983). *Lifting Capacity Determination as a Function of Task Variables*. Texas Tech University.

Balogun, J. A., Robertson, R. J., Goss, F. L., Edwards, M. A., Cox, R. C., & Metz, K. F. (1986). Metabolic and Perceptual Responses while Carrying External Loads on the Head and by Yoke. *Ergonomics*, 29(12), 1623–1636.

Baxter, C. S., Stalhammar, H., & Troup, J. D. (1986). A Psychophysical Study of Heaviness for Box Lifting and Lowering. *Ergonomics*, 29(9), 1056–1062.

Borg, G. (1962). Physical Performance and Perceived Exertion. *Studia Psychologica et Paedagogica, Series altera, Investigationes XI*. Lund, Sweden: Gleerup.

Borg, G. (1970). Perceived Exertion as an Indicator of Somatic Stress. *Scandinavian Journal of Rehabilitation Medicine*, 92–98.

Borg, G. (1998). *Borg's Perceived Exertion and Pain Scales*. Human Kinetics.

Davies, C. T., & Sargent, A. J. (1979). The Effects of Atropine and Practolol on the Perception of Exertion During Treadmill Exercise. *Ergonomics*, 1141–1146.

Davies, C. T., & White, M. J. (1982). Muscle Weakness Following Dynamic Exercise in Humans. *Journal of Applied Physiology*, 53(1), 236–241.

Dryden, R. (1973). *A Predictive Model for the Maximum Permissible Weight*. Texas Tech University.

Elkblom, B., & Goldberg, A. N. (1971). The Influence of Physical Training and Other Factors on the Subjective Rating of Perceived Exertion. *Acta Physiologica Scandinavica*, 83, 399–406.

Emanuel, I., Chafee, J., & Wing, J. (1956). *A Study of Human Weight Lifting Capabilities for Loading Ammunition into the F-86 Aircraft*. U.S. Air Force: WADC-TR 066-367.

Fernandez, F. (1986). *Psychophysical Lifting Capacity over Extended Periods*. Texas Tech University.

Fleet Marine Force. (n.d.). Retrieved August 30, 2011, from Wikipedia: http://en.wikipedia.org/wiki/Fleet_Marine_Force.

Foreman, T. K., Baxter, C. E., & Troup, J. D. (1984). Ratings of Acceptable Load and Maximal Isometric Lifting Strengths: The Effects of Repetition. *Ergonomics*, 27, 1283–1288.

Garg, A. & Ayoub, M.M. (1980). What criteria exist for determining how much load can be lifted safely?. *Human Factors*, 22, 475–486.

Hafez, H. (1984). *An Investigation of Biomechanical, Physiological, and Environmental Heat Stresses Associated with Manual Lifting in Hot Environments*. Texas Tech University.

Houghton, F. C., & Yagloglou, C. P. (1923). Determination of the Comfort Zone. *Journal of the American Society of Heating and Ventilation Engineers*, 29, 515–536.

Jiang, B. (1984). *Psychophysical Capacity Modeling of Individual and Combined Material Handling Activities*. Texas Tech University.

Karwowski, W. (1982). *A Fuzzy Sets Based Model on the Interaction Between Stresses Involved in Manual Lifting Tasks*. Texas Tech University.

Knipfer, R. (1974). *Predictive Models for the Maximum Acceptable Weight of Lift*. Texas Tech University.

Legg, S. J., & Myles, W. S. (1981). Maximum Acceptable Repetitive Lifting Workloads for an 8-hour Work day Using Psychophysical and Subjective Rating Methods. *Ergonomics*, 24(12), 907–916.

McConville, J. T., & Hertzberg, H. T. (1966). A Study of One Hand Lifting: Final Report. Wright-Patterson AFB, OH: Aerospace Medical Research Laboratory, Technical Report, AMRL-TR-66-17.

McDaniel, J. (1972). *Prediction of Acceptable Lift Capability*. Texas Tech University.

Mital, A. (1986). Comparison of Lifting Capabilities of Industrial and Non-industrial Populations. *Human Factors Proceedings. Dayton*, 239–241.

Mital, A. (1983). The Psychophysical Approach in Manual Lifting – A Verification Study. *Human Factors*, 485–491.

Mital, A., & Aghazadeh, F. (1987). Psychophysical Lifting Capabilities for Overreach Heights. *Ergonomics*, 30(6), 901–909.

Mital, A., & Ayoub, M. M. (1980). Modeling of Isometric Strength and Lifting Capacity. *Human Factor*, 22(3), 285–290.

NIOSH. (1981). *Work Practices Guide for Manual Lifting*. Cincinnati, OH: NIOSH.

Nordin, M., Ortengren, R., Envall, L., & Andersson, G. (1984). The Influence of Paint Characteristics on the Workload Produced. *Ergonomics*, 27(4), 409–423.

Poulson, E. (1970). Prediction of Maximum Loads in Lifting from Measurement of Back Muscle Strength. *Progressive Physical Therapy*, 1, 146–149.

Snook, S. (1978). The Design of Manual Handling Tasks. *Ergonomics*, 21(2), 963–985.

Snook, S. H., & Irvine, C. H. (1967). Maximum Acceptable Weight of Lift. *American Industrial Hygiene Association Journal*, 28, 322–329.

Snook, S. H., Irvine, C. H., & Bass, S. F. (1970). Maximum Weights and Work Loads Acceptable to Male Industrial Workers. *American Industrial Hygiene Association Journal*, 31, 579–586.

Stevens, S. (1960). The Psychophysics of Sensory Function. *American Scientist*, 48, 226–253.

Stevens, S. (1976). *Psychophysics: Introduction to its Perceptual Neural and Social Prospects*. New York, NY: John Wiley and Sons, Inc.

Switzer, S. A. (1962). Weight – Lifting Capabilities of a Selected Sample of Human Males. Wright Patterson Air Force Base, OH: Aerospace Research Lab Report No. AD-284054.

Weather Information. (n.d.). Retrieved August 30, 2011, from USMC: http://www.quantico.usmc.mil/weather.asp.

ADDITIONAL SOURCES

Agahazadeh, F. (1974). Lifting Capacity as a Function of Operator and Task Variables. Texas Tech University: Master of Science Thesis, Lubbock, TX.

Asmussen, E., & Heeboll-Nielsen, K. (1961). Isometric Muscle Strength of Adult Men and Women, *Community Testing Observation Institute of the Danish National Association for Infantile Paralysis*, E. Asmussen, A. Fredsted & E. Ryge (eds.) Institute of the Danish National Association for Infantile Paralysis, vol. 11, pp. 1–44.

Asmussen, E., & Heeboll-Nielsen, K. (1962). Isometric Muscle Strength in Relation to Age in Men and Women, *Ergonomics*. 5, 167–169.

Ayoub, M. A. (1976). Optimum Design for Containers for Manual Materials Handling Tasks. Paper presented at International Symposium on Safety in Manual Materials Handling. Cincinnati, OH.

Ayoub, M. M., J. D. Denardo, J. L. Smith, N. J. Bethea, B. K. Lambert, L. R. Alley, & Duran, B. S. (1982). Establishing Physical criteria for Assigning Personnel to Air Force Jobs – Final Report, Air Force Office of Scientific Research. Contract No. F49620-79-C-0006, September.

Ayoub, M. M., R. D. Dryden, & Knipfer, R. E. (1976). Psychophysical based Models for Prediction of Lifting Capacity of the Industrial Worker. Proceedings of Society of Automotive Engineers. Pennsylvania, Warrendale.

Ayoub, M.M., R.D. Dryden, J.W. McDaniel, R.E. Knipfer, & Dixon, D. (1979), Predicting Lifting Capacity. *American Industrial Hygiene Association Journal*. 40, 1076–1084.

Ayoub, M.M., & LoPresti, P. (1971). The Determination of an Optimum Size Cylindrical Handle by Use of Electromyography, *Ergonomics*. 14(4), 609–618.

Belding, H. S. (1971). Ergonomics Guide to Assessment of Metabolic and Cardiac Costs of Physical Work, *American Industrial Hygiene Association Journal*. 32(8), 660–664.

Ciriello, V. M. & Snook, S. (1978). The Effects of Size, Distance, Height, and Frequency on Manual Handling Performance, Proceedings of the Human Factors Society. Detroit, MI.

Das, R.K. (1951). Energy Expenditure in Weight Lifting by Different Methods, Ph.D. Dissertation, University of London.

Davies, B. T. (1972). Moving Loads Manually, *Applied Ergonomics*. 3(4), 190–194.

Drury, C. G. & Deeb, J. M. (1986a). Handle Positions and Angles in a Dynamic Lifting Task: Part 1. Biomechanical Considerations, *Ergonomics*. 29(6), 743–768.

Drury, C. G. & Deeb, J. M. (1986b). Handle Positions and Angles in a Dynamic Lifting Task: Part 2. Psychophysical Measures and Heart Rate. *Ergonomics*. 29(6), 769–777.

Drury, C.G., Law, C.H. & Pawenski, C.S. (1982). A Survey of Industrial Box Handling. *Human Factors*. 24(5), 553–565.

Ekblom, B., Astrand, P. O., Saltin, B., Stenberg, J., & Wallstrom, B. (1968). Effects of Training on Circulatory Response to Exercise, *Journal of Applied Physiology* 24, 618–627.

Frederik, W. S. (1959). Human Energy in Manual Lifting. *Modern Material Handling*, pp. 74–76.

Jiang, B. (1981). A Manual Materials Handling Study of Bag Lifting, M.S. Thesis, Texas Tech University.

Jiang, B. & Ayoub, M. M. (1987). Modelling of Maximum Acceptable Load of Lifting by Physical Factors, *Ergonomics*. 30(3), 529–538.

Jiang, B., Smith, J. L., & Ayoub, M. M. (1986). Psychophysical Modeling for Combined Manual Materials-Handling Activities. *Ergonomics*. 29(10), 1173–1190.

Kroemer, K. H. E. (1976). The Assessment of Human Strength. Symposium of Safety in Manual Material Handling.

Kroemer, K. H. E. (1983). An Isoinertial Technique to Assess Individual Lifting Capacity. *Human Factors*. 26(5), 493–606.

Kroemer, K. H. E., & Howard, J. M. (1970). Towards Standardization of Muscle Strength Testing. *Medicine & Science in Sports* 2, 224–226.

Mital, A. (1985). A Comparison between Psychophysical and Physiological Approaches Across High Frequency Ranges, *Human Ergology*. Submitted for publication.

Mital, A. (1987). Patterns of Difference Between the Maximum Weights of Lift Acceptable to Experienced and Inexperienced Materials Handlers. *Ergonomics*. 30(8), 1137–1147.

Mital, A. & Ayoub, M.M. (1981). Effects of Task Variables and Their Interactions in Lifting and Lowering Loads. *American Industrial Hygiene Association Journal*. 42(2), 134–142.

Mital, A. & Shell, R.L. (1984). A Comprehensive Metabolic Energy Model for Determining Rest Allowances for Physical Tasks. *Journal of Methods-Time Measurements*. XI(2).

Muth, M. B., Ayoub, M.A., & Gruver, W.A. (1978). A Non-Linear Programming Model for the Design and Evaluation of Lifting Tasks. *Safety in Manual Material Handling*, C.G. Drury (ed.) Cincinnati, OH: NIOSH, DHEW(NIOSH), Publication 78-185, pp. 96–109.

NIOSH. (1986). *Consumer Products Safety Commission*, National Electronic Surveillance System.

Poulson, E. & Jorgenson, K. (1971). Back Muscle Strength, Lifting and Stoop Working Postures. *Applied Ergonomics*. 1, 133–137.

Selan, J. L. (1986). Relation Between the Type A Behavior Pattern and Social Facilitation in the Expression of the Psychophysical Maximum Acceptable Weight of Lift, Ph.D. Dissertation, Texas Tech University.

Snook, S. H. & Ciriello, V. M. (1974). The Effects of Heat Stress on Manual Handling Tasks. *American Industrial Hygiene Association Journal*. 36, 681–695.

Snook, S. H. & Irvine, C. H. (1966). The Evaluation of Physical Tasks in Industry. *American Industrial Hygiene Association Journal*. 27, 228–233.

9

HAND TOOLS

LEARNING OBJECTIVE

The students will learn about hand tool designs and be able to describe the key principles of hand tool selection and how they interact to affect performance and safety. They will also be able to inform workers of the risk factors associated with powered hand tool use and methods to control or limit exposure.

BRIEF HISTORY OF TOOL MAKING

In the centuries prior to 1800, craftsmen made their own tools. The craftsmen made these tools to fit their hands, and many tools had a specialized purpose. The journeymen would teach their apprentices how to make tools in the same manner they were taught by their masters. This tradition was millennia old. Around the year 1800, the concept of interchangeable parts was developed by Eli Whitney of the Cotton Gin fame. Before that time firearms, like everything else, were manufactured one at a time and, even if the gunsmith was following a pattern, one musket might vary enough from another so that parts were not interchangeable. In fact, if a musket became broken the owner would need to find a gunsmith to fix it, instead of just replacing the broken parts. Eli Whitney won a contract to manufacture between 10,000 and 15,000 muskets in 1798. During the course of delivering on the contract, he developed the concept of interchangeable parts, though the concept was not brought to fruition until after his death in 1825. Other manufacturers truly succeeded with the concept before Whitney (Hounshell, 1984).

Occupational Ergonomics: A Practical Approach, First Edition.
Theresa Stack, Lee T. Ostrom and Cheryl A. Wilhelmsen.

Making tools and manufacturing equipment standard began with Whitney, but then in the late 1800s, Fredrick Taylor took it many steps further. Frederick Taylor was the father of the concept of Scientific Management and was one of the first management consultants. Taylor felt that control of a manufacturing process should be with management and not workers. He also sought to truly standardize most all aspects of manufacturing process. Tools, workbench sizes, and door openings were all standardized based on Taylor's principles (Hughes, 1989).

These ideas of standardization are in direct conflict with ergonomics. Of course, some of Taylor's ideas were appropriate. He was the first, as obvious as it seems now, to suggest workers use a big shovel for shoveling coke used in steel making because coke is light, and a small shovel for shoveling iron ore because it is heavy. Also, he developed the concept of tool cribs and supplying the correct tool for a job for the workers instead of them bringing tools from home. However, standardizing the tool handle length and bench heights is counter intuitive to the idea of ergonomics.

The reasonable approaches outlined in this document can be directly applied to challenges such as the following:

- Deciding whether to stay with traditional tool designs or opt for new designs
- Evaluating the effectiveness of different designs
- Choosing a tool of the right size and shape for the task and the user.

This document also contains an easy-to-use checklist for comparing tools against several design characteristics that have been shown to reduce physical stresses on the user (NIOSH, 2004).

INTRODUCTION

Nonpowered hand tools are widely used in a variety of industries including construction, manufacturing, and agriculture. National data suggests that a large number of injuries, known as musculoskeletal disorders, are attributable to hand tool use in occupational settings, resulting in unnecessary suffering, lost workdays, and economic costs. Prevention of work-related musculoskeletal disorders is a high priority.

To the untrained eye, however, it may be difficult to evaluate tools from an ergonomic point of view. The purpose of this document is to demystify the process and help employers and workers identify nonpowered hand tools that are less likely to cause injury – those that can be used effectively with less force, less repeated movement, and less awkward positioning of the body. Presented here are the ergonomic basics of hand tool use. These principles are meant to complement the ordinary process of deciding on what tool to select by knowing how it is used and the task to which it will be applied.

Some tools are advertised as "ergonomic" or are designed with ergonomic features. A tool becomes "ergonomic" only when it fits the task you are performing, and it fits your hand without causing awkward postures, harmful contact pressures, or other safety and health risks. If you use a tool that does not fit your hand or use the tool in a way it was not intended, you might develop an injury, such as carpal tunnel syndrome, tendonitis, or muscle strain. These injuries do not happen because of a single event, such as a fall. Instead, they result from repetitive movements that are performed over time or for a long period of time, which may result in damage to muscles, tendons, nerves, ligaments, joints, cartilage, spinal discs, or blood vessels.

DEFINITIONS

Figures 9.1–9.3 show some general tool definitions.

The cost of an injury can be high, especially if the injury prevents them from doing their job.

The best tool is one that meets the following requirements:

- Fits the job one is doing
- Fits the work space available
- Reduces the force one need to apply
- Fits their hand.

Guidelines for Selecting Hand Tools to Reduce Your Risk of Injury

A. Know the job.
B. Look at the work space.
C. Improve the work posture.
D. Select the tool.

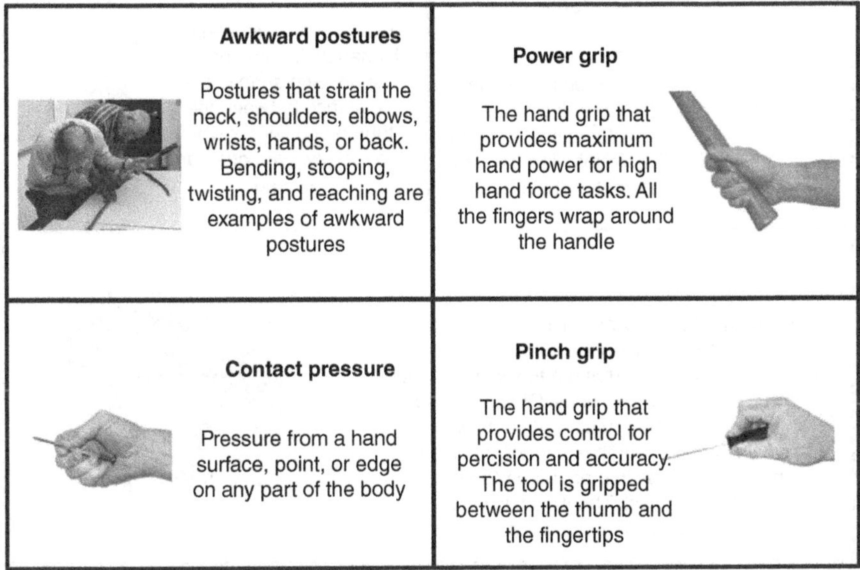

Figure 9.1 Tool definitions (Combination of graphics adapted from the Washington State Department of Labor and Industries guide to selecting non-powered hand tools and original Ostrom photos and graphics)

Hook grip

Characterized by a flat hand, curled fingers, and thumb used passively to stabilize the load, for example, auto steering wheels. Load is supported by fingers. This grip is most effective when the arms are down at the side of the body

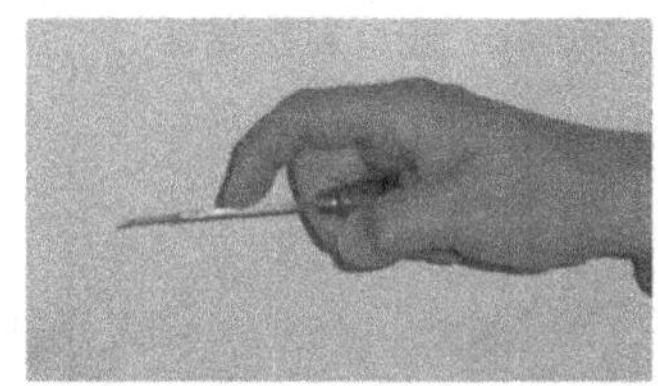

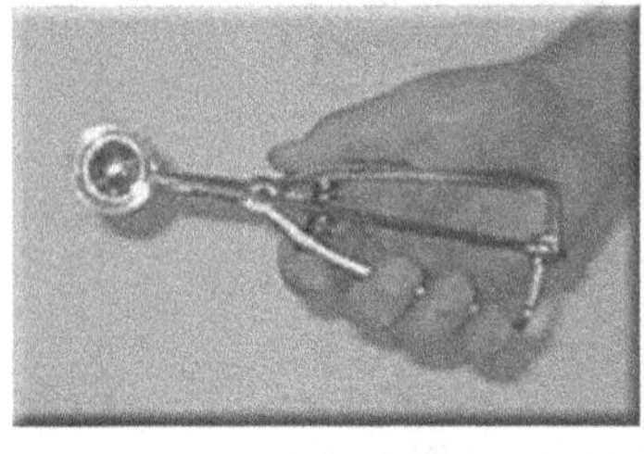

Ambidextrous tools

Tools that can be used with either hand

Figure 9.2 More tool definitions (Combination of graphics adapted from the Washington State Department of Labor and Industries guide to selecting non-powered hand tools and original Ostrom photos and graphics)

Single-handed tools

Tube-like tools measured by handle length and diameter. Diameter is the length of the straight line through the center of the handle

Double-handle tools

Plier-like tools measured by handle length and grip span. Grip span is the distance between the thumb and fingers when the tool jaws are open and closed

Oblique grip

A variant of the power grasp characterized by gripping across the surface of an object, for example, carrying a tray with handles-lift up and power grip ends. Oblique grasp is around 65% grip strength of power grasp. Grasp strength is strongly affected by hand span. A grip of around 2-2.5 in. is the strongest. The thumb is very important in this grasp

Figure 9.3 Still more tool definitions (Combination of graphics adapted from the Washington State Department of Labor and Industries guide to selecting non-powered hand tools and original Ostrom photos and graphics)

Know the Job Before selecting a tool, think about the job you will be doing. Tools are designed for specific purposes. Using a tool for something other than it's intended purpose often damages the tool and could cause pain, discomfort, or injury. One reduce chances of being injured when they select a tool that fits the job one will be doing.

The list of tools in each category shows a few examples of tools that are most frequently used (Figure 9.4).

Next, consider whether they need the tool to provide power or precision. Then select the tool with the correct handle diameter or grip span. Figure 9.5 shows the power grip style and the appropriate tool. Figure 9.6 shows tools for precision tasks.

Look at the Work Space Now look at the work space. Awkward postures may cause them to use more force. Select a tool that can be used within the space available. For example, if they work in a cramped area and high force is required, select a tool that is held with a power grip (Figure 9.7). A pinch grip will produce much less power than a power grip. Exerting force with a pinch grip means one will work harder to get the job done.

If they work in a cramped space, they may not be able to use a long-handle tool. Use of a long-handle tool may cause awkward postures or harmful contact pressure on the hand as they use more force. Instead, use a tool that fits within the work space. A short-handle tool can help them reach their target directly as they keep their wrist straight (Figure 9.8).

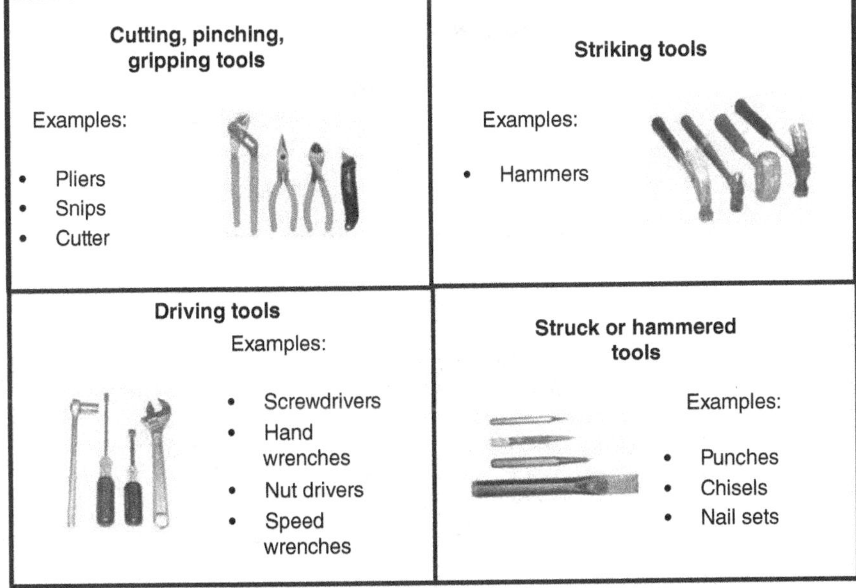

Figure 9.4 Tool categories (Combination of graphics adapted from the Washington State Department of Labor and Industries guide to selecting non-powered hand tools and original Ostrom photos and graphics)

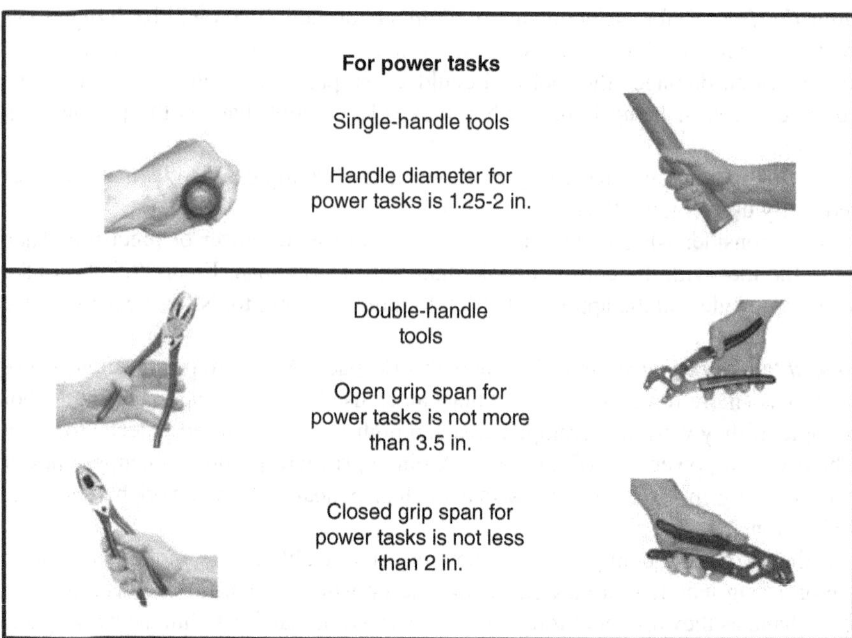

Figure 9.5 Power grip and oblique grip (Combination of graphics adapted from the Washington State Department of Labor and Industries guide to selecting non-powered hand tools and original Ostrom photos and graphics)

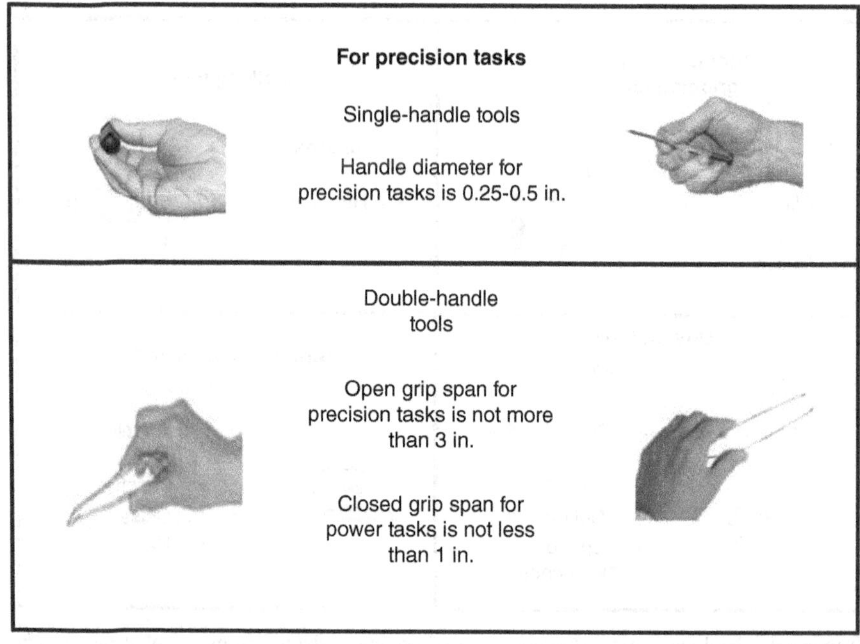

Figure 9.6 Precision grip (Combination of graphics adapted from the Washington State Department of Labor and Industries guide to selecting non-powered hand tools and original Ostrom photos and graphics)

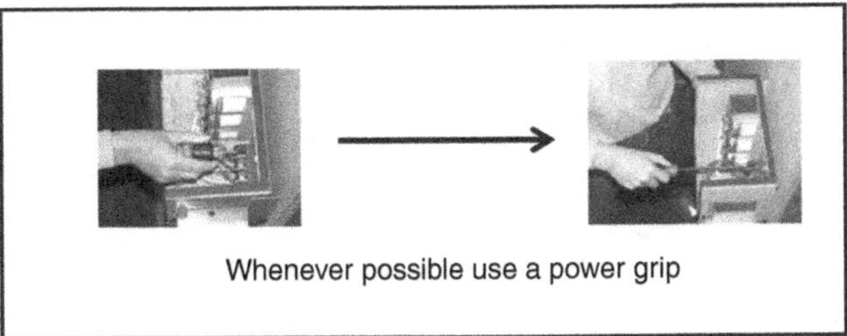

Figure 9.7 Change the tool for the task (Combination of graphics adapted from the Washington State Department of Labor and Industries guide to selecting non-powered hand tools and original Ostrom photos and graphics)

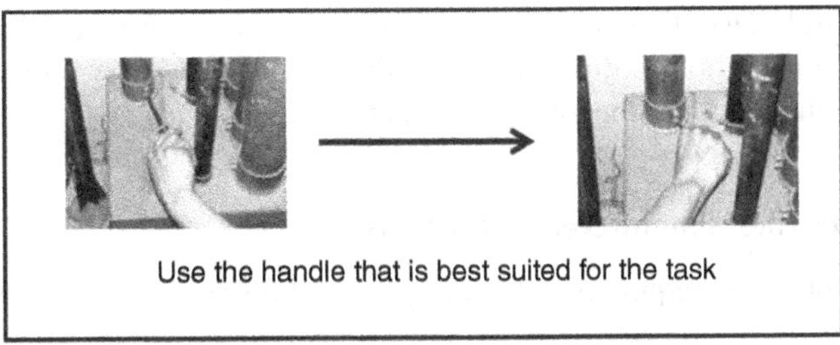

Use the handle that is best suited for the task

Figure 9.8 Select the correct tool for the task (Combination of graphics adapted from the Washington State Department of Labor and Industries guide to selecting non-powered hand tools and original Ostrom photos and graphics)

Improve the Work Posture Awkward postures make more demands on the body. In some cases, the placement of the work piece will affect the shoulder, elbow, wrist, hand, or back posture. Whenever possible, choose a tool that requires the least continuous force and can be used without awkward postures. The right tool will help to minimize pain and fatigue by keeping the neck, shoulders, and back relaxed and arms at sides (Figures 9.9 and 9.10).

For example, avoid raising shoulders and elbows. Relaxed shoulders and elbows are more comfortable and will make it easier to drive downward force.

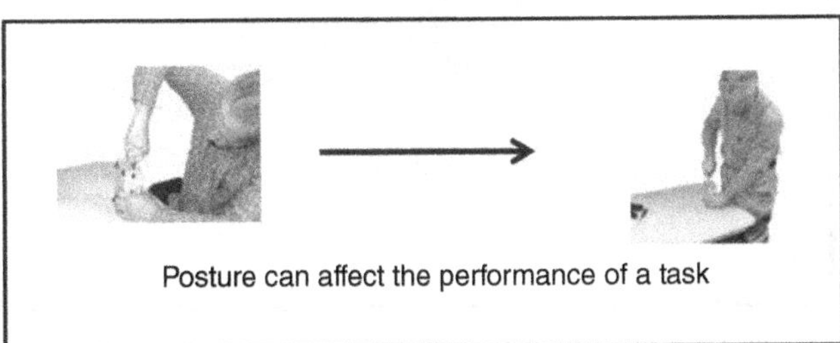

Posture can affect the performance of a task

Figure 9.9 Use the correct posture for the tool being used (Combination of graphics adapted from the Washington State Department of Labor and Industries guide to selecting non-powered hand tools and original Ostrom photos and graphics)

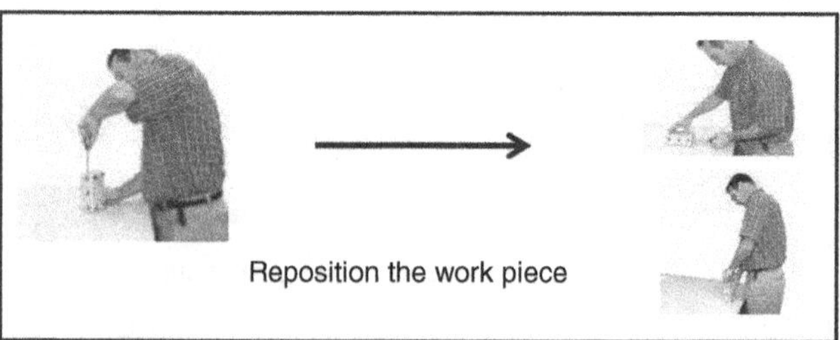

Figure 9.10 Change the part location to obtain the correct posture for the tool being used (Combination of graphics adapted from the Washington State Department of Labor and Industries guide to selecting non-powered hand tools and original Ostrom photos and graphics)

Select the Tool Over time, exposure to awkward postures or harmful contact pressures can contribute to an injury. One can reduce the risk of injury if they select hand tools that fit the hand and the job you are doing.

Tips for Selecting Nonpowered Hand Tools

Tools used for power require high force. Tools used for precision or accuracy require low force.

1. Power grip: Select a tool that feels comfortable with a handle diameter in the range of 1.25–2 in. One can increase the diameter by adding a sleeve to the handle (Figure 9.11).
2. For double-handle tools (plier-like) used for power tasks: Select a tool with a grip span that is at least 2 in. when fully closed and no more than 3.5 in. when

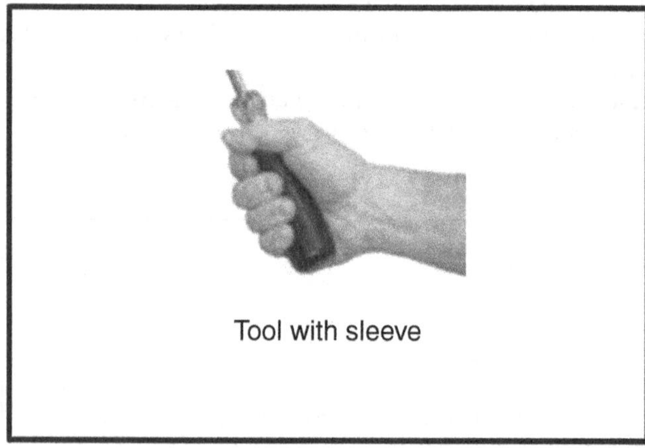

Figure 9.11 Tool with sleeve (Combination of graphics adapted from the Washington State Department of Labor and Industries guide to selecting non-powered hand tools and original Ostrom photos and graphics)

fully open. When continuous force is required, consider using a clamp, a grip, or locking pliers (Figure 9.12).

3. For double-handle tools (plier-like) used for precision tasks: Select a tool with a grip span that is not less than 1 in. when fully closed and no more than 3 in. when fully open.
4. For double-handled pinching, gripping, or cutting tools: Select a tool with handles that are spring-loaded to return the handles to the open position.
5. Select a tool without sharp edges or finger grooves on the handle (Figure 9.13).
6. Select a tool that is coated with soft material. Adding a sleeve to the tool handle pads the surface but also increases the diameter or the grip span of the handle (Figure 9.11).

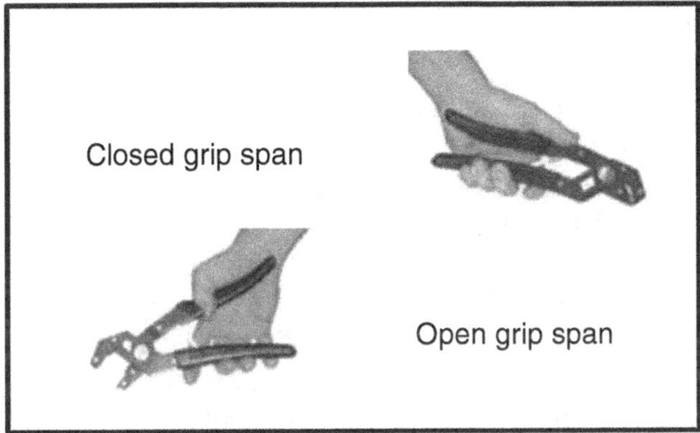

Figure 9.12 Open grip tool (Combination of graphics adapted from the Washington State Department of Labor and Industries guide to selecting non-powered hand tools and original Ostrom photos and graphics)

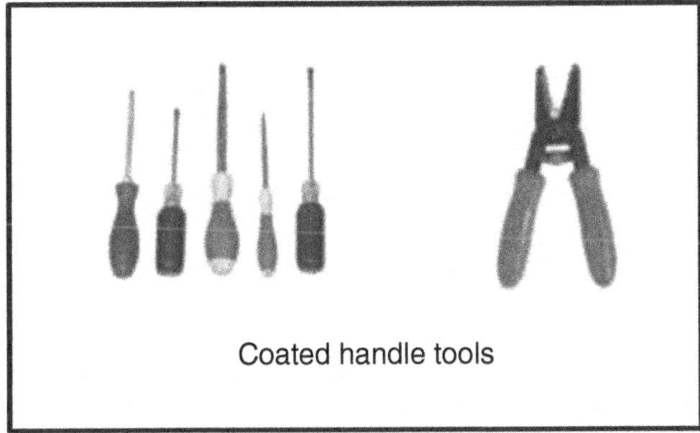

Figure 9.13 Tools with coated handles (Combination of graphics adapted from the Washington State Department of Labor and Industries guide to selecting non-powered hand tools and original Ostrom photos and graphics)

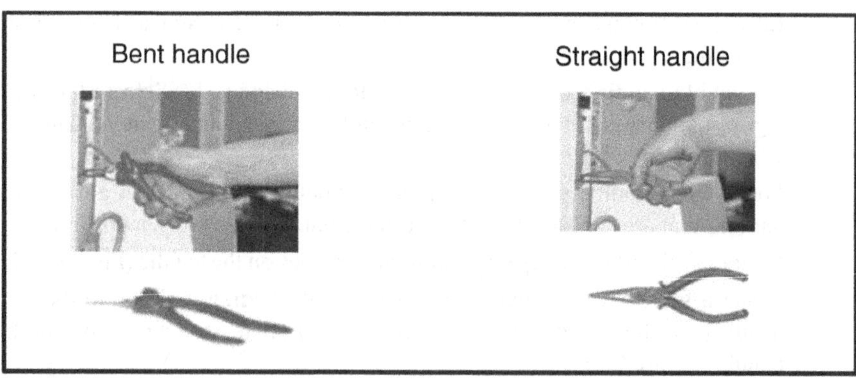

Figure 9.14 Tool with bent handle (Combination of graphics adapted from the Washington State Department of Labor and Industries guide to selecting non-powered hand tools and original Ostrom photos and graphics)

7. Select a tool with an angle that allows them to work with a straight wrist.

   Tools with bent handles are better than those with straight handles when the force is applied horizontally (in the same direction as their straight forearm and wrist) (Figure 9.14).

   Tools with straight handles are better than those with bent handles when the force is applied vertically.
8. Select a tool that can be used with their dominant hand or with either hand (Figure 9.15).
9. For tasks requiring high force: select a tool with a handle length longer than the widest part of the hand – usually 4–6 in. (Figure 9.16).

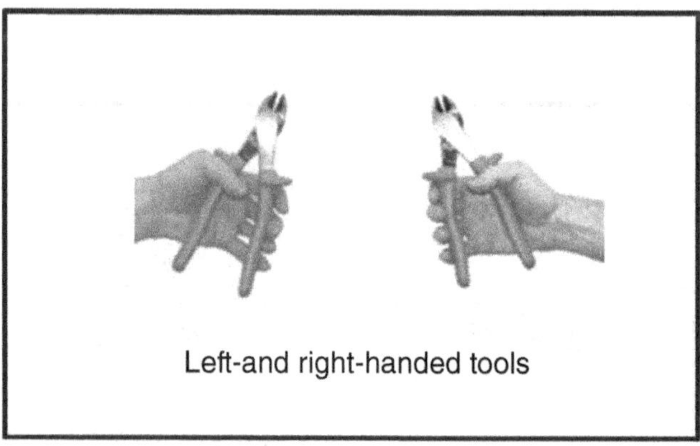

Figure 9.15 Choice of right- and left-handed tools (Combination of graphics adapted from the Washington State Department of Labor and Industries guide to selecting non-powered hand tools and original Ostrom photos and graphics)

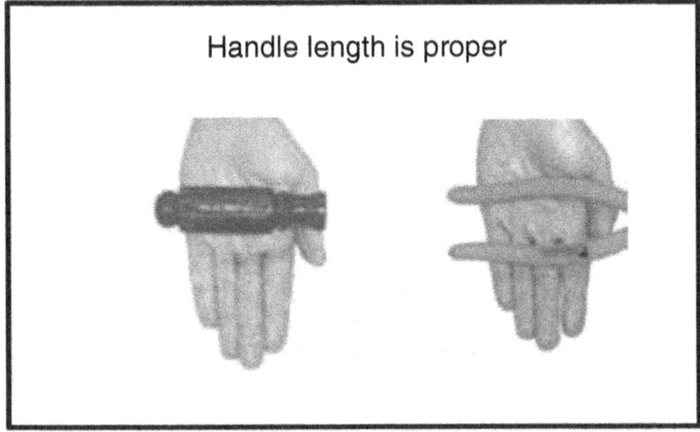
Handle length is proper

Figure 9.16 Handle length (Combination of graphics adapted from the Washington State Department of Labor and Industries guide to selecting non-powered hand tools and original Ostrom photos and graphics)

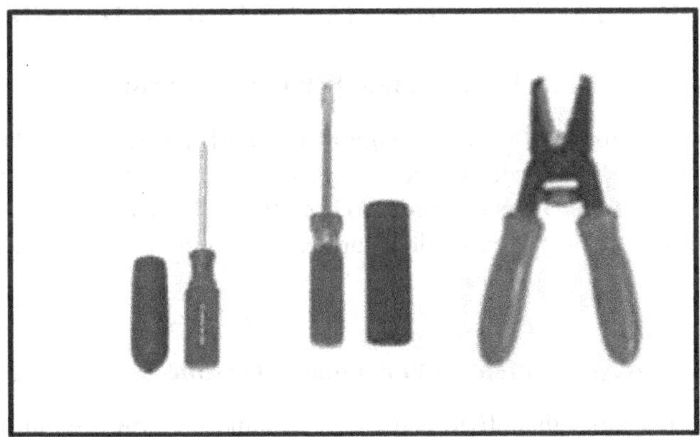

Figure 9.17 Grip sleeves and proper handles (Combination of graphics adapted from the Washington State Department of Labor and Industries guide to selecting non-powered hand tools and original Ostrom photos and graphics)

Prevent contact pressure by making sure the end of the handle does not press on the nerves and blood vessels in the palm of the hand.

10. Select a tool that has a nonslip surface for a better grip. Adding a sleeve to the tool improves the surface texture of the handle. To prevent tool slippage within the sleeve, make sure that the sleeve fits snugly during use.

    Remember: A sleeve always increases the diameter or the grip span of the handle (Figure 9.17).

Figure 9.18 shows tools with spring returns for aiding in opening the tools.

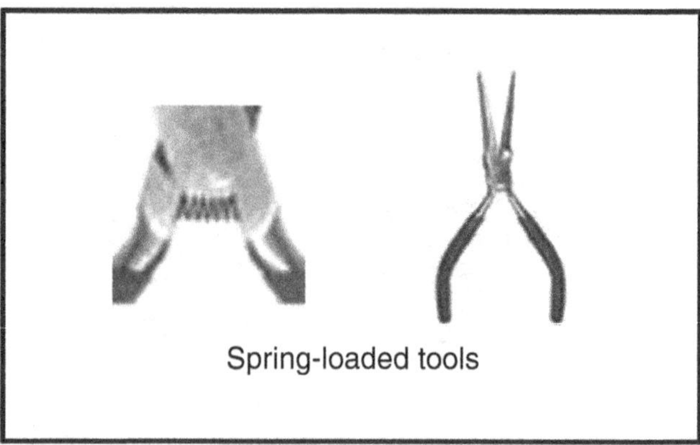

Figure 9.18 Spring return tools (Combination of graphics adapted from the Washington State Department of Labor and Industries guide to selecting non-powered hand tools and original Ostrom photos and graphics)

ERGONOMIC SELECTION CRITERIA FOR POWERED HAND TOOLS

Select Tools That Can Be Used Without Bending the Wrist

Use the right tool for the job. The design of the workstation and the layout of the work piece will influence their handle choice. Work surfaces may need to be angled to match the tool, or vice versa, in order to keep the body in a neutral posture. Pistol grip tools are best for work on vertical surfaces to maintain a neutral wrist posture (Figure 9.19).

Select Tools That Are as Light as Functionally Possible

Tools that weigh more than 10 lb can cause extreme forearm discomfort in a few minutes. For tools heavier than 4 lb, a second handle can help disperse the weight. Tools that are used frequently and weigh more than 1 lb should be counterbalanced.

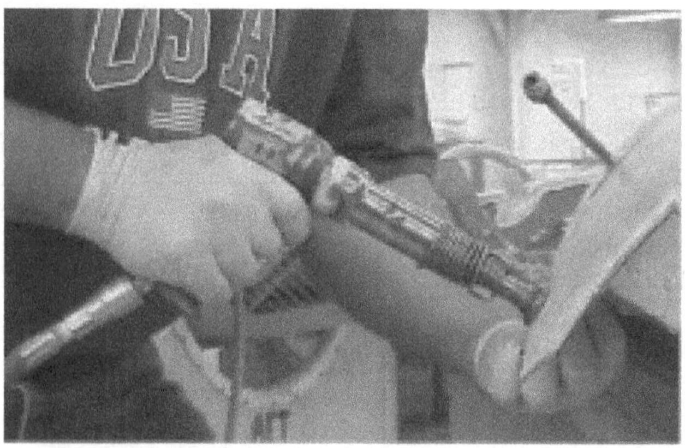

Figure 9.19 Use of pistol grip tools

Select the right tool for the particular job to keep the person's wrist in a neutral posture.

Consider the Design of the Handle

Choose tools with vibration-damping handles (i.e., rubber, plastic, or cork). Choose handles that are located close to or below the center of gravity of the tool. Select tools with rounded and smooth handles to aid grip and with a trigger strip instead of a trigger button.

Select Tools That Will Minimize Vibration Exposure

Vibrating tools can cause vascular spasms or a constriction of blood vessels in the fingers, which then appear white or pale (Figure 9.20). Vascular constriction may lead to numbness and swelling of hand tissue, with a loss of grip strength. Vibration-induced white finger, also known as VWF or "Reynaud's phenomenon," and hand-arm vibration syndrome (HAVS) cause tingling, numbness, or pain that can be brought on or intensified by exposure to cold.

There are preventive actions that can be taken to reduce the impact of vibration:

- Reduce the number of hours or days vibrating tools are used in accordance with the American Conference of Governmental Industrial Hygiene Threshold Limit Values.
- Arrange tasks to alternate use of vibrating and nonvibrating tools.

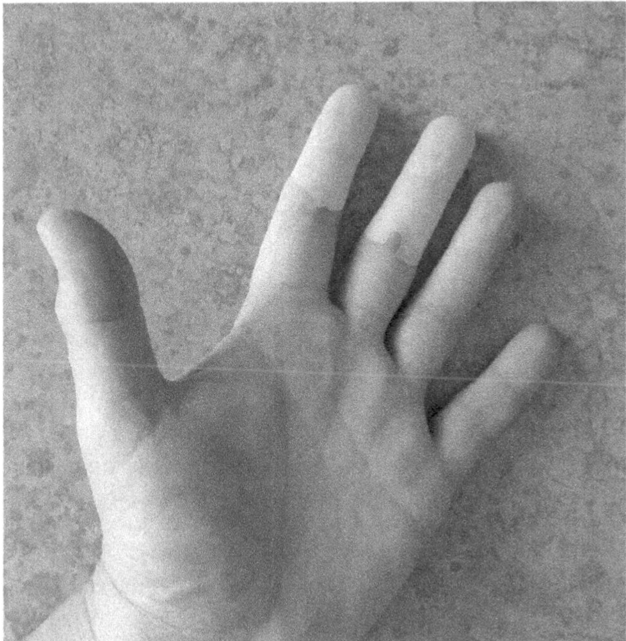

Figure 9.20 Vibration-induced white finger

- Schedule tool maintenance so tools remain sharp, lubricated, and properly tuned.
- Select tools that perform satisfactorily with the least vibration. Ask tool manufacturers to furnish vibration and frequency data on their tools.
- Use gloves with vibration-damping materials in the palms and fingers. Ensure workers keep warm at work, especially their hands. To be effective, gloves should fit, be full finger, and comply with ANSI S3.40-2002 or ISO 10819 standards.
- Use tools with vibration-damping handles.
- Keep hands warm and dry.
- Avoid using tobacco or stimulant drugs that may restrict blood flow to the skin by as much as 40%.

KEY POINTS

The key points of this chapter are as follows:

There are four types of hand grips:

- Power – best for heavy tasks
- Pinch – best for precise tasks
- Oblique
- Hook

Ergonomic tools:

- Fits the job one is doing
- Fits the work space available
- Reduces the force one needs to apply
- Fits the hand

When using tools you should do the following:

- Know the job.
- Look at the work space.
- Improve the work posture.
- Select the tool.

HAND TOOL CASE STUDY

Abstract

This report documents the ergonomic assessments performed on the traditional method of splicing on 4/0 electrical cable and the Melni connector to perform a splice. A master electrician performed three (3) splices using each methodology. The tasks were recorded using digital video and still photography. The data were analyzed using two methods:

1. Observational and video analysis was used to detail the steps in the procedure and to identify the ergonomic risk factors associated with each of the two (2) sets of tasks.

2. The Rapid Upper Limb Assessment (RULA) tool was also used to perform a postural analysis of the two (2) sets of tasks (Ergonomic Concepts, n.d.).

The ergonomic risk factors associated with the traditional splice method were as follows:

- Leaning forward back postures in excess of 20° for several aspects of the task
- Twisted back postures for several aspects of the task
- Repetition associated with crimping the splice
- High forces and duration when crimping the splice
- Compression of the thigh and hands when crimping the splice.

The ergonomic risk factors associated with the Melni connector splice method were as follows:

- High hand forces and twisted wrist when inserting the wire into the Melni connector to ensure it is seated
- Awkward neck posture when inserting the wire into the Melni connector
- Awkward back posture when using the channel locks to tighten the couplers
- Awkward wrist postures when tightening the gripper/seal ring on the Melni connector.

The RULA score for the traditional splice method was seven (7), indicating a person is working in the worst posture with an immediate risk of injury from their work posture and changes should obscure immediately to prevent injury. The RULA score for the most stressful aspects of the Melni connector splice method was five (5) or less, indicating a less stressful task.

The traditional splicing method took approximately 9 min to perform, while the Melni connector method took approximately 50 s to perform. Also, the Melni connector method used fewer tools and had seven (7) fewer basic steps.

The overall conclusion is that the Melni connector splice method has significant ergonomic benefit over the traditional method of performing splices on 4/0 electrical cable. The Melni method should replace the traditional method whenever feasible.

Introduction and Background

Here is an ergonomic assessment comparison of the traditional method of performing a splice connection on 4/0 gauge aluminum cable, with using the Melni connector to perform the same sort of splice. The data collection consisted of a master electrician performing three (3) cable splices using the traditional splice method and three (3) splices using the Melni connector. The Melni connector is shown in Figure 9.21. The two tasks were observed and recorded with video and with still photographs. These tasks were performed in a laboratory setting.

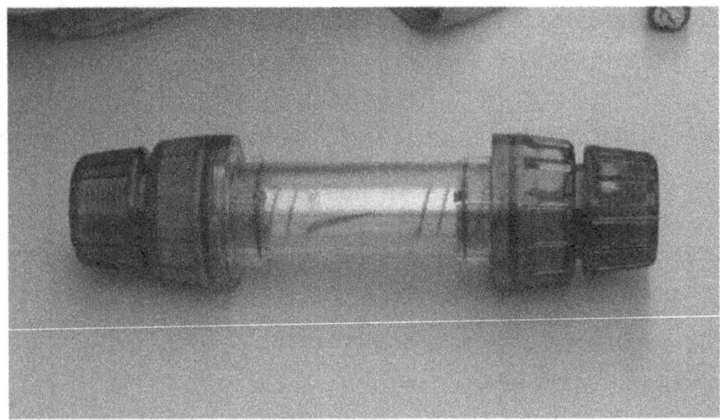

Figure 9.21 Melni connector

The data were analyzed using two (2) methodologies. These were as follows:

1. Using the observations, pictures, and videos of the task to identify the steps in the procedure and to identify the ergonomic risk factors associated with each of the two (2) sets of tasks. The following are traditional ergonomic risk factors(Ergonomic Concepts, n.d.):
   a. Force
   b. Posture
   c. Repetition
   d. Duration
   e. Vibration
   f. Compression
2. Using the RULA evaluation tool to perform a postural analysis of the two (2) sets of tasks (McAtamney & Corlett, 1993).

This report discusses the results of the assessments, the analysis of the traditional splicing task, and the analysis of using the Melni connector to perform the splicing task.

Analysis of the Traditional Splicing Task

The equipment used for the traditional splicing task is as follows:

- Wire
- Butt splice connectors
- Crimping tool
- Utility knife or utility knife and wire stripper
- Heat shrink wrap
- Heat source (propane torch or heat gun)
- Wire cutter.

The basic steps for performing the traditional splice of 4/0 gauge wire are as follows:

1. Retrieve tools.
2. Provide access to the wire that requires splicing.
3. Adjust the crimping tool to the correct splice.
4. Cut the wire.
5. Strip the insulation back approximately 2 inches.
6. Ensure wire is clean of dirt or other contaminants.
7. Slide a shrink wrap sleeve on the wire.
8. Insert an end of the wire into the end of the butt splice connector
9. Ensure the crimping tool is adjusted correctly for the butt splice connector
10. Crimp the butt splice connector using the crimping tool.
11. Rotate the wire 90°.
12. Crimp the butt splice connector.
13. Insert the other end of the wire into the end of the butt splice connector.
14. Crimp the butt splice connector using the crimping tool.
15. Rotate the wire 90°.
16. Crimp the butt splice connector.
17. Slide the shrink wrap sleeve over the splice.
18. Use the heat source to shrink the sleeve.
19. Store tools.

The total operation of these steps took approximately 9 min to perform.

The ergonomic risk factors identified in this procedure were as follows:

- Leaning forward back postures in excess of 20° for several aspects of the task
- Twisted back postures for several aspects of the task
- Repetition associated with crimping the splice
- High forces and duration of high forces when crimping the splice
- Compression of the thigh and hands when crimping the splice.

In reality, these tasks were being performed in an ideal environment to include being indoors, good lighting, and ambient temperatures in the range of 70–75 °F. In addition, the splices were performed on a table. When performed in the field, the task could be even more stressful. For instance, if the tasks were performed in a ditch or on overhead wires the postures and forces would vary greatly.

The steps of the process with ergonomic risk factors were 9, 10, 12, 13, 14, and 16. The crimping steps (10, 12, 14, and 16) appeared to be the most stressful and are discussed next.

Figure 9.22 shows a photo at the start of the crimping step. The master electrician is in an awkward posture. His torso is leaning forward, one handle of the tool is

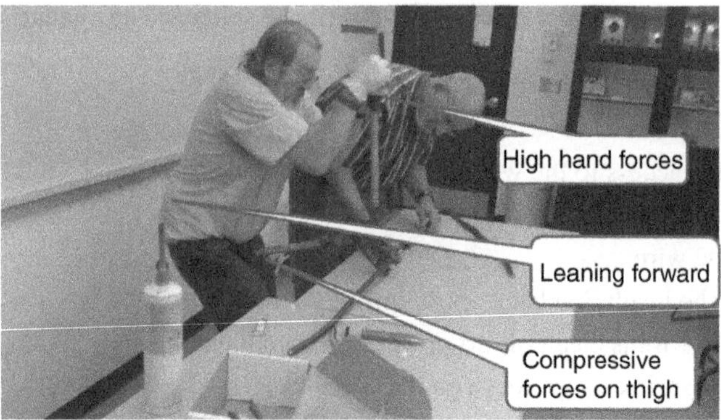

Figure 9.22 Start of crimping step

compressing his thigh, his upper arm approaching shoulder height, and he is putting a great amount of force on the handles. He must attain this posture four (4) times each time he splices one (1) 4/0 cable. So, it is performed four (4) times in a 9-min sequence of steps. Each crimping step took approximately 9 s to perform.

At the midpoint of the crimping step, as shown in Figure 9.23, the master electrician is in a very awkward posture. His back is bent forward and twisted, his neck is twisted, and he is continuing to apply a great amount of force using his body weight and his hands. His left upper arm is above shoulder height and abducted (away from the body), and the handle of the tool is in a position where it could slip.

Figure 9.24 shows the end point of the crimping step. As the figure shows, the master electrician is in a very awkward posture with his back leaning forward almost the end of a supportable range and twisted, his neck is in a twisted posture; handle of the tool is still unsupported and could slip, and he is still exerting a great amount of force.

The RULA analysis was peroformed on three of the most stressful postures associated with this task. Figure 9.22 shows the start of the crimping task. A RULA analysis was performed on this posture. The completed RULA form is shown in Figure 9.25.

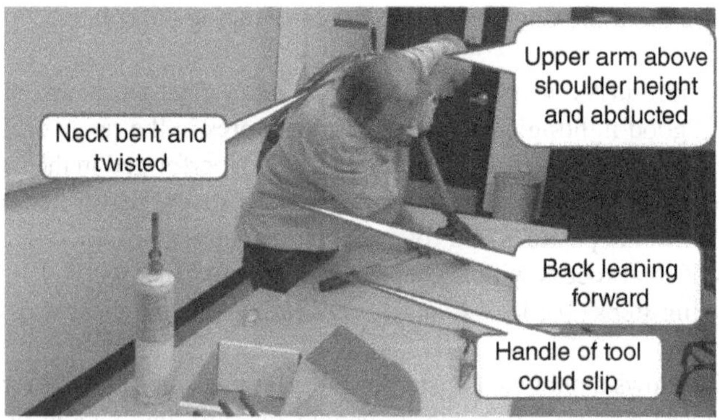

Figure 9.23 Midpoint crimping step

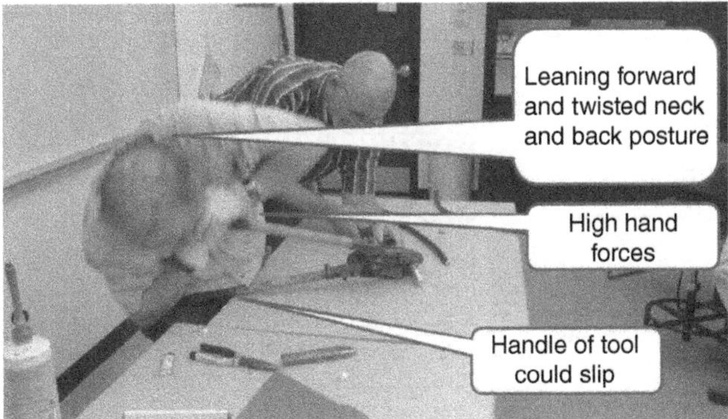

Figure 9.24 End point of crimping step

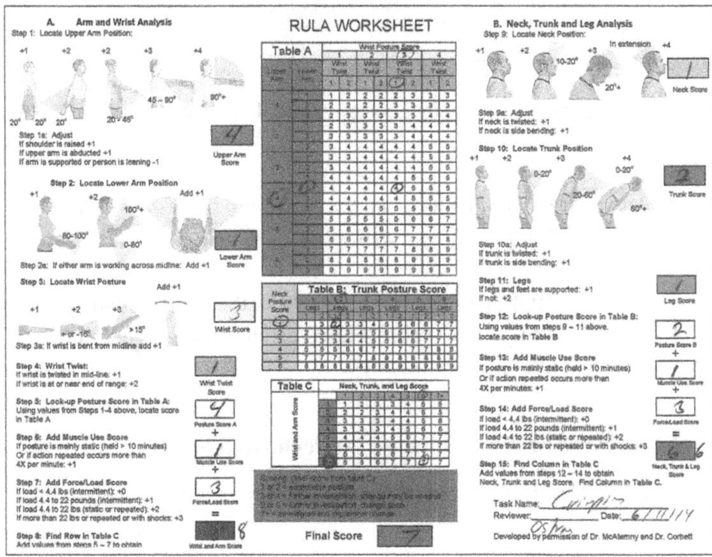

RULA WORKSHEET

A. Arm and Wrist Analysis

Step 1: Locate Upper Arm Position:

+1 +2 +2 +3 +4

20° 20° 20° 20°-45° 45–90° 90°+

Step 1a: Adjust
If shoulder is raised +1
If upper arm is abducted +1
If arm is supported or person is leaning -1

Upper Arm Score: 4

Step 2: Locate Lower Arm Position

+1 60-100° +2 100°+ 0-60° Add +1

Step 2a: If either arm is working across midline: Add +1

Lower Arm Score: 1

Step 3: Locate Wrist Posture

+1 +2 + or -15° +3 >15° Add +1

Step 3a: If wrist is bent from midline add +1

Wrist Score: 3

Step 4: Wrist Twist:
If wrist is twisted in mid-line: +1
If wrist is at or near end of range: +2

Wrist Twist Score: 1

Step 5: Look-up Posture Score in Table A:
Using values from Steps 1-4 above, locate score in Table A

Posture Score A: 4
+

Step 6: Add Muscle Use Score
If posture is mainly static (held > 10 minutes)
Or if action repeated occurs more than 4X per minute: +1

Muscle Use Score: 1
+

Step 7: Add Force/Load Score
If load < 4.4 lbs (intermittent): +0
If load 4.4 to 22 pounds (intermittent): +1
If load 4.4 to 22 lbs (static or repeated): +2
If more than 22 lbs or repeated or with shocks: +3

Force/Load Score: 3
=

Step 8: Find Row in Table C
Add values from steps 5 – 7 to obtain

Wrist and Arm Score: 8

Table A

Table B: Trunk Posture Score

Table C

Final Score: 7

B. Neck, Trunk and Leg Analysis

Step 9: Locate Neck Position:

+1 +2 10-20° +3 20°+ In extension +4

Neck Score: 1

Step 9a: Adjust
If neck is twisted: +1
If neck is side bending: +1

Step 10: Locate Trunk Position

+1 +2 0-20° +3 20-60° +4 0-20° 60°+

Trunk Score: 2

Step 10a: Adjust
If trunk is twisted: +1
If trunk is side bending: +1

Step 11: Legs
If legs and feet are supported: +1
If not: +2

Leg Score: 1

Step 12: Look-up Posture Score in Table B:
Using values from steps 9 – 11 above, locate score in Table B

Posture Score B: 2
+

Step 13: Add Muscle Use Score
If posture is mainly static (held > 10 minutes)
Or if action repeated occurs more than 4X per minutes: +1

Muscle Use Score: 1
+

Step 14: Add Force/Load Score
If load < 4.4 lbs (intermittent): +0
If load 4.4 to 22 pounds (intermittent): +1
If load 4.4 to 22 lbs (static or repeated): +2
If more than 22 lbs or repeated or with shocks: +3

Force/Load Score: 3
=

Step 15: Find Column in Table C
Add values from steps 12 – 14 to obtain Neck, Trunk and Leg Score. Find Column in Table C.

Neck, Trunk & Leg Score: 6

Task Name: Crimping
Reviewer: Osborn Date: 6/11/14

Developed by permission of Dr. McAtamny and Dr. Corbett

Figure 9.25 Example of a completed RULA form for crimping

The RULA score developed from this analysis was seven (7). A seven (7) represents the worst postural score under the RULA technique, indicating a change needs to be made to the task. This step is performed four (4) times for each splice. So, this action is performed four (4) times in the 9-min task, allowing very little time for muscle recovery.

Figure 9.23 shows the midpoint in the crimping task. The completed RULA form is shown in Figure 9.26. The RULA score developed from this analysis was seven (7). A seven (7) again represents the worst postural score under the RULA technique, indicating a change needs to be made to the task. This step is also performed four (4) times for each splice. So, this action is performed four (4) times in the 9-min task.

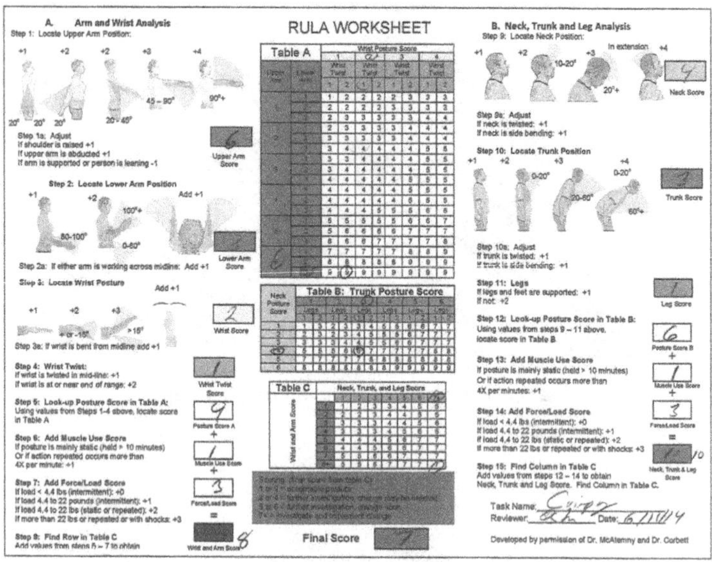

A. Arm and Wrist Analysis
Step 1: Locate Upper Arm Position:
+1 +2 +2 +3 +4
20° 20° 20° 20–45° 45–90° 90°+
Step 1a: Adjust
If shoulder is raised +1
If upper arm is abducted +1
If arm is supported or person is leaning -1
Upper Arm Score
Step 2: Locate Lower Arm Position
+1 +2 Add +1
100°+ 60-100° 0-60°
Lower Arm Score
Step 2a: If either arm is working across midline: Add +1
Step 3: Locate Wrist Posture
+1 +2 +3 Add +1
+ or -15° >15°
Wrist Score
Step 3a: If wrist is bent from midline add +1
Step 4: Wrist Twist:
If wrist is twisted in mid-line: +1
If wrist is at or near end of range: +2
Wrist Twist Score
Step 5: Look-up Posture Score in Table A:
Using values from Steps 1-4 above, locate score in Table A
Posture Score A
+
Step 6: Add Muscle Use Score
If posture is mainly static (held > 10 minutes)
Or if action repeated occurs more than 4X per minute: +1
Muscle Use Score
+
Step 7: Add Force/Load Score
If load < 4.4 lbs (intermittent): +0
If load 4.4 to 22 pounds (intermittent): +1
If load 4.4 to 22 lbs (static or repeated): +2
If more than 22 lbs or repeated or with shocks: +3
Force/Load Score
=
Step 8: Find Row in Table C
Add values from steps 5 – 7 to obtain
Wrist and Arm Score

RULA WORKSHEET
Table A
Table B: Trunk Posture Score
Table C
Final Score

B. Neck, Trunk and Leg Analysis
Step 9: Locate Neck Position:
+1 +2 +3 In extension +4
10-20° 20°+
Neck Score
Step 9a: Adjust
If neck is twisted: +1
If neck is side bending: +1
Step 10: Locate Trunk Position
+1 +2 +3 +4
0-20° 20-60° 60°+
Trunk Score
Step 10a: Adjust
If trunk is twisted: +1
If trunk is side bending: +1
Step 11: Legs
If legs and feet are supported: +1
If not: +2
Leg Score
Step 12: Look-up Posture Score in Table B:
Using values from steps 9 – 11 above, locate score in Table B
Posture Score B
+
Step 13: Add Muscle Use Score
If posture is mainly static (held > 10 minutes)
Or if action repeated occurs more than 4X per minutes: +1
Muscle Use Score
+
Step 14: Add Force/Load Score
If load < 4.4 lbs (intermittent): +0
If load 4.4 to 22 pounds (intermittent): +1
If load 4.4 to 22 lbs (static or repeated): +2
If more than 22 lbs or repeated or with shocks: +3
Force/Load Score
=
Step 15: Find Column in Table C
Add values from steps 12 – 14 to obtain Neck, Trunk and Leg Score. Find Column in Table C.
Neck, Trunk & Leg Score
Task Name:
Reviewer: Date:
Developed by permission of Dr. McAtemny and Dr. Corbett

Figure 9.26 Another completed RULA form example

Figure 9.24 shows the final phase of the crimping step and Figure 9.27 shows the RULA analysis. This part of the crimping task also scored a RULA score of seven (7).

As stated above, a RULA score of seven (7) is the worst postural score possible using this methodology. This score indicates that the task should be modified immediately to avoid injury.

Analysis of Using the Melni Connector to Perform a Splice

The equipment used for the Melni connector splicing task is as follows:

- Wire
- Melni connector
- Two (2) pairs of channel lock pliers.

The basic steps for performing the splicing task with the Melni connector for 4/0 wire are as follows:

1. Retrieving tools
2. Providing access to the wire that requires splicing
3. Cutting the wire
4. Stripping the insulation back approximately 2 in.

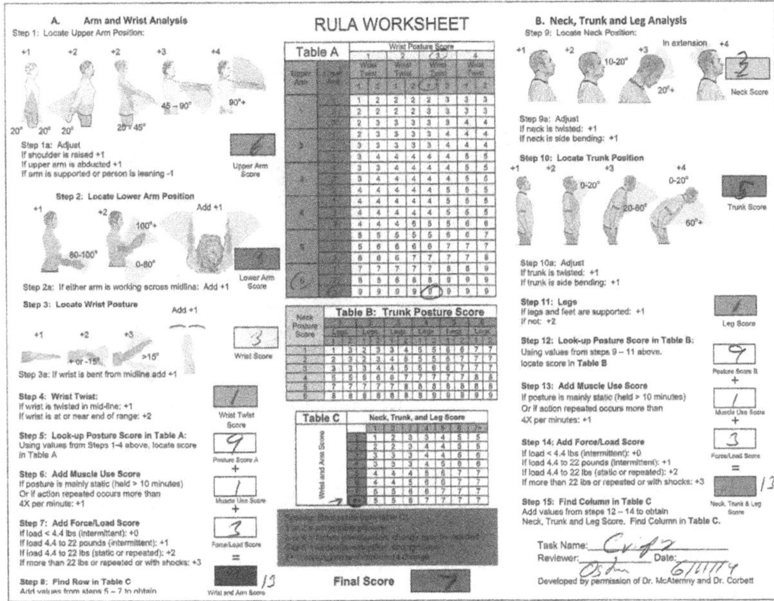

RULA WORKSHEET

A. Arm and Wrist Analysis

Step 1: Locate Upper Arm Position:

+1 +2 +2 +3 +4

20° 20° 20° 20–45° 45 – 90° 90°+

Step 1a: Adjust
If shoulder is raised +1
If upper arm is abducted +1
If arm is supported or person is leaning -1

Upper Arm Score: 6

Step 2: Locate Lower Arm Position

+1 60-100° +2 100°+ 0-60° Add +1

Lower Arm Score

Step 2a: If either arm is working across midline: Add +1

Step 3: Locate Wrist Posture

+1 +2 + or -15° +3 >15° Add +1

Wrist Score: 3

Step 3a: If wrist is bent from midline add +1

Step 4: Wrist Twist:
If wrist is twisted in mid-line: +1
If wrist is at or near end of range: +2

Wrist Twist Score: 1

Step 5: Look-up Posture Score in Table A:
Using values from Steps 1-4 above, locate score in Table A

Posture Score A: 9

+

Step 6: Add Muscle Use Score
If posture is mainly static (held > 10 minutes)
Or if action repeated occurs more than 4X per minute: +1

Muscle Use Score: 1

+

Step 7: Add Force/Load Score
If load < 4.4 lbs (intermittent): +0
If load 4.4 to 22 pounds (intermittent): +1
If load 4.4 to 22 lbs (static or repeated): +2
If more than 22 lbs or repeated or with shocks: +3

Force/Load Score: 3

=

Wrist and Arm Score: 13

Step 8: Find Row in Table C
Add values from steps 5 – 7 to obtain

Table A — Wrist Posture Score

Table B: Trunk Posture Score

Table C — Neck, Trunk, and Leg Score

Final Score: 7

B. Neck, Trunk and Leg Analysis

Step 9: Locate Neck Position:

+1 +2 10-20° +3 20°+ In extension +4

Neck Score: 3

Step 9a: Adjust
If neck is twisted: +1
If neck is side bending: +1

Step 10: Locate Trunk Position

+1 +2 0-20° +3 20-60° +4 60°+ 0-20°

Trunk Score: 5

Step 10a: Adjust
If trunk is twisted: +1
If trunk is side bending: +1

Step 11: Legs
If legs and feet are supported: +1
If not: +2

Leg Score: 1

Step 12: Look-up Posture Score in Table B:
Using values from steps 9 – 11 above.
locate score in Table B

Posture Score B: 9

+

Step 13: Add Muscle Use Score
If posture is mainly static (held > 10 minutes)
Or if action repeated occurs more than 4X per minutes: +1

Muscle Use Score: 1

+

Step 14: Add Force/Load Score
If load < 4.4 lbs (intermittent): +0
If load 4.4 to 22 pounds (intermittent): +1
If load 4.4 to 22 lbs (static or repeated): +2
If more than 22 lbs or repeated or with shocks: +3

Force/Load Score: 3

=

Neck, Trunk & Leg Score: 13

Step 15: Find Column in Table C
Add values from steps 12 – 14 to obtain
Neck, Trunk and Leg Score. Find Column in Table C.

Task Name: Crif 2
Reviewer: Os Jim Date: 6/11/14

Developed by permission of Dr. McAtemny and Dr. Corbett

Figure 9.27 Completed RULA analysis form

5. Ensuring the wire is clean of dirt or other contaminants
6. Inserting an end of the wire into the end of the Melni connector
7. Hand tightening the gripper/seal ring on the Melni connector
8. Inserting the other end of the wire into the end of the Melni connector
9. Hand tightening the gripper/seal ring on the Melni connector
10. Twisting the couplers on one end of the Melni connectors with the channel lock pliers
11. Ensuring the coupler on the other side of the connector is tight
12. Storing tools.

The procedure for using the Melni connector has seven (7) fewer basic steps, and the steps have many fewer ergonomic risk factors. This process took approximately 50 s to perform, meaning duration is not a factor as it is with the traditional method. From observing the task and analyzing the videos, the steps with ergonomic risk factors were steps 6–10. The ergonomic risk factors for the steps were as follows:

- High hand forces and twisted wrist when inserting the wire into the Melni connector to ensure it is seated (steps 6 and 8)
- Awkward neck posture when inserting the wire into the Melni connector (steps 6 and 8)

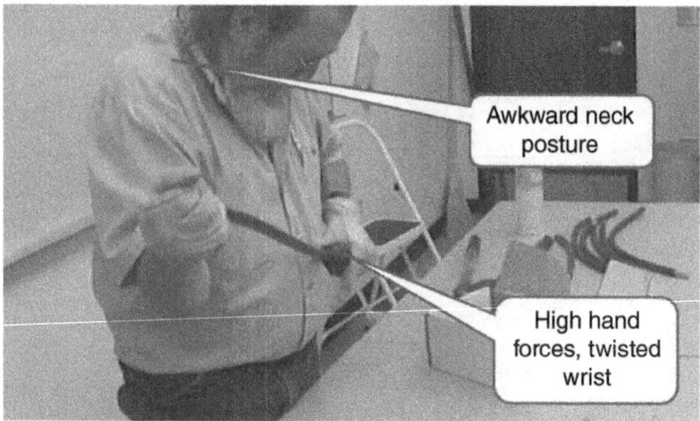

Figure 9.28 Inserting wire into Melni connector

- Awkward back posture when using the channel locks to tighten the couplers (step 10)
- Awkward wrist postures when tighten the gripper/seal ring on the Melni connector (step 9).

As with the traditional method, the postural issues will also depend on the environment where the splice is being performed. The conditions of this test were considered ideal.

Figure 9.28 shows the most stressful step using the Melni connector. This figure shows that the master electrician is slightly bent forward, and he is using force to push the wire into the Melni connector.

Figure 9.29 shows the posture when using the channel locks to tighten the Melni connector couplings. Figure 9.30 shows the posture when the master electrician tightens the gripper/seal ring on the Melni connector.

The resulting RULA scores were 5, 4, and 4, respectively, indicating that the stress should be further investigated and changed at some point in time but is not considered imminent as with the traditional method.

Figure 9.31 shows the RULA analysis for wire insertion step.

KEY POINTS

The results from the analysis showed the following points:

- There were seven (7) fewer steps associated with the Melni connector for performing a splice of 4/0 cable.
- The time to perform the splicing task was approximately 8 min less for the Melni connector compared with the traditional splicing method.
- The ergonomic hazards associated with using the Melni connector were fewer in number and, according to the RULA analyses, less stressful. The RULA scores

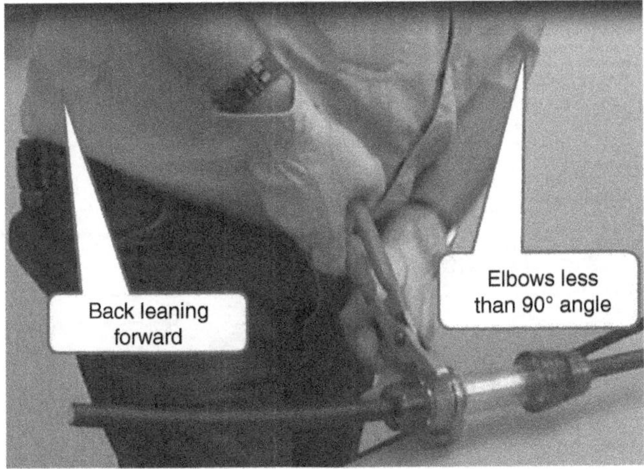

Figure 9.29 Tightening coupler step for Melni connector

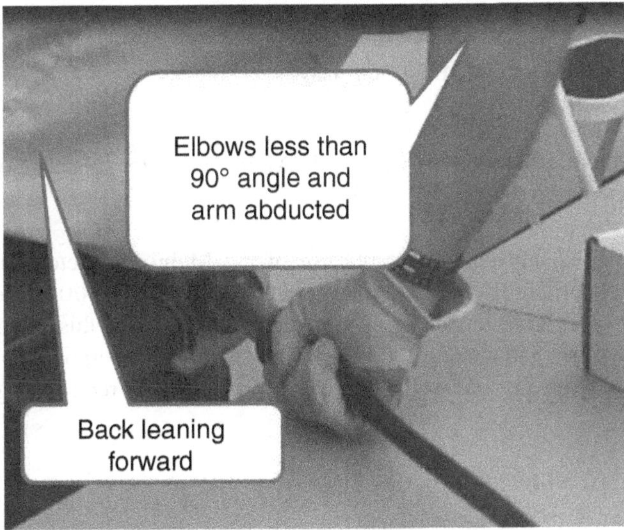

Figure 9.30 Tightening gripper/seal ring step for Melni connector

for the three (3) parts of the crimping steps for the traditional splicing method all were seven (7), whereas the highest RULA score for any of the steps associated with the Melni connector was five (5).

- The risk of the crimping tool slipping while performing the crimping steps for a traditional splice appears to be great and could lead to an acute injury.

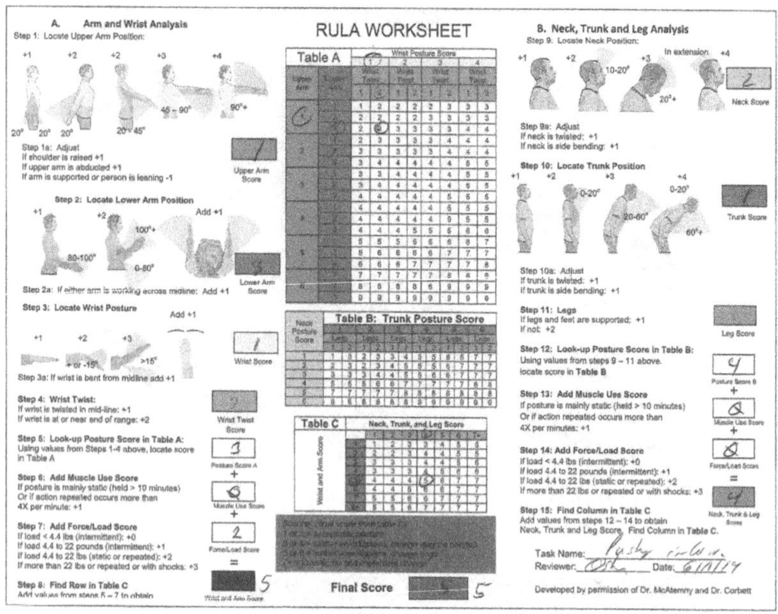

Figure 9.31 RULA analysis for wire insertion step

- The only tools required to perform the Melni connector splice are two (2) channel locks.

The overall conclusion is that the use of the Melni connector provides great ergonomic benefit over the traditional method of splicing 4/0 cable and takes considerably less time to perform. It is recommended that this method be used whenever possible and electricians should be trained on the ergonomic risk factors associated with this tasks as well as stretching techniques to reduce fatigue.

REVIEW QUESTIONS

1. What is the best grip for tasks requiring high forces?
2. What is the best grip for precision tasks?
3. Are low-power tools the answer to ergonomic issues?

REFERENCES

Ergonomic Concepts. (n.d.). Retrieved 2015, from Ergoweb: http://ergoweb.com/knowledge/ergonomics-101/concepts/.

Hounshell, D. A. (1984). *From the American System to Mass Production*. Baltimore: Johns Hopkins University.

Hughes, T. P. (1989). *American Genesis: A Century of Invention and Technological Enthusiasm*. Penguin.

McAtamney, L., & Corlett, E. N. (1993). RULA: A Survey Method for the Investigation of Work-Related Upper Limb Disorders. *Applied Ergonomics*, 24(2), 91–99.

NIOSH. (2004). *A Guide to Selecting Non-Powered Hand Tools*. NIOSH, p. 164.

10

VIBRATION

LEARNING OBJECTIVE

The students will be able to identify different sources of vibration, as well as the risks of permanent damage to the body, if exposed to high levels of vibration. The students will be able to identify the signs and symptoms involved in exposure to vibration and learn how to reduce vibration exposure.

INTRODUCTION

The topic of vibration and occupational exposure to vibration is complex, and this chapter provides a simplified overview of the topics.

Vibration can occur in any possible plane, and at any point a source of vibration can be generating vibration waves in more than one plane. Quite honestly, everyone living in a modern society, even those living in remote areas, experience some level of vibration almost constantly. For instance, the Earth vibrates at approximately 7.83 Hz (Schumann & Konig, 1954). Sound waves are vibration waves that are transmitted through air and oscillate our ear drums. We only perceive sound waves in range in frequency from relatively low frequencies (20 Hz) up to 20,000 Hz in young children. As people age, they generally lose the ability to hear higher frequency sounds. Vibrating equipment also generates sound waves, though, for this chapter, occupational exposure to the sound generated by the vibration is not the focus.

Occupational exposure to vibration can result from a wide range of equipment including grinders, drills, generators, motors, vehicles, compressors, fans, and vibration waves generated from flowing fluids and air.

Occupational Ergonomics: A Practical Approach, First Edition.
Theresa Stack, Lee T. Ostrom and Cheryl A. Wilhelmsen.

DEFINITIONS

Vibration – The oscillating, reciprocating, or other periodic motion of a rigid or elastic body or medium forced from a position or state of equilibrium.

Frequency (*F*) – The frequency of a vibration, measured in hertz (Hz), is simply the number of to and fro movements made in each second. 100 Hz would be 100 complete cycles in 1 s.

Amplitude – The amplitude of vibration is the magnitude of vibration. A vibrating object moves to a certain maximum distance on either side of its stationary position. Amplitude is the distance from the stationary position to the extreme position on either side and is measured in meters (m). The intensity of vibration depends on amplitude.

Acceleration (*A*) – Acceleration is a measure of how quickly speed changes with time. The measure of acceleration is expressed in units of measure (meters or feet per second) per second or meters or feet per second squared (m/s^2). The magnitude of acceleration changes from zero to a maximum during each cycle of vibration. It increases as a vibrating object moves further from its normal stationary position.

Speed/Velocity (*V*) – The speed of a vibrating object varies from zero to a maximum during each cycle of vibration. It moves fastest as it passes through its natural stationary position to an extreme position. The vibrating object slows down as it approaches the extreme, where it stops and then moves in the opposite direction through the stationary position toward the other extreme. Speed of vibration is expressed in units of measure per second (m/s). For instance, 1 ft/s or 1 m/s.

Wave length (λ) – The measure of peak to peak of one vibration cycle.

Resonance – Every object tends to vibrate at one particular frequency called the natural frequency. The measure of natural frequency depends on the composition of the object, its size, structure, weight, and shape. If we apply a vibrating force on the object with its frequency equal to the natural frequency, it is a resonance condition. A vibrating machine transfers the maximum amount of energy to the object when the machine vibrates at the object's resonant frequency.

The human body has natural resonance frequencies. Figure 10.1 shows some of the resonance frequencies of various parts of the human body.

Figures 10.2 and 10.3 show diagrams of cyclical vibration waves and how these various vibration elements tie together.

Two forms of occupational exposure may be distinguished: **whole-body vibration** (WBV), which is transmitted by mobile or fixed machines where the operator is standing or seated. Trucks, aircraft, heavy equipment, and boats are examples of things that cause WBV.

Hand-arm vibration (HAV) is transmitted from things in the hand to the hands and arms.

Whole-body vibration (WBV) is the mechanical vibration that, when transmitted to the whole body, produces risk to the health and safety of workers, in par-

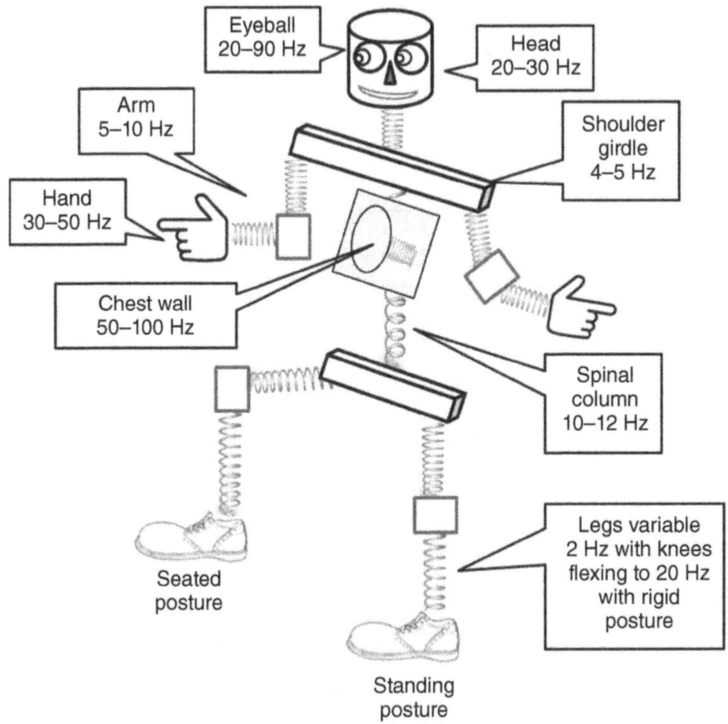

Figure 10.1 Resonance frequencies of the human body

ticular, lower back morbidity and trauma of the spine. Figure 10.4 shows a common military helicopter seat. These types of seats have no vibration dampening properties.

HAV is the mechanical vibration that, when transmitted to the human hand-arm system, entails risks to the health and safety of workers. In particular, vascular, bone or joint, neurological, or muscular disorders.

Table 10.1 shows a listing of the common sources of vibration and the areas they affect.

Potential Injuries Resulting from Vibration Exposure

Raynaud's syndrome or phenomenon was first described as "a condition, a local syncope (loss of blood circulation) to the fingers and hand" (Raynaud, 1888). A more modern term for Raynaud's syndrome is Hand Arm Vibration (HAVS). These terms, along with vibration-induced white finger describe the same condition. Early stages of HAVS are characterized by tingling or numbness in the fingers. Temporary tingling or numbness during or soon after use of a vibrating hand tool is not considered HAVS. To be diagnosed as HAVS, these neurologic symptoms must be more persistent and occur without provocation by immediate exposure to vibration. Other symptoms of HAVS include blanching, pain, and flushing. The symptoms usually appear

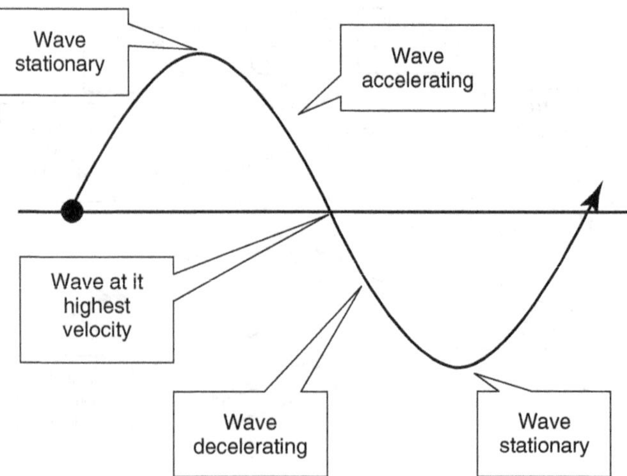

Figure 10.2 Sigmoidal vibration wave

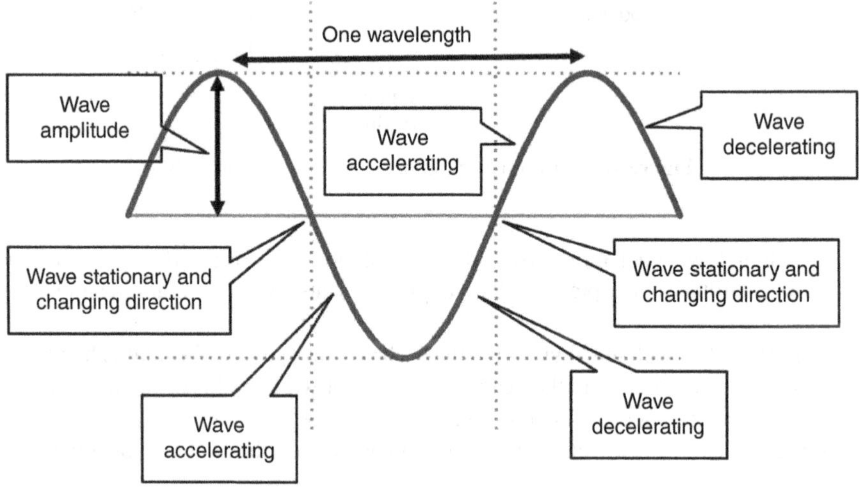

Figure 10.3 Changing velocity in a vibration wave

suddenly and are precipitated by exposure to cold. With continuing exposure to vibration, the signs and symptoms become more severe, and the pathology may become irreversible.

The severity of HAVS can be measured using a grading system developed by Taylor (1974). After a clinical observation and an interview, a worker can be placed into one of the categories shown in Table 10.2.

Another classification scheme is from the Stockholm Workshop and is shown in Table 10.3 (Gemne et al., 1987).

Figure 10.4 Helicopter seat

There are several other factors that can affect the development of HAVS and are listed in Table 10.4.

Whole-body vibration (WBV) can contribute fatigue, insomnia, stomach problems, headache, and "shakiness" shortly after or during exposure. These symptoms are similar to those that many people experience after a long car or boat trip. After daily exposure over a number of years, WBV can affect the entire body and result in a number of health disorders such as back injuries. Sea, air, or land vehicles cause motion sickness when the vibration exposure occurs in the 0.1–0.6 Hz frequency range. Bus and truck drivers found that occupational exposure to WBV could have

TABLE 10.1 Sources of Vibration

Industry	Type of Vibration	Common Source of Vibration
Agriculture	Whole body	Tractors
Military, commercial, and general aviation	Whole body	Aircraft
Military and others	Whole body and hand-arm	Boats
Boiler making	Hand-arm	Pneumatic tools
Construction	Whole body	Heavy equipment vehicles
	Hand-arm	Pneumatic tools, jackhammers
Diamond cutting	Hand-arm	Vibrating hand tools
Forestry	Whole body	Tractors
	Hand-arm	Chain saws
Foundries	Hand-arm	Vibrating cleavers
Furniture manufacture	Hand-arm	Pneumatic chisels
Iron and steel	Hand-arm	Vibrating hand tools
Lumber	Hand-arm	Chain saws
Machine tools	Hand-arm	Vibrating hand tools
Mining	Whole body	Vehicle operation
	Hand-arm	Rock drills
Riveting	Hand-arm	Hand tools
Rubber	Hand-arm	Pneumatic stripping tools
Sheet metal	Hand-arm	Stamping equipment
Shipyards	Hand-arm	Pneumatic hand tools
Shoe-making	Hand-arm	Pounding machine
Stone dressing	Hand-arm	Pneumatic hand tools
Textile	Hand-arm	Sewing machines, looms
Transportation	Whole body	Vehicles

contributed to a number of circulatory, intestinal, respiratory, muscular, and back disorders. The combined effects of body posture, postural fatigue, dietary habits, and WBV are the possible contributors for these disorders. Studies show that WBV can increase heart rate, oxygen uptake, and respiratory rate and can produce changes in blood and urine. Many studies have reported decreased performance from workers exposed to WBV.

VIBRATION MEASUREMENT

Vibration measurement is complex due to its many components – displacement, velocity, acceleration, and frequencies. In addition, vibration can occur in any possible plane. Also, each of these components can be measured in different ways – peak-to-peak, peak, average, RMS (Root Mean Square), each of which can be measured in the time domain (real-time, instantaneous measurements with an oscilloscope or data acquisition system) or frequency domain (vibration magnitude at different frequencies across a frequency spectrum), or just a single number for "total vibration".

TABLE 10.2 Stages of Hand Arm Vibration Syndrome

Stage	Condition of Fingers	Work and Social Interference
00	No tingling, numbness, or blanching of fingers	No complaints
OT	Intermittent tingling	No interference with activities
ON	Intermittent numbness	No interference with activities
TN	Intermittent tingling and numbness	No interference with activities
01	Blanching of a fingertip with or without tingling and/or numbness	No interference with activities
02	Blanching of one or more fingers beyond tips, usually during winter	Possible interference with nonwork activities; no interference at work
03	Extensive blanching of fingers; during summer and winter	Definite interference at work, at home, and with social activities; restriction of hobbies
04	Extensive blanching of most fingers; during summer and winter	Occupation usually changed because of severity of signs and symptoms

Source: Adapted from Taylor (1974).

TABLE 10.3 Stockholm Workshop Classification of Stages of Raynaud's Syndrome

The Stockholm Workshop Classification Scale for Sensorineural Changes in Fingers Due to Hand-Arm HAVS

Stage	Symptoms
OSN	Exposed to vibration but no symptoms
1SN	Intermittent numbness, with or without tingling
2SN	Intermittent or persistent numbness, reduced sensory perception

Source: Adapted from Gemne et al. (1987).

VIBRATION MONITORS

The following provides brief discussion of various types of measurements that are of concern when measuring vibration. This discussion is adapted from information obtained from the Centers for Disease Control (CDC) website (CDC, 2012). Human response to vibration is dependent on several factors including frequency, amplitude, direction, point of application, time of exposure, clothing and equipment, body size, body posture, body tension, and composition. A complete assessment of exposure to vibration requires the measurement of acceleration in well-defined directions, frequencies, and duration of exposure. The vibration will generally be measured along three (x, y, and z) axis.

A typical vibration measurement system includes a device (accelerometer) to sense vibration, a recorder, a frequency analyzer, a frequency-weighting network, and a display such as a meter, printer, or recorder. The accelerometer produces an electrical signal in response to vibration. The size of this signal is proportional to

TABLE 10.4 Factors That Can Influence the Development of Hand Arm Vibration Syndrome

Factors That Influence the Effect of Vibration on the Hand		
Physical Factors	Biodynamic Factors	Individual Factors
Acceleration of vibration	Grip forces – how hard the worker grasps the vibrating equipment	Operator's control of tool
Frequency of vibration	Surface area, location, and mass of parts of the hand in contact with the source of vibration	Machine work rate
Duration of exposure each workday	Hardness of the material being contacted by the handheld tools, for example, metal in grinding and chipping	Skill and productivity
Years of employment involving vibration exposure	Position of the hand and arm relative to the body	Individual susceptibility to vibration
State of tool maintenance	Texture of handle-soft and compliant versus rigid material	Smoking and use of drugs Exposure to other physical and chemical agents
Protective practices and equipment including gloves, boots, work-rest periods	Medical history of injury to fingers and hands, particularly frostbite	Disease or prior injury to the fingers or hands

acceleration applied to it. The frequency analyzer determines the distribution of acceleration in different frequency bands. The frequency-weighting network mimics the human sensitivity to vibration at different frequencies. The use of weighting networks gives a single number as a measure of vibration exposure (i.e., units of vibration) and is expressed in meters per second squared (m/s^2).

Measurement for Hand-Arm Vibration

Exposure measurement for HAV will generally be conducted for workers using handheld power tools such as drills, grinders, needle guns, and jackhammers. The first step is to determine the type of vibration that will be encountered because a different accelerometer will be used depending on whether an impact (e.g., jackhammer or chipper) or nonimpact (e.g., chain saws or grinders) tool is being used. The accelerometer will be attached to the tool (or held in contact with the tool by the user) so the axis are measured while the worker grasps the tool handle. The z-axis is generally from the wrist to the middle knuckle, the x-axis is from the top of the hand down through the bottom of the hand and wrapped fingers, and the y-axis runs from right to left across the knuckles of the hand. The measurement should be made as close as possible to the point where the vibration enters the hand.

The frequency-weighting network for HAV is given in the International Organization for Standardization (ISO) standard ISO 5349-1 (Mechanical

Vibration – Measurement and Evaluation of Human Exposure to Hand-Transmitted Vibration – Part 1: General Requirements). The human hand does not appear to be equally sensitive to vibration energy at all frequencies. The sensitivity appears to be highest around 8–16 Hz (hertz or cycles per second), so the weighting networks will generally emphasize this range. Vibration amplitudes, whether measured as frequency-weighted or frequency-independent acceleration levels (m/s^2), are generally used to describe vibration stress (American National Standards Institute, American Conference of Governmental Industrial Hygienists (ACGIH), ISO, and the British Standards Institution). These numbers can generally be read directly from the human vibration meter used. The recommendations of most advisory bodies are based on an exposure level likely to cause the first signs of Stage II Hand-Arm HAVS (white finger) in workers.

OSHA does not have standards concerning vibration exposure. The ACGIH has developed threshold limit values (TLVs) for vibration exposure to handheld tools. The exposure limits are given as frequency-weighted acceleration. The frequency weighting is based on a scheme recommended in ISO 5349-1. Vibration-measuring instruments have a frequency-weighting network as an option. The networks list acceleration levels and exposure durations to which, ACGIH has determined, most workers can be exposed repeatedly without severe damage to the fingers. The ACGIH advises that these values be applied in conjunction with other protective measures, including vibration control.

The ACGIH recommendations are based on exposure levels that should be safe for repeated exposure, with minimal risk of adverse effects (including pain) to the back and the ability to operate a land-based vehicle. Table 10.5 shows the ACGIH vibration exposure TLVs.

Whole-Body Vibration

The measurement of WBV is important when measuring vibration from large pieces of machinery that are operated in a seated, standing, or reclined posture. WBV is measured across three (x, y, and z) axis. The orientation of each axis is as follows: z is from head to toe, x is from front to back, and y is from shoulder to shoulder. The accelerometer must be placed at the point where the body comes in contact with the vibrating surface, generally on the seat or against the back of the operator.

TABLE 10.5 ACGIH Vibration Exposure Limits

Threshold Limit Values for Exposure to Hand Arm Vibration in Any of the *X*, *Y*, or *Z* Planes		
Daily Exposure (h)	Dominant Frequency of Weighted-Component of Acceleration Which Shall Not Be Exceeded (RMS)	
	m/s^2	G's
4–8h	4	0.40
2–4	6	0.61
1–2	8	0.81
Less than 1	12	1.22

Source: Used with permission from ACGIH.

The measurement device is generally an accelerometer mounted in a hard rubber disc. This disc is placed in the seat between the operator and the machinery. Care should be taken to ensure that the weight of the disc does not exceed more than about 10% of the weight of the person being measured.

One of the important issues with vibration measurement is ensuring the measuring equipment is properly calibrated. Vibration equipment will not generally be calibrated by the user. These devices will generally be sent back to the manufacturer for calibration on an annual basis.

Special Considerations The ISO standard suggests three different types of exposure limits for whole body vibration, of which only the third type is generally used occupationally and is the basis for the ACGIH TLVs:

1. The **reduced-comfort boundary** is for the comfort of passengers in airplanes, boats, and trains. Exceeding these exposure limits makes it difficult for passengers to eat, read, or write when traveling.
2. The **fatigue-decreased proficiency boundary** is a limit for time-dependent effects that impair performance. For example, fatigue impairs performance in flying, driving, and operating heavy vehicles.
3. The **exposure limit** is used to assess the maximum exposure allowed for WBV. There are two separate tables for exposures. One table is for longitudinal (foot to head; z-axis) exposures, with the lowest exposure limit at 4–8 Hz based on human body sensitivity. The second table is for transverse (back to chest and side to side; x and y axes) exposures, with the lowest exposure limit at 1–2 Hz based on human body sensitivity. A separate set of "severe discomfort boundaries" is given for 8-h, 2-h, and 30-min exposures to WBV in the 0.1–0.63 Hz range.

REDUCING VIBRATION EXPOSURE

Engineering Controls

The engineering controls related to controlling vibration include the following:

- Installing equipment that vibrates less
- Proper maintaining the equipment
- Remotely operating the vibrating equipment
- Changing the process. For instance, protecting steel to prevent rusting; therefore, reducing or eliminating the need to sand or grind it before finishing.

Vibration-Dampened Tools

Tools can be designed or mounted in ways that help reduce the vibration level. For example, using vibration-dampened chain saws reduces acceleration levels by a factor of about 10. These types of chain saws must be well maintained to ensure the dampened qualities. If a tool is dropped or otherwise damaged, it needs to be inspected, repaired, or discarded. Maintenance must include periodic replacement of shock absorbers. Some pneumatic tool companies manufacture antivibration tools such as antivibration pneumatic chipping hammers, pavement breakers, and vibration-dampened pneumatic riveting guns.

Most reparable tool manufacturers list the level a tool vibrates as it comes out of the box. Tools that vibrate less should be purchased, if they have enough power to do the job.

Vibration Absorbing Gloves

Conventional protective gloves (e.g., cotton, leather), commonly used by workers, do not reduce vibration that is transferred to workers' hands when they are using vibrating tools or equipment. Vibration absorbing gloves are made of a layer of viscoelastic material. Actual measurements have shown that such gloves have limited effectiveness in absorbing low-frequency vibration, the major contributor to vibration-related disorders. Therefore, they offer little protection against developing vibration-induced white finger syndrome. However, gloves do provide protection from typical industrial hazards (e.g., cuts, abrasions) and from cold temperatures that, in turn, may reduce the initial sensation of white finger attacks. Also, some of these types of gloves require the user to grip tools with a stronger grip that might lead to other occupational injuries.

Safe Work Practices

Along with using vibration-dampened tools and vibration absorbing gloves, workers can reduce the risk of HAVS by the following safe work practices:

- Employ a minimum hand grip consistent with safe operation of the tool or process.
- Wear sufficient clothing, including gloves, to keep warm.
- Avoid continuous exposure by taking rest periods.
- Rest the tool on the work piece whenever practical.
- Refrain from using faulty tools.
- Maintain properly sharpened cutting tools.
- Consult a doctor at the first sign of vibration disease and ask about the possibility of changing to a job with less exposure.

Employee Education

Training programs are an effective means of heightening the awareness of HAVS in the work place. Training should include proper use and maintenance of vibrating tools to avoid unnecessary exposure to vibration. Vibrating machines and equipment often produce loud noise as well. Therefore, training and education in controlling vibration should also address concerns about noise control.

Whole-Body Vibration

The following precautions help to reduce WBV exposure:

- Limit the time spent by workers on a vibrating surface.
- Mechanically isolate the vibrating source or surface to reduce exposure.
- Ensure that equipment is well maintained to avoid excessive vibration.
- Install vibration-dampening seats.

The vibration control design is an intricate engineering problem and must be set up by qualified professionals. Many factors specific to the individual work station govern the choice of the vibration isolation material and the machine mounting methods.

KEY POINTS

- HAV can cause various health issues, including HAVS.
- Vibration is difficult to measure, and the measuring equipment can be very costly.
- The specification sheet for good quality tools contains information about the vibration level of tools.
- Vibration absorbing gloves can reduce the transmission from the tool to the hands.
- Internally dampened tools can help reduce the potential for exposure to vibration.

REVIEW QUESTIONS

1. A worker is grinding on a pipe. The tool's specifications state that the tool vibrates at $10\,m/s^2$. How many hours during the day can the worker operate the tool under ACGIH guidelines?
2. What types of injuries could a helicopter pilot experience from exposure to whole body vibration?
3. How might a company prevent all vibration-related hand-arm injuries?
4. Vibration exposure can occur off the job as well as on the job. What types of off-the-job activities could contribute to vibration exposure?
5. A company is planning to purchase a new piece of industrial equipment. What things could you suggest that would help reduce vibration exposure?

REFERENCES

Gemne, G. et al. (1987). Environment and Health. *Scandinavian Journal of Work*, 13, 4, 275–278.

Raynaud, M. (1888). *Local Asphyxia and Symmetrical Gangrene of the Extremities*. London: New Sydenham Society.

Schumann, W. O. and Konig, H. (1954). The Observation of Atmospherics at the Lowest Frequencies. Retrieved from The Healers Journal: http://www.thehealersjournal.com/2012/05/21/the-schumann-resonance-earths-powerful-natural-vibration/#sthash.qrXSOhWM.dpuf

Taylor, W. (1974). *The Vibration White Finger*. London: Academic Press.

11

INDUSTRIAL WORKSTATION DESIGN

LEARNING OBJECTIVE

Students will be able to recognize the best work surface height for a given task and identify the basic principles behind seated, standing, and leaning workstations. They will be able to identify common workstation solutions and how to apply work space envelopes.

Most all workers perform their tasks at some sort of workstation. Whether the worker is a truck driver in a vehicle, an office worker at a desk, or a healthcare worker in a hospital, they work at some sort of workstation. This chapter discusses specific industrial types of workstations that involve assembly repair and fabrication types of tasks (Ostrom, 1994).

INTRODUCTION

A workstation is a location where a person performs one or more tasks that are required as part of his/her job. The design of the workstation can have a profound impact on the person's ability to safely and effectively perform the required tasks. Reach capability, body size, muscle strength, and visual capabilities are just a few of the factors that should be considered in workstation design. The design guidelines to be discussed in this section include the following:

- Accommodate people with a range of body sizes or anthropometric dimensions.

Occupational Ergonomics: A Practical Approach, First Edition.
Theresa Stack, Lee T. Ostrom and Cheryl A. Wilhelmsen.

- Permit several working positions/postures to promote better blood flow and muscle movement.
- Design workstations from the working point of the hands. People work with their hands so we want their working height to be relative to their hand height.
- Place tools, controls, and materials between the shoulder and waist height, where they have the greatest mechanical advantage.
- Provide higher work surfaces should be provided for precision work, and lower work surfaces for heavy work.
- Round or pad work surface edges should be rounded or padded to reduce compressive forces.
- Provide well-designed chairs in order to support the worker.

As each of the design guidelines is discussed, keep in mind that some of these principles may not be applicable to designs for individuals with special needs. Different tasks may also require different design guidelines, such as a shorter bus driver who may need a brake pedal extension.

GENERAL WORKSTATION GUIDELINES

Workstations should be designed to accommodate the anthropometric characteristics of a range of workers. In general, a well-designed workstation should be able to accommodate 90–95% of a worker population. This helps to ensure that workers can perform their job tasks comfortably and in a safe manner. There are several different sources of data that can assist with ergonomic design of the workplace.

Anthropometric tables contain data on human body dimensions and can be extremely useful in designing a work environment. When designing a new workstation, you should first identify the target audience, determine the appropriate anthropometric measurements, and ideally, create a mock-up for trial.

Figure 11.1 is an example of a height-adjustable table with a hand crank. This is an example of how you can accommodate workers with a range of body dimensions.

For operations where highly repetitive actions are required, where existing tools are not optimal, and where the workplace cannot be adjusted, a specially designed tool should be considered. Examples of special purpose tools include the following:

- Spring-loaded scissors that prevent irritation of the backs and sides of the fingers caused by opening conventional shears
- Bent pliers to maintain the wrist in a neutral posture
- Adjustable position (straight to pistol) screwdriver for multiple uses that encourages the use of neutral postures (Refer to the Hand Tool Section for more information on tools and neutral postures).

Situations when special tool designs are not recommended include the following:

- If the tool is used for a short part of a multitask cycle and thus becomes an extra tool to pick up and set down frequently.

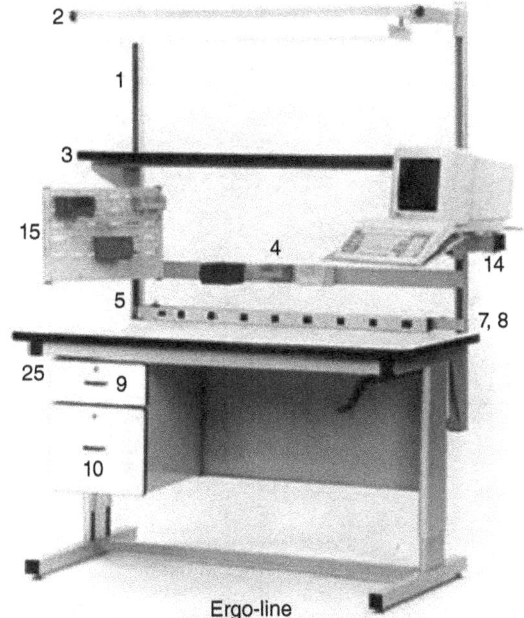

Figure 11.1 Hand crank workstation (Photo with permission from Pro-Line)

- If the operation is not a continuous one and occurs only occasionally in the work shift
- If the workspace around the operator is limited, and there is no place to set tools between uses (Eastman-Kodak, 2003).

MOCK-UP OR FITTED TRIALS

The following illustrates the advantages of using a mock-up before actually purchasing equipment. In this example, aircraft access doors were being repaired. The workers were bending over into awkward, stressful postures when working on the access doors. One of the workers created a fixture for holding the access doors, which would allow them to rotate 180°. Figure 11.2 is the original design developed by a worker. As you can see, the heavy-duty locking pin and open support structure allowed almost 180° of rotation. The large diameter casters created low push forces, even when the door was attached to the fixture.

Navy Hazard Abatement funding was used to improve and mass produce these fixtures. Prior to mass production of the fixture, a prototype (mock-up) was fabricated and put into use by the workers. The first prototype (Figure 11.3) was inferior to the one made by the workers, and needed improvements were identified during this trial use. The wheels were too small, which required high forces to move the fixture and access door. The locking mechanism was unstable, and the support structure actually

Figure 11.2 Original fixture

Figure 11.3 Prototype fixture

reduced the amount of rotation. A great amount of information was gained from creating and testing the mock-up that helped in designing the final product (Figure 11.4).

The final fixture design, shown on the left, permits 180° of rotation, and is height adjustable and has a heavy-duty locking pin and large diameter casters to reduce push/pull forces.

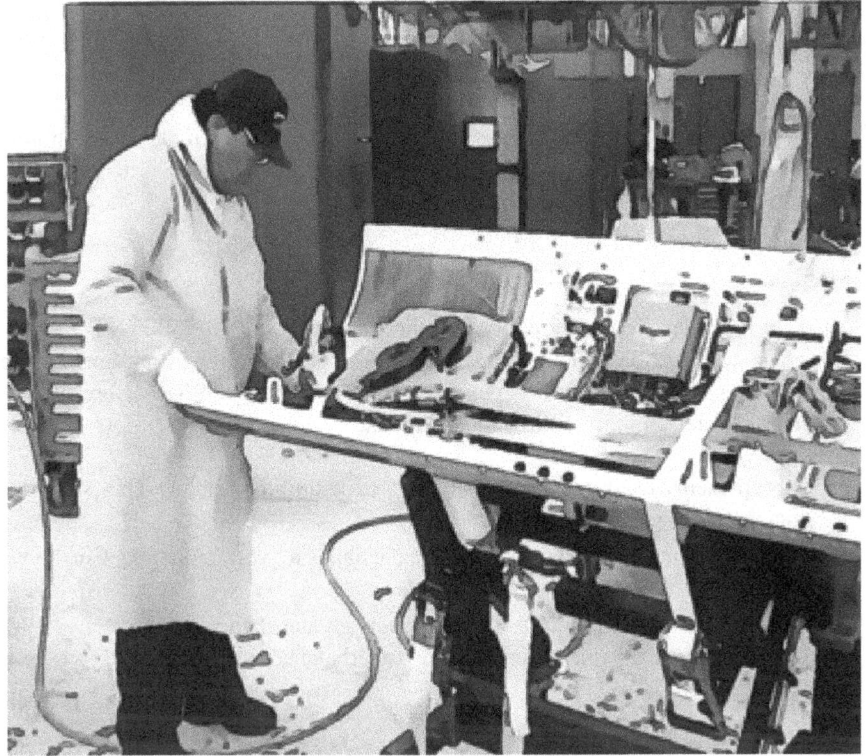

Figure 11.4 Final fixture

The original fixture design was built in-house; we borrowed the design and expanded on it. A great amount of worker pride was shown by using the original design. The new fixture can be shared with other activities, and they save 3–5 days of labor per door now.

GENERAL WORKSTATION GUIDELINES PERMIT SEVERAL WORKING POSITIONS/POSTURES

One of the tenets of biomechanics is that static loading of muscles and joints is something that should be avoided. Static loading is considered a potential risk factor for WMSD development. To avoid static loading, it is recommended that workers be allowed to periodically change their working posture. When changing posture, the individual utilizes different muscle groups, increases blood flow, and allows the muscles to rest.

There are many possible postures for working. Each has benefits and disadvantages to be considered. We will discuss a few of them in detail:

- Sitting
- Reclined

- Standing
- Sit/stand
- Leaning.

Each of the postures a worker uses should be equally safe and free from ergonomic stresses. One approach for permitting different working postures is to allow the worker to both sit and stand while performing job tasks.

Seated Posture

Seated workstations are very common, but there are guidelines to follow when determining if a workstation should be seated. A seated workstation is not appropriate when working with heavy items, repetitive extended reaching, or heavy forces.

Sitting is preferred in the following situations:

- All repeatedly used items are easily located within the seated workspace (i.e., "reach envelope" shown in Figure 11.5)
- The items being handled do not require the hands to work more than 6 in. above the work surface on average.
- No large forces are required (heavy objects greater than approximately 10 lb). Using mechanical assists, such as a counterbalance, may eliminate or reduce large forces.

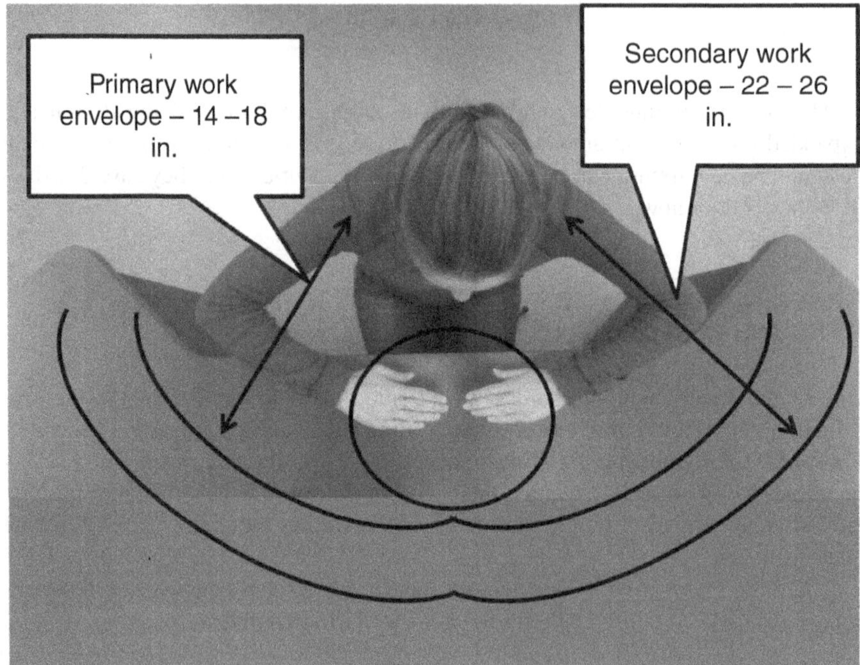

Figure 11.5 Horizontal reach envelope

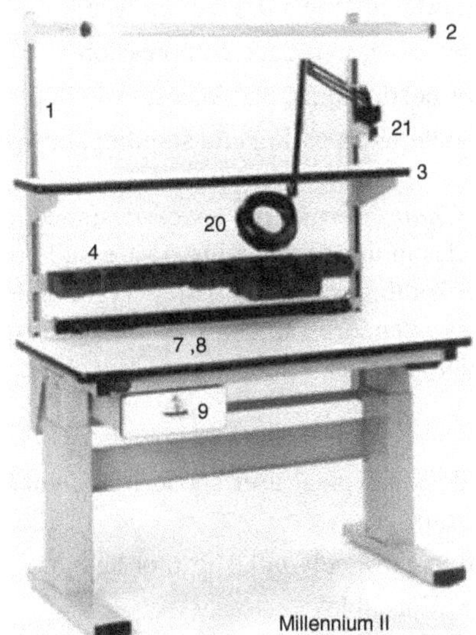

Figure 11.6 Adjustable workbench (Photo with permission from Pro-Line)

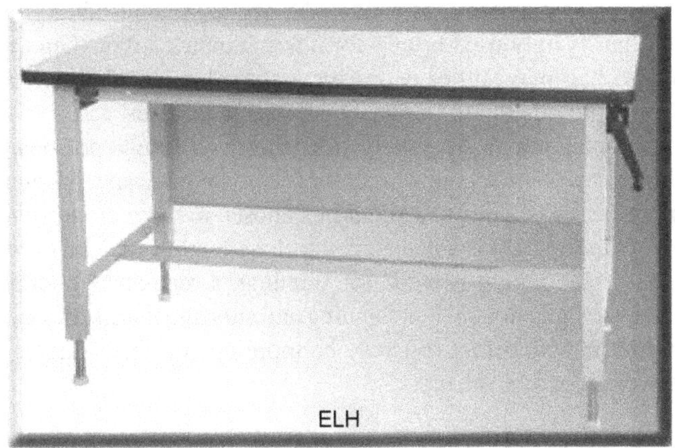

Figure 11.7 Manual adjustable workbench (Photo with permission from Pro-Line)

- Fine assembly or writing tasks are done for a majority of the shift (e.g., precision work or visual inspection).

Note: Workers should not handle more than 10 lb from a seated position.

If the workstation is a seated workstation, they need to provide the following:

- All tools, parts, and controls to be within easy reach.
- Appropriate height work surface for the individual/task, height adjustable preferred. Figures 11.6 and 11.7 illustrate adjustable work benches.
- Chair providing good support
- Foot support if necessary (solutions for this will be shown later in this module)

- Part/task should be oriented toward the employee.
- Padded work surface edge, to reduce compression
- Allow clearance for head, thighs, and knees.
- If feasible, alternate between sitting and standing, or leaning.

Provide Well-Designed Chairs Providing a well-designed, ergonomic chair can be extremely important in designing or choosing a safe and comfortable workstation. Many manufacturers market their chairs as being "ergonomic". What makes a chair ergonomic? With regard to vendors and chair manufacturers, the term "ergonomic" is often overused. In order to select a good ergonomic chair, a few issues need to be considered:

- How many hours does a typical user spend in a chair? (24-h chairs are more durable than 8-h chairs)
- Is the user population extremely tall, short, or heavy?
- Is the work surface adjustable?
- What is the height of the work surface?

The duration of time a user spends in a chair will affect the type of cushion and padding that the chair may need, as well as the degree of adjustability required. For example, if a chair is only used briefly for a few minutes a day, then a very simple and rudimentary chair may suffice. However, a chair that a worker uses for an entire work shift should provide the full range of adjustable features.

If the user population is extremely tall, short, or heavy persons, a standard ergonomic chair may not meet the workers' needs. There are special chairs designed for tall, short, or heavy persons that should be chosen in these circumstances.

If the work surface height is adjustable, the chair height should also be adjustable.

If the work surface height is made for standing, a regular ergonomic chair will probably not adjust high enough. For standing and sit/stand workstations, a prop stool or an elevated chair with a foot rest may be more appropriate. Figures 11.8–11.10 show a variety of ergonomic chairs.

Solutions for Workers Who Are Required to Sit Awkward neck/back angles are common issues for seated tasks and compression on the forearms/elbows is also common.

Figures 11.11 and 11.12 show a before and after example of a seated workstation. The before photo shows a worker with neck and back bent forward to see her work and compression from the edge of the desk on her elbows. In the second photo you can see task lighting, the work piece in a fixture which angles it toward the worker, and a padded forearm rest to reduce compression.

It is common for workers to rest their elbows on the surface of a table or work bench to stabilize the hands and rest the shoulders. However, the tissues of the forearm can be placed under pressure where work surfaces have sharp edges. These edges can

Figure 11.8 Traditional ergonomic chair (Photo with permission from IzzyPlus-www.izzyplus.com)

dig into the forearm, causing discomfort and potential nerve damage. These pressures can be alleviated by using pads and rounded corners on working surfaces.

General Workstation Guidelines – Design From the Working Point of the Hands
Working Envelope keep everything within easy reach. An ergonomically designed workstation should keep most of a worker's tasks within a certain maximum distance (referred to as a "working envelope"). The working envelope for a specific worker is defined by the sweep radius of the arms, with the hands in a grasping or reaching position. In general, forward reaches more than 20 in. (50 cm) in from the body when standing, or 15 in. (38 cm) in from the body when sitting, should be avoided. An occasional reach beyond this range is permissible, since the momentary effect on the shoulders and back is transient; however, frequent reaching of this type can quickly become fatiguing.

The primary seated work envelope is 14–18 in. where the most frequently used items should be kept. The secondary reach zone for occasional reaches is 22–26 in. Two-handed reaches are generally shorter than one-handed reaches.

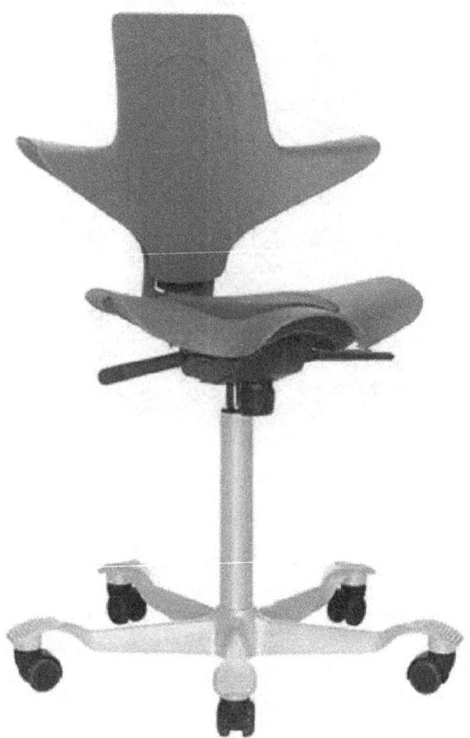

Figure 11.9 Saddle style of chair (Photo with permission from IzzyPlus-www.izzyplus.com)

Figure 11.10 Reversible chair (Photo with permission from IzzyPlus-www.izzyplus.com)

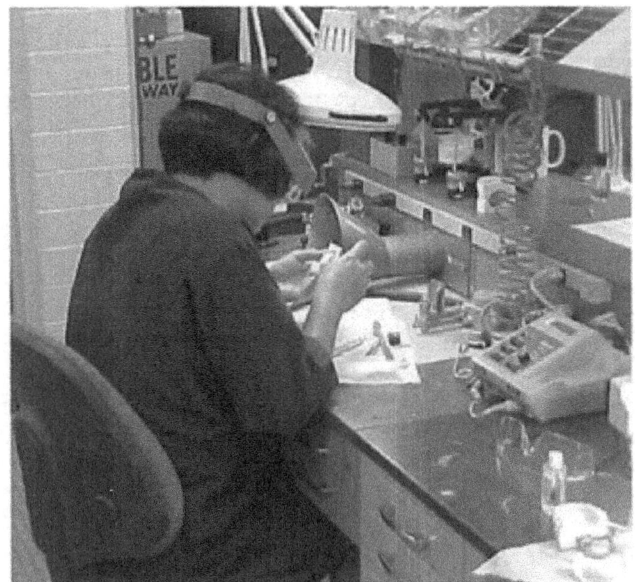

Figure 11.11 Worker in awkward posture

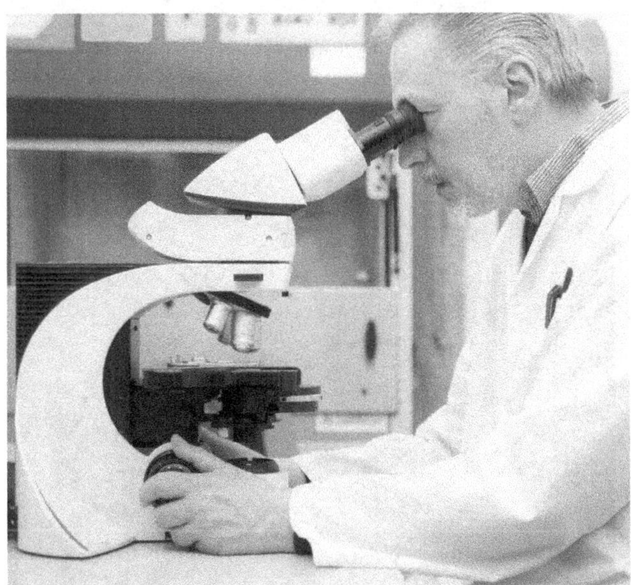

Figure 11.12 Worker illustrates good posture

Figure 11.5 shows the horizontal reach envelope and Figure 11.13 shows the vertical reach envelope (Grandjean & Kroemer, 1997)
Figure 11.14 shows how a chair that allows backward sitting can provide postural support to a worker.

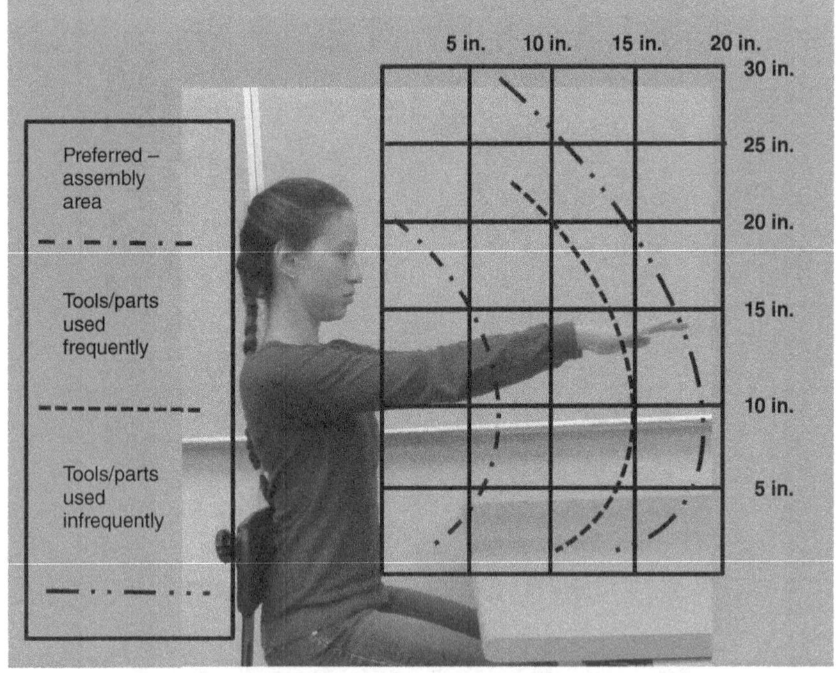

Figure 11.13 Vertical reach envelope

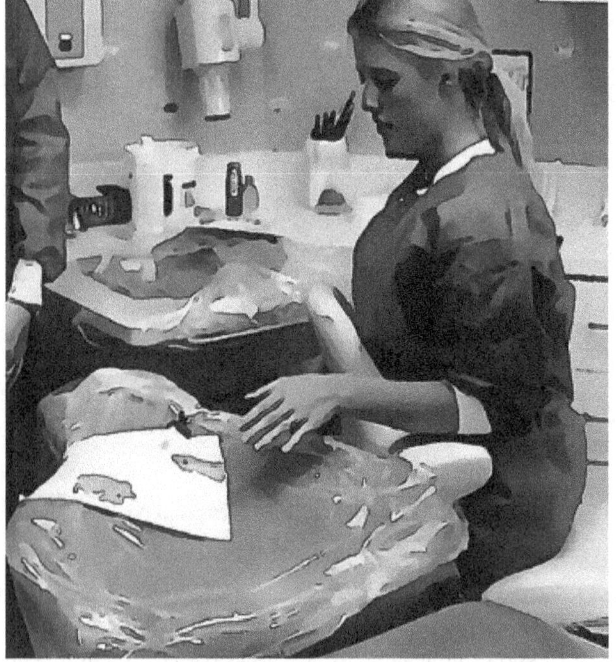

Figure 11.14 Alternative seated posture

Figure 11.15 Poor posture

Laboratories Common Issues We see a lot of seated workstations in laboratories and the solutions are similar for different work environments. Figure 11.15 shows a woman using a microscope with her back severely bent and compressive force on her arms.

Other common laboratory problems include a lack of knee space, pipette use, eye strain, and fatigue and extensive standing.

Laboratories Common Solutions

A few simple solutions for laboratory use include the following:

- Tilting, padded forearm supports
- Ample storage space to allow unencumbered access to tools
- Risers or adjustable work surface
- Mats and foot rests
- Adjustable chair with foot ring
- Shorter tubes, electric pipettes, multipipettes
- Sit/stand work stations
- Video microscope – with a screen instead of eye tubes.

Figure 11.16 illustrates the posture when provided a chair with back support, a padded surface for the arms, and a device to bring the work closer to the individual so they are not leaning forward in an awkward posture.

GENERAL WORKSTATION GUIDELINES PERMIT SEVERAL WORKING POSITIONS/POSTURES

People do not usually prefer to work standing, they feel it is tiring. The truth is that standing produces the least amount of stress on the spine-intervertebral disc between

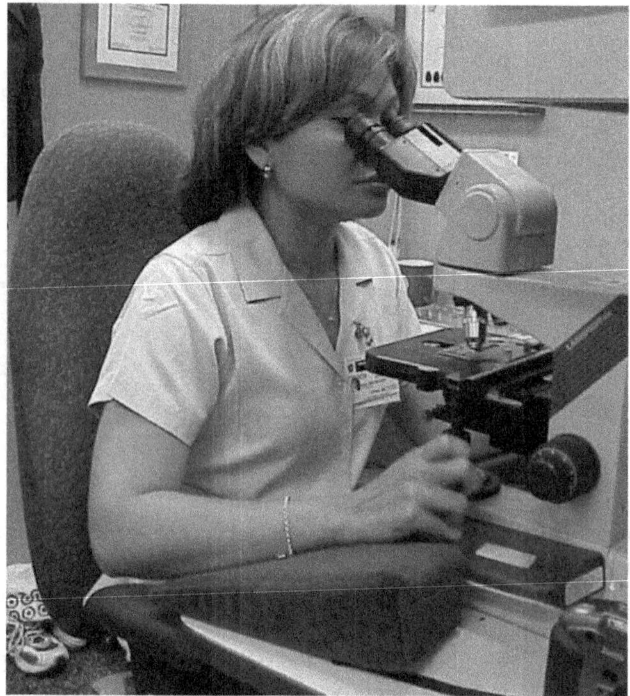

Figure 11.16 Good back support

the third and fourth vertebrae. Even standing bent over is less stressful to the spine than sitting in a poor posture, especially if the seats are not supportive.

Note: Laying down produces ~75% less disc pressure than standing at ease.

Reclined Posture

When employees are working overhead (such as aircraft mechanics), a recliner allows them to get closer to their work and reduces the static loading associated with holding their arms, neck, and back off the ground. Figure 11.17 shows the use of a creeper to avoid static loading in an awkward posture for an aircraft mechanic.

Standing Workstations

Standing workstations are appropriate when the task deals with heavy objects and/or frequent extended reaches, or the worker has to move around a lot. Standing gives us greater mobility, our reaching capacity is greater and it is faster for us to walk to another area since they don't have to get up out of a chair.

Standing is also preferred if the workplace or workstation does not have knee clearances for seated operation, or downward forces must be exerted, such as in wrapping and packing operations. If feasible, provide a sit/stand or lean chair and ensure there is suitable seating available for breaks and meals.

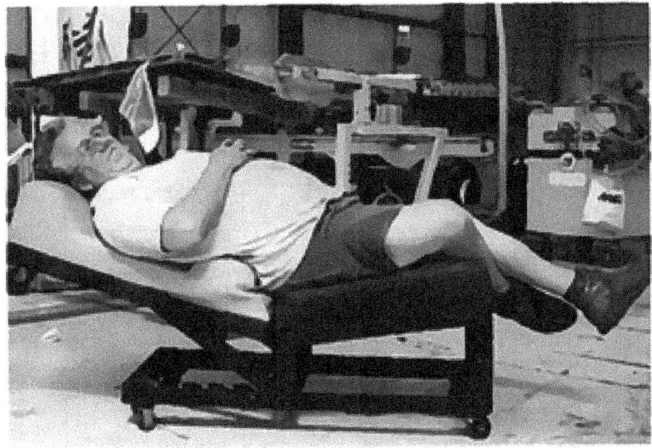

Figure 11.17 Industrial creeper

Venous pooling increases the heart rate as the heart tries to maintain constant cardiac output. Walking actually helps the heart by providing a "milking action" in the leg muscles that aids in moving the blood from the legs back to the heart. Venous pooling causes swelling of the legs (edema) and varicose veins in addition to increasing the load on the heart.

The veins are the body's blood storage location. If the legs do not move, the blood from the heart tends to come down the legs but not go back up; this is called venous pooling (Konz et al., 1990). Venous pooling is caused by standing with no leg movement. This type of inactive standing causes more discomfort than active standing, where the individual walks 2–4 min every 15 min (Mital, 1989). Shoes that are worn unequally may also present a problem. Worn shoes require the wearer to stand on a curved surface rather than a flat surface. Figure 11.18 shows attributes of a good standing workstation.

There are many ways to make standing more comfortable for a worker. One can provide a support to lean against.

Figure 11.18 shows a foot rail. Foot rails should be at least 1 in. in diameter, 6 in. off ground. Foot rails allow the user to shift their weight from 1 ft to another and relieve pressure. This is the same concept behind foot rails in bars – if the patrons are more comfortable, they are more likely to stay.

Hard floors should be avoided or covered with padding in areas where standing is required. Plastic, cork tile, wood, and carpet are preferable to concrete or metal grating. However, these types of flooring are not always practical for an entire work area. In these cases, individuals may modify their standing area with carpet, rubber mats, old rugs, or wooden platforms.

Figures 11.6 and 11.7 show there should be adjustable height work surface (hand crank and pneumatic) platforms that can be stacked to create steps or taller platforms and a height-adjustable platform.

When selecting a work surface consider the following criteria:

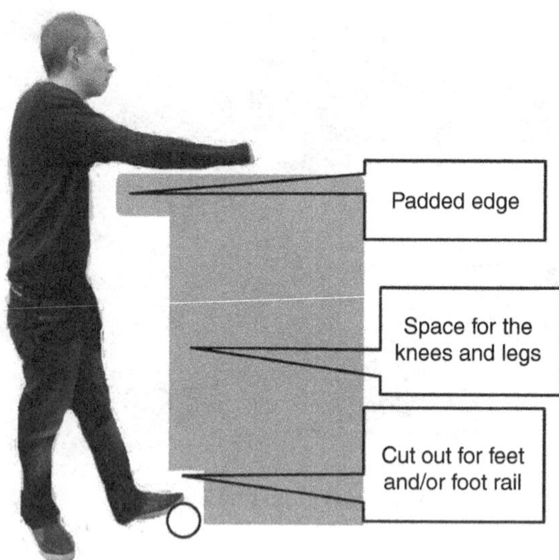

Figure 11.18 Attributes of a good standing workstation

- If the work surface is used by multiple users during different shifts, strongly consider making the work surface user adjustable.
- If the user of work surfaces changes regularly (for instance, work surfaces in locations that have a high turnover rate or office buildings where people change offices regularly) consider user-adjustable work surfaces. If user-adjustable work surfaces are not an alternative, the work surface should be maintenance adjustable.
- If adjustability of the work surface is not feasible, consider adjusting the height of the individual with platforms or person lifts as shown in Figure 11.19.
- When no adjustability is possible, ensure that the "working height" is at or just below the elbow rest height of the 50th percentile of the typical population that works at that particular job (This should be used as the last alternative and is not recommended).

Leaning Workstations

Leaning is preferred in the following situations (commonly referred to as sit/stand workstations):

- Repetitive operations are done with **frequent reaches**
  - more than **16 in. forward** and/or
  - more than **6 in. above** the work surface
- **Multiple tasks** are performed, some best-done sitting and others best done standing.
  - Employee has to **travel** away from work area.

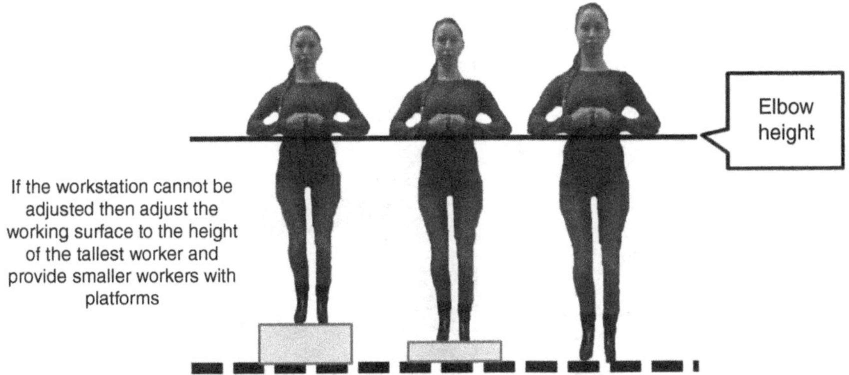

Figure 11.19 Height-adjustable platform

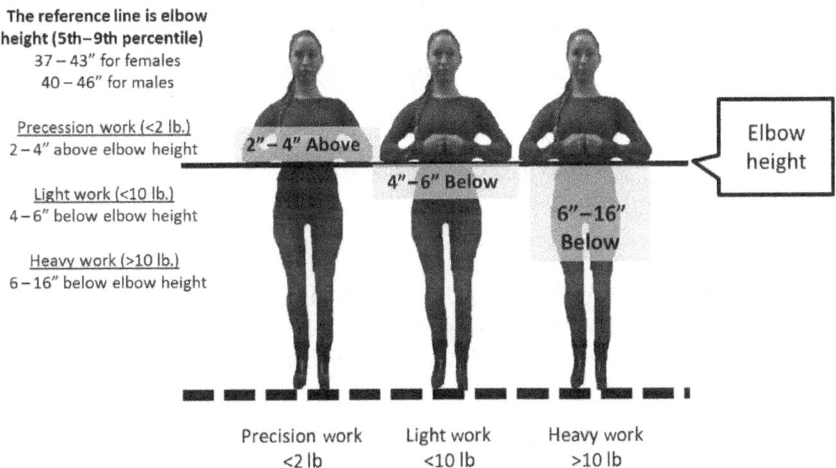

Figure 11.20 Standing work height guidelines

- There is not enough room or adjustability to alternate sitting and standing.

Lean chairs support two-thirds, of the user's body weight while maintaining upright posture. The semisitting posture relieves stress on the spine and the muscles and increases circulation in the lower extremities by providing relief from prolonged standing. This also allows the user to easily get up.

Standing Work Height Figure 11.20 shows the recommended work heights for standing workstations. These heights are based on whether the work to be performed is considered light, medium, or heavy. The rule of thumb is that the work station should be lower for heavier work.

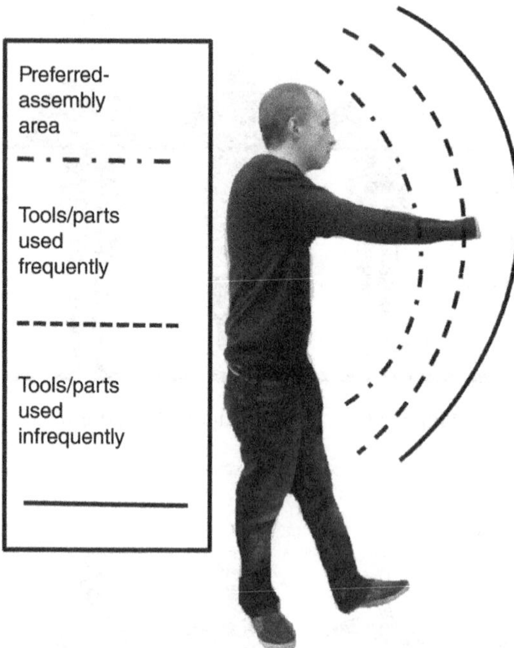

Figure 11.21 Vertical reach envelopes

Whether the workstation is designed to be standing, sitting, or a sit/stand workstation, the vertical reach envelope of the individual needs to be considered. Figure 11.21 shows the vertical reach envelopes that need to be considered.

General Workstation Guidelines

- Present parts upright and forward facing.
- Avoid extended reaches or twisting.
- Allow for several equally safe postures/positions.
- Conserve momentum and use gravity.
- Alternate arms (or legs) for repetitive work – consider "mirror image" assembly lines.
- Use two hands for power.
- Avoid static loads.
- Design for a small woman's reach; a large man's fit.
- Encourage neutral postures.
- Utilize large muscles for force, small muscles for precision.
- Provide adjustable seating with supportive backrest and arm supports as necessary.
- Ensure the feet are supported and proper illumination provided.

Table 11.1 provides a summary of the criteria that should be applied when deciding whether to design a workstation to be sitting, standing, or sit/stand.

KEY POINTS

Summary of seated workstation design guidelines

If a job does not require a great deal of physical strength and can be done in a limited space, then the work should be done in a sitting position.
Note: Sitting all day is not good for the body, especially for the back. Therefore, there should be some variety in the job tasks performed so a worker is not required to do sitting work only. A good chair is essential for sitting work. The chair should allow the worker to change the general working positions easily.
The following are some ergonomic guidelines for seated work:

- The worker needs to be able to reach the entire work area without stretching or twisting unnecessarily.
- A good sitting position means that the individual is sitting straight in front of and close to the work.
- The work table and the chair should be designed so that the work surface is approximately at the same level as the elbows.
- The back should be straight and the shoulders relaxed.
- If possible, there should be some form of adjustable support for the elbows, forearms, or hands.

Summary of standing workstation design guidelines

Standing for long periods of time to perform a job should be avoided whenever possible. Long periods of standing work can cause back pain, leg swelling, problems with blood circulation, sore feet, and tired muscles. Here are some guidelines to follow when standing work cannot be avoided:

- The worker needs to be able to reach the entire work area without stretching or twisting unnecessarily.
- A good sitting position means that the individual is sitting straight in front of and close to the work.
- The work table and the chair should be designed so that the work surface is approximately at the same level as the elbows.
- The back should be straight and the shoulders relaxed.
- If possible, there should be some form of adjustable support for the elbows, forearms, or hands.
- If a job must be done in a standing position, a chair or stool should be provided for the worker, and he or she should be able to sit down at regular intervals.
- Workers should be able to work with their upper arms at their sides and without excessive bending or twisting of the back.

TABLE 11.1 Criteria for Designing a Workstation

Parameters	Standing Workstation	Sit–Stand/Leaning Workstation	Sit to Stand Workstation	Sitting Workstation	Special Considerations
Heavy load and/or with high forces – greater than 10 lb	Work surface height 6–16 in. below elbow height	Not recommended	Not recommended	Not recommended	Provide a place for short breaks. A lean station or a chair
Intermittent work with moderately high forces – 2–10 lb	Work surface height 4–6 in. below elbow height	Work surface height 4–6 in. below elbow height	Not recommended	Not recommended	Provide a place for short breaks. A lean station or a chair
Extended reach envelope	If variable tasks are required	If variable tasks are required	If moderately fine work is required and/or forces are less than 10 lb while sitting	If fine or precise work is required and/or forces are less than 2 lb	Ensure worker can reach items required without attaining awkward postures
Variable work surface heights	If forces greater than 10 lb are required. Work surface height 6–16 in. below elbow height	If moderately high forces are required – 2–10 lb	If moderately fine work is required and/or forces are less than 10 lb while sitting	If work can be performed at approximately elbow height and forces are less than 2 lb	Ensure worker does not attain sustained awkward postures

Repetitive movements	If forces greater than 10 lb are required. Work surface height 6–16 in. below elbow height	If moderately high forces are required – 2–10 lb	If moderately fine work is required and/or forces are less than 10 lb while sitting	If work can be performed at approximately elbow height and forces are less than 2 lb	Ensure worker has adequate recovery time from repetitious movements
Repetitive tasks requiring moderately high visual attention – production inspection	If extended reaches are required	If worker can remain in the leaning posture and perform all tasks required	If moderately fine work is required and/or forces are less than 10lb while sitting	If work can be performed at approximately elbow height and forces are less than 2 lb	Ensure adequate illumination and worker has adequate time from repetitious movements
Precise work and/or requiring demanding visual attention	Not recommended	Not recommended	Not recommended	Work surface height 2–4 in. above elbow height	Ensure adequate illumination and worker has adequate recovery time from visually demanding tasks

- The work surface should be adjustable for workers of different heights and for different job tasks.
- If the work surface is not adjustable, then provide a pedestal to raise the work surface for taller workers. For shorter workers, provide a platform to raise their working height.
- A footrest should be provided to help reduce the strain on the back and to allow the worker to change positions. Shifting weight from time to time reduces the strain on the legs and back.
- There should be a mat on the floor so the worker does not have to stand on a hard surface. A concrete or metal floor can be covered to absorb shock. The floor should be clean, level, and not slippery.
- Workers should wear shoes with arch support and low heels when performing standing work.
- There should be adequate space and knee room to allow the worker to change body position while working.
- The worker should not have to reach to do the job tasks. Therefore, the work should be performed 8–12 in. (20–30 cm) in front of the body.

REVIEW QUESTIONS

1. List those items that are important when considering setting up a new workstation.
2. A workstation should be setup from__________________________.
3. What is one thing that can be done to reduce the amount of stress on a worker's back who is in a seated workstation?
4. Why is alternating from one posture to another so important?
5. When setting up a workstation the items that are used seldom should be placed ___________?
6. Why would a seated workstation be selected over a standing workstation?
7. Conversely, why would a standing workstation be selected over a seated workstation?
8. What is the ergonomic risk factor associated with a sharp edge on a workbench?
9. Why an Irish bar? Lol why not an Italian bar?
10. A generator has to be repaired. There are multiple areas on the generator that have to be adjusted. How could one set up a workstation so that all these access points can be worked on?

REFERENCES

Eastman-Kodak. (2003). *Kodak's Ergonomic Design for People at Work*. Eastman Kodak Company.

Grandjean, E., and Kroemer, K. (1997). Fitting the Task to the Human. *A Textbook of Occupational Ergonomics* 5th edn. CRC Press.

Konz, S., Bandla, V., Rys, M., and Sambasvian, J., Eds. (1990). Standing on Concrete Versus Floor Mats. *Advances in Ergonomics and Safety II*. Taylor and Francis, pp. 991–998.

Mital, A. (1989). Footprints. In S. Konz, & S. Subramanian, *Advances in Ergonomics and Safety I*. Taylor and Francis, pp. 203–205.

Ostrom, L. (1994). *Creating the Ergonomically Sound Work place*. Jossey-Bass.

ADDITIONAL SOURCE

Konz, S., Bandla, V., Rys, M., and Sambasvian, J., Standing on Concrete vs Floor Mats, In *Advances in Industrial Ergonomics and Safety II*, B. Das (ed.) London: Taylor and Francis, pp. 991-998 1990.

12

MANUAL MATERIALS HANDLING

LEARNING OBJECTIVES

At the end of the module, students will be able to apply aspects of manual material handling. Students will review and identify the factors that affect manual material handling methods, and apply common manual material handling solutions.

INTRODUCTION

Few industries are able to avoid the need for products or raw materials to be handled manually (Marras, 2006). Manual handling includes any tasks that require a person to lift, lower, push, pull, hold, or carry any object, animal, or person (see Figure 12.1).

MANUAL MATERIAL HANDLING AND INJURIES

Manual materials handling (MMH) has long been recognized as a contributor to back pain. The National Institute of Occupational Safety and Health (NIOSH) found evidence for a causal relationship between "heavy physical work" and back pain and strong evidence for a causal relationship between lifting and forceful movement and back pain. Other notable injuries associated with MMH include slips, trips and falls, and crushing injuries.

Occupational Ergonomics: A Practical Approach, First Edition.
Theresa Stack, Lee T. Ostrom and Cheryl A. Wilhelmsen.

Figure 12.1 Heavy lifting over long distances and benefit from a platform cart

Strategies for Reducing MMH Injures and Increasing Productivity

Several approaches can be used to reduce all types of injuries associated with MMH:

- A system approach to material handling to reduce the amount of product handled and the use of automated transfer devices
- Teaching people to lift safely
- Selecting people for specific tasks according to their strength and endurance
- Developing the workers' strength and endurance capacity through industrial fitness and total health programs
- Redesigning the workplace, containers, and jobs to make the handling suitable for more people (The Ergonomics Group, Health and Environmental Laboratories, Eastman Kodak Company, 1989).

This module addresses the systems approach and redesigning the workplace (providing aids, containers, and jobs) to make the handling suitable for more people (Figures 12.2–12.4).

THE SYSTEMS APPROACH TO MMH

The systems approach to MMH can best be accomplished in the planning phase of a facility or operation. As much as feasible, automated movement of bulk,

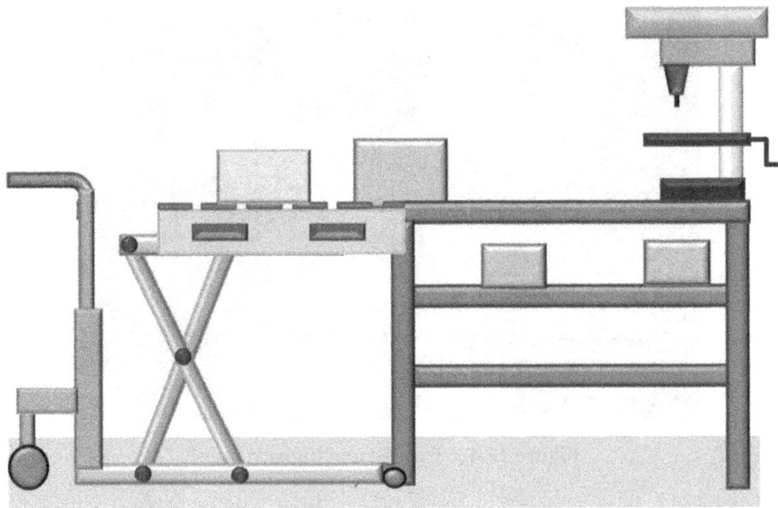

Figure 12.2 Pallet scissor l

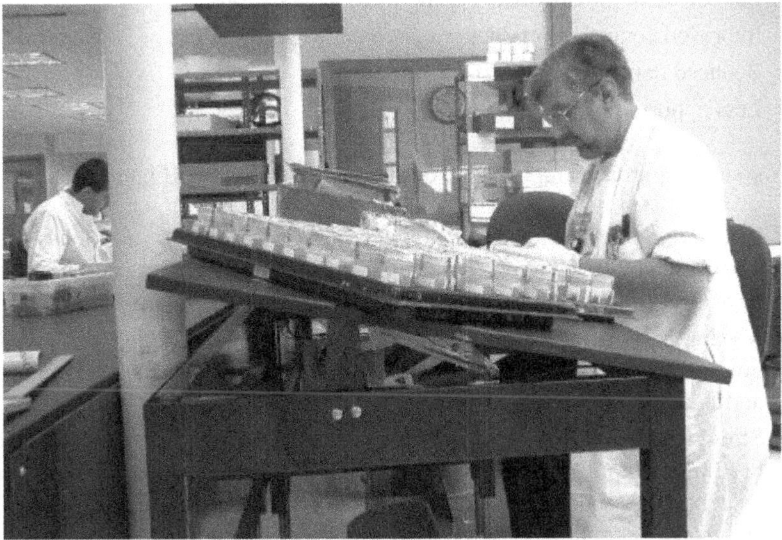

Figure 12.3 Tilting lift table orients work to the worker so they do not need to bend forward

Figure 12.4 Two-person lift made easy

intermediate, and final-stage products are incorporated into the process. A number of design principles, even in an established operation, can be beneficial.

Advantages of a systems approach:

- Reduction of in-process inventory
- Elimination of multiple remaining products
- Less delay in moving materials between workstations
- Improved control of production
- Reduced damage to product
- Less opportunity for human error.

A systems approach employs one or more of the following (Figure 12.5):

- **Unit size or load:** Increase the quantity, size, or weight of unit loads so they must be handled with powered equipment. A large drum instead of individual bags is an example of this principle.
- **Mechanization:** Mechanize operations with motorized conveyors, overhead hoists between workplaces, vacuum assists, and motorized carts (Figures 12.6 and 12.7).
- **Standardization:** Standardize handling methods as well as types and sizes of handling equipment. People are more likely to use equipment they are familiar with. Use equipment that is compatible across systems. For example, an automated stair climber can be used in place of a lift gate.
- **Adaptable:** Choose equipment that can perform a variety of tasks (Figure 12.8).

Figure 12.5 Awkward posture could benefit from a lift table

Figure 12.6 Rotating lift table used for inspection or building

- **Dead weight:** Reduce the ratio of the dead weight to the load being moved. Heavy metallic trays or cans can be replaced with a lighter suitable material.
- **Gravity:** Use gravity to move material. Gravity feed conveyors or canted storage units (so material flows forward). Lower materials instead of lifting them.
- **Automation:** Provide automation to include production handling and storage functions such as an automated stacker–retriever.

Figure 12.7 Vacuum assist for lifting heavy items (Courtesy of http://www.anver.com)

Figure 12.8 Walk behind fork truck or stacker

Redesign of Work Places and Jobs

Redesigning a workplace can eliminate the hazard or greatly reduce the risk of injury. Not all tasks lend themselves to redesign; for example, the environment in which a firefighter works cannot be changed, or others may simply be cost prohibitive to redesign. Some strategies are included below:

- Provide ways to adjust material to be handled so that less lifting and more sliding can be done. For example, a scissors lift table.
- Provide comfortable, appropriately shaped hand holds or handles on containers or objects.
- Rotate people between heavy and lighter jobs.
- Alternate jobs so that the same body regions (muscle groups) are not over exposed.
- Ensure two people lift objects over a certain weight and are trained in proper lifting techniques. These objects can be tagged stating a two-person lift is required.
- Locate materials within the workers' power zone (knuckles to shoulders).
- Provide carts and handling aids to support the weight of objects carried more than a few feet.
- Provide tools to help reduce forces in the hands.

MMH EQUIPMENT AND SOLUTIONS

(Department of Health and Human Services, 2007)

Easier Ways to Manually Lift, Fill, or Empty Containers

Use a scissors lift, load lifter, or pneumatic lifter to raise or lower the load so that it is level with the work surface. Then slide the load instead of lifting.

Options available are as follows:

- Portable (wheels)/fixed (attached to the floor)
- Zero-clearance (can use a pallet jack or fork truck to drive on top of them)
- Powered/spring lift mechanism
- Use as workbench
- Tops:
  - Lazy-Susan (top rotates 360°)
  - Conveyer top
  - Tilting top
  - Specialized: pallet, drum, roll

Use a dolly to make things easier to push and pull.

Figure 12.9 Manual material handling lift table

Raise the workers so they are close to the surface they need to reach with platforms, catwalks, portable stairs, or person lifts. Work within the power zone by raising the product or the worker, or providing various working height surfaces. Figure 12.9 shows a simple solution for keeping the work within the workers' power zone.

Store heavier or bulkier containers so that they can be handled within the power zone where the workers have the greatest strength. Angle the work toward the worker, also angling the product uses gravity to bring product forward. Tugging and towing equipment can be purchased and can be mated to anything that needs to be moved. Figures 12.10 and 12.11 highlight the use of equipment to reduce the stress and strain on a worker found in a common repeatable task.

SUMMARY

Manual material handling takes many forms and is present in all industries. Back injuries as well as caught-in or -between injuries are commonly linked to manual material handling tasks. Anticipation of overexertion, heavy, awkward lifting, and repeated lifting can lead to the proper abatement method. Manual material handling solutions do not need to be costly to be effective in accident aversion. Simple means to bring the work within the workers' power zone can reduce physical as well as physiological demands (Figure 12.12). Reference the survey tools found in the Appendix C and Chapter 14 for qualitative methods of manual material handling evaluation.

Figure 12.10 Before: Worker climbs and stretches while balancing on the ledge of a van to reach heavy ladder risking falls and soft tissue injuries

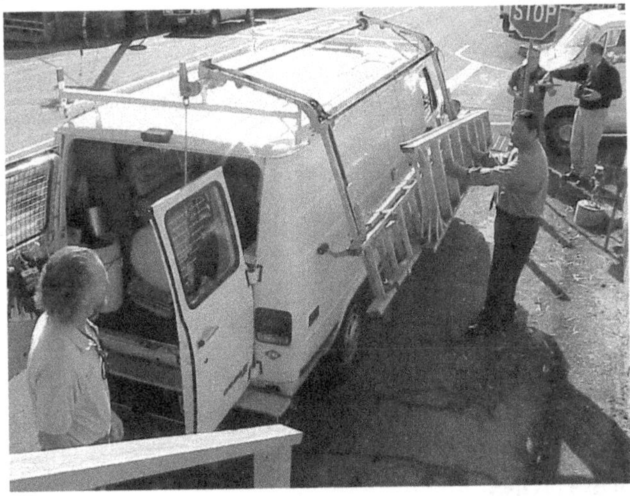

Figure 12.11 After: Worker safely removes ladder from curb side

KEY POINTS

- Eliminate lifting from below the knees by focusing on bringing the work to the elbow height of the workers. Lifting assists such as scissor lift tables and pallet lifts are useful solutions.

Figure 12.12 Arm lifts help the worker to maintain a neutral upright posture and eliminate compression on the hands (Adapted from www.forearmforklift.com)

- Reduce forceful pushing, pulling, and lifting by using simple transport devices such as carts or dollies.
- Use more sophisticated equipment such as powered stackers, hoists, cranes, or vacuum-assist devices for tasks that occur frequently or have a high potential for injury.

REVIEW QUESTIONS

1. What can be done to prevent manual material handling injuries?
2. What types of injuries are common in manual material handling environments?
3. What is one use for a portable scissor lift table in a manual material handling environment?

EXERCISES

See the exercise appendix.

REFERENCES

Marras, W. (2006). *Fundamentals and Assessment Tools for Occupational Ergonomics* 2nd edn. CRC Press.

The Ergonomics Group, Health and Environmental Laboratories, Eastman Kodak Company. (1989). *Ergonomic Design for People at Work*, Vol. 2. Wiley.

13

WORK-RELATED MUSCULOSKELETAL DISORDERS

LEARNING OBJECTIVE

At the end of the module, the students will have a basic understanding of various musculoskeletal disorders that are caused by occupational exposure to physical workplace risk factors. Anatomy of the muscular and skeletal systems is covered as well as injury and disorder prevention techniques.

INTRODUCTION

Musculoskeletal disorders are a broad class of disorders involving damage to muscles, tendons, ligaments, peripheral nerves, joints, cartilage, vertebral discs, bones, and/or supporting blood vessels. Work-related musculoskeletal disorder (WMSD) is a subcategory; these are injuries and illnesses that are caused or aggravated by working conditions. MSDs are not typically caused by acute events but occur slowly over time due to repeated wear and tear or microtraumas.

WMSDs are also known as cumulative trauma disorders (CTDs), repetitive strain injuries (RSIs), repetitive motion trauma (RMT), or occupational overuse syndrome. Examples of WMSDs include herniated disc, epicondylitis (tennis elbow), tendinitis, de Quervain's disease (tenosynovitis of the thumb), trigger finger, and Reynaud's syndrome (vibration white finger).

Researchers have identified specific physical workplace risk factors for the development of WMSDs: force, posture, compression, repetition, duration, vibration, and

Occupational Ergonomics: A Practical Approach, First Edition.
Theresa Stack, Lee T. Ostrom and Cheryl A. Wilhelmsen.

temperature. Exposure to these risk factors can result in decreased blood flow, elongation, compression, tears or strains to muscles, tendons, ligaments and nerves as well as disc or joint damage. When present for sufficient duration, frequency, or magnitude, physical workplace risk factors may cause WMSDs. In addition, personal risk factors, such as physical conditioning, existing health problems, gender, age, work technique, hobbies and organizational factors (e.g., job autonomy, quotas, deadlines) contribute to, but do not cause, the development of WMSDs.

Applying ergonomics principles to reduce a worker's exposure to the physical workplace risk factors decreases the chance of injury.

This unit is not designed to impart skills in the diagnoses of WMSDs, only occupational healthcare providers do so. This unit is a background in some of the more common WMSDs and explores anatomic features that are affected by WMSDs. It is written from the view point of an ergonomist and not a healthcare provider. For convenience, this chapter is divided into four sections:

- The musculoskeletal system
- Disorders of the
  - spine
  - upper extremities
  - lower extremities.

THE MUSCULOSKELETAL SYSTEM

The musculoskeletal system's primary functions include supporting the body, allowing motion, and protecting vital organs. The musculoskeletal system is an organ system that gives humans the ability to move using their muscular + skeletal systems (musculoskeletal). The musculoskeletal system provides form, support, stability, and movement to the body. It is made up of the bones of the skeleton, cartilage (framework), muscles (movement), tendons, ligaments, and other connective tissue (support–connectivity) that supports and binds tissues and organs together (Figure 13.1).

The skeleton is the framework of the human anatomy (Figure 13.2), supporting the body and protecting its internal organs. Two hundred and six bones compose the skeleton, about half of which are in the hands and feet. Most of the bones are connected to other bones at flexible joints, which lend the framework a high degree of flexibility (see Figure 13.3). Only one bone, the hyoid, is not directly connected to another bone in such an articulation.

Humans have over 650 muscles that differ in size according to the jobs they do. These muscles constitute 40% of body weight. The special function of muscle tissue is contraction, making the muscle shorter and thicker. There are three kinds of muscle tissue: striated muscle, smooth muscle, and cardiac muscle. Most of the body's muscle consists of striated muscle, which is the skeletal muscle. It is also called a voluntary muscle because it can be consciously controlled via the central nervous system, unlike the cardiac muscles.

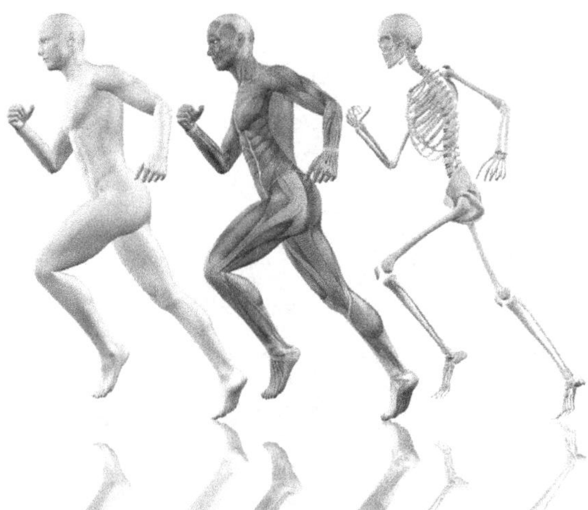

Figure 13.1 Parts of the musculoskeletal system (Used with permission Clip Art LLC; http://www.clipartof.com/cgi-bin/download_image.pl?q=1299636_SMJPG_7GN82358 YB166203F.jpg)

Muscles are named according to their shape, location, or a combination (see Figure 13.4). They are further categorized according to functions such as flexion, extension, or rotation (see Figure 13.6). For example, the subscapularis is in interior scapula muscle, the biceps brachii is a branched (three-headed) biceps muscle. Muscles, tendons, and ligaments work together to support the body, hold it upright, and control movement during rest and activity. Skeletal muscles help move the skeleton. Skeletal muscles contract when stimulated by a nerve. They can pull on the skeleton in the direction of the **contraction**. Muscles can only pull so they need to be arranged in antagonistic (or opposing) pairs; that is, one muscle in the pair contracts, while the other relaxes. The two muscles work together to produce movement of a joint, to steady a joint, and to prevent movement. For example, the biceps muscles contract, and your forearm is pulled toward your upper arm (flexion). Then your triceps muscles contract to pull your forearm away from your upper arm (extension) (Figures 13.5 and 13.6). This is an example of an antagonistic pair, one causes flexion (angle gets smaller) and the other causes extension (angle gets larger). Tendons and ligaments bear a burden of forces during muscular contraction and, therefore, are common injury sites for WMSD when the physical workplace risk factors go uncontrolled.

Muscles are attached to bone either directly or indirectly (via tendons). The muscle + skeleton system can be thought of as a machine. The nerves provide the energy to power the muscles. Muscles shorten and pull on a tendon (a cable). The levers are the bones that results in movement around a joint. Ligaments typically, but not always, provide the tunnels for the tendons to pass under and stabilize bones by crossing joints (Figure 13.7).

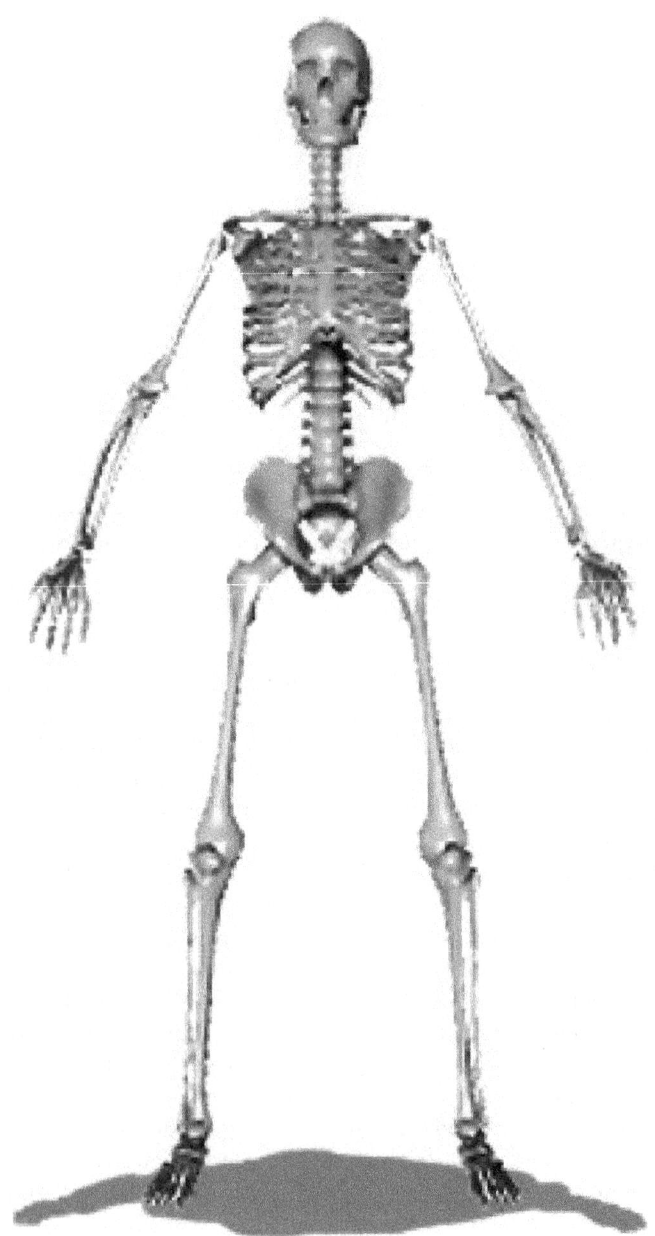

Figure 13.2 Skeleton (Adapted with permission from Ergonomics Image Gallery)

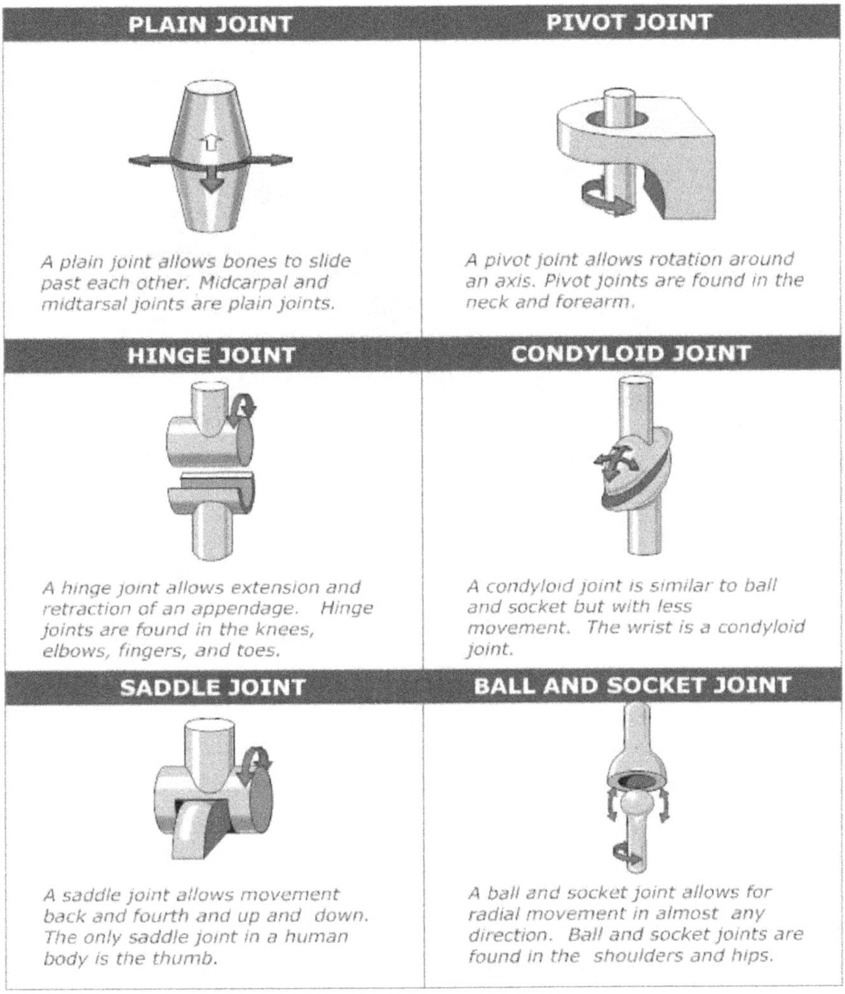

PLAIN JOINT	PIVOT JOINT
A plain joint allows bones to slide past each other. Midcarpal and midtarsal joints are plain joints.	A pivot joint allows rotation around an axis. Pivot joints are found in the neck and forearm.
HINGE JOINT	**CONDYLOID JOINT**
A hinge joint allows extension and retraction of an appendage. Hinge joints are found in the knees, elbows, fingers, and toes.	A condyloid joint is similar to ball and socket but with less movement. The wrist is a condyloid joint.
SADDLE JOINT	**BALL AND SOCKET JOINT**
A saddle joint allows movement back and fourth and up and down. The only saddle joint in a human body is the thumb.	A ball and socket joint allows for radial movement in almost any direction. Ball and socket joints are found in the shoulders and hips.

Figure 13.3 Types of joints (Adapted with permission Lippincott Williams & Wilkins, Baltimore, MD)

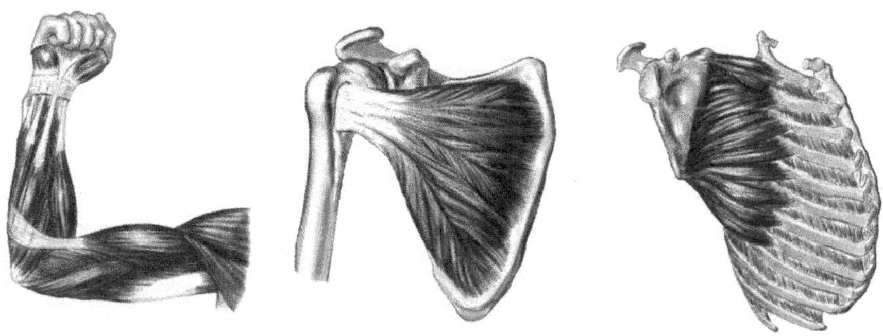

Figure 13.4 Different muscle shapes (Adapted from Clip Art Explosion)

Figure 13.5 Muscles in motion (Adapted from http://navylive.dodlive.mil/2011/12/22/u-s-navy-seabees-building-bridges/)

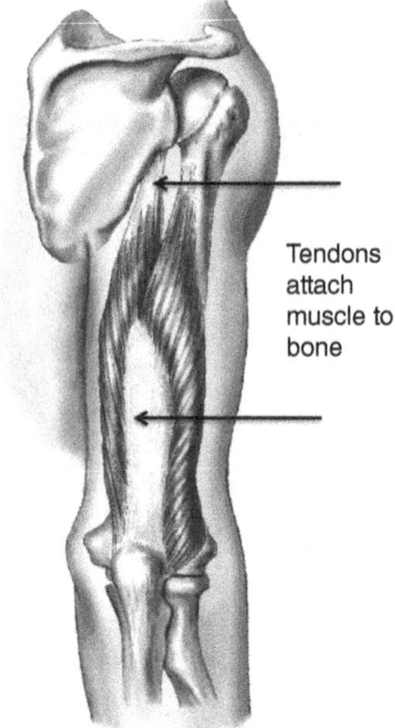
Tendons
attach
muscle to
bone

Figure 13.6 Biceps and triceps muscles (Adapted from Clip Art Explosion)

Tendons

Tendons are an integral part of the musculoskeletal system. These sturdy bands of fibrous tissue attach muscles to bones; they pretty much hold it all together. Tendons are like cables, their length and thickness depending on their function; Figure 13.7, for example, shows wide, thick tendons supporting the hip versus long, thin tendons for fast, repetitive finger movements. Tendons transfer forces for movement from

Muscles have one action, contraction. The shortening of the muscle moves two bones closer together. Two or more bones meet at a joint. As seen here the humerus meets the ulna and radius bones at the elbow joint

Muscles are attached to bone via a tendon, so when a muscle contracts, it pulls on the tendon that moves the bones closer together.
Tendons and muscles pass through tunnels or pulleys typically oriented in the opposite or crisscross configuration to the tendons. These tunnels represent the ligaments that provide stability to a joint and hold tendons onto bones

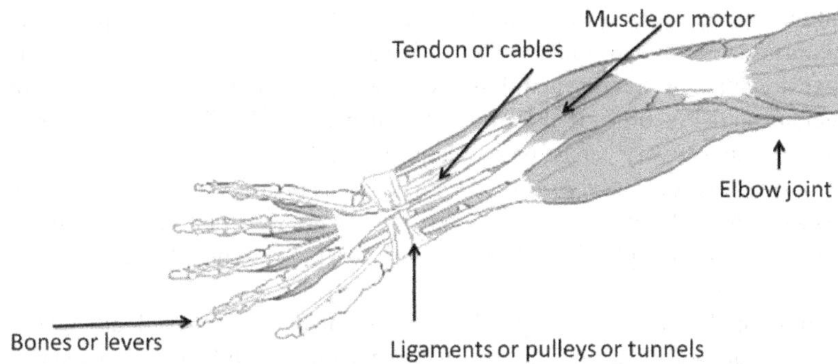

Figure 13.7 Bones, tendons and muscles and levers, cables and pulleys (Adapted with permission from Lippincott Williams & Wilkins, Baltimore, MD; http://www.clipartof.com/download?txn=577e6995e66d8d9c7f0dc5d21e5f9aa2)

the muscles to the skeletal system (bones) and are common sites for WMSD injury. Tendon disorders generally occur at or near the joint where, surrounded by synovial sheaths, the tendon rubs on adjacent ligaments or bones. Repetitive movements or awkward posture can cause friction between the tendon and the bone, resulting in heat or inflammation at that site as well as wear. Think of a wire cable rubbing against a seized pulley.

To understand how tendons become diseased, one must understand tendon function and repair mechanisms. As muscles contract, tendons are subjected to mechanical loading and viscoelastic deformation. Tendons must have both tensile resistance to loading (to move attached bones) and elastic properties (to enable bones to move around turns, as in the hand). When placed under tension, tendons first elongate without significant increase in stress. With increased tension, tendons become stiffer in response to this further loading. If the load on these structures exceeds the elastic limit of the tissue (its ability to recoil to its original configuration), permanent changes occur. During subsequent loading of the damaged tendon, less stiffness is observed (i.e., more creep). In addition, if recovery time between contractions is too short, deformation can result in pathologic changes that decrease the tendon's ultimate strength.

- Tensile strength is resistance of a material to forces tending to tear it apart.
- Shear stress results when a force tends to make part of the body or one side of a plane slide past another.

- Stress is the internal force exerted by one part of an elastic body upon the adjoining part.
- Strain is the deformation or change in dimension occasioned by stress.
- Stress causes strain.

Tendons and ligaments are elastic and will "creep" (i.e., stretch) in response to tensile loading. Creeping involves progressive fiber recruitment and loss of the natural waviness of collagen fibers. That is, when a tendon is subjected to prolonged elongation and loading, the magnitude of the tensile force will gradually decrease (relaxation) and the length of the tendon will gradually increase (creep) to a level of equilibrium. Picture an outstretched arm carrying a 5-lb paint bucket. During repetitive loading, the tendon exhibits these properties and then recovers if there is sufficient recovery time. If the time interval between loadings does not permit restoration, then recovery can be incomplete, even if the elastic limit is not exceeded. The natural recovery property of a tendon (ligament or muscle) can be maintained or enhanced through administrative controls such as job rotation or engineering controls, which make a physical change to the workplace to match it to the workers capabilities (Figure 13.8).

Tendons are also subject to perpendicularly oriented compressive loading. This is seen when tendons are loaded as they turn corners, around pulleys, or bony surfaces. Friction is generated at these locations as the tendon slides against adjacent surfaces, causing a shearing force. This friction is significant in the hand and wrist. Higher levels of muscle tension are required to achieve a specific level of strength at the fingertip during nonneutral wrist postures, and those tendons are subject to greater shear stress with nonneutral wrist postures. Tendon friction is proportional to the axial tension of the tendon, the coefficient of friction between the tendon and its adjacent surface, and the angle of the tendon as it turns about a pulley. This may be a cause of surface degeneration in tendons. Internal degeneration may be the result of friction-induced internal heat generation.

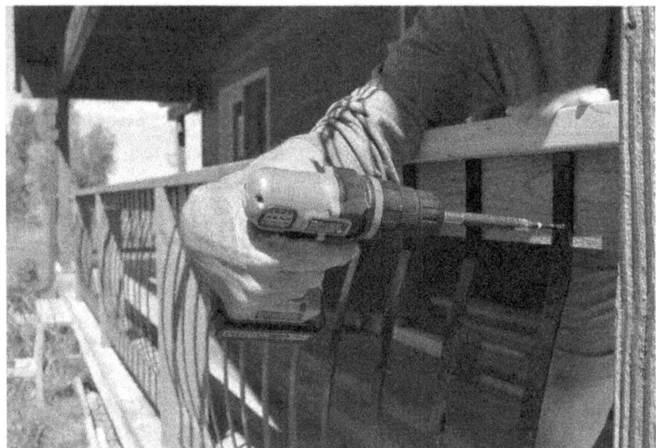

Figure 13.8 Worker in one neutral posture is likely to develop a WMSD of the upper extremities or spine

Tendons are surrounded by a protective slippery sheath called a synovial sheath. A synovial sheath is a tubular bursa surrounding a tendon. Tendon tissue slides or glides in the sheath due to synovial fluid within the sheath. Passage of blood vessels and nerves into the tendon tissue is through the sheath. Damage or compression of the sheath can reduce blood flow to the tendon and eventually contribute to degradation.

Initial symptoms of tendon MSD include impaired motion, tenderness, pain on resisted contraction or passive stretch, swelling, or crepitation. With time, swelling and thickening of the tendon may occur from fibril disruption, partial laceration, impairment of blood flow and diffusion of metabolites, and the localized repair process. Ultimately, this limits the normal smooth passage of the tendon through its fibro-osseous canal. These chronic tissue changes are recognized as triggering.

Ligaments

Ligaments are fibrous bands or sheets of connective tissue linking two or more bones, cartilages, or structures together. One or more ligaments provide stability to a joint. Excessive movements such as hyperextension or hyperflexion may be restricted or prevented by ligaments (Figure 13.9).

Movement of the body occurs about joints and between adjacent bones. Ligaments connect and hold all bones together (bone to bone). The ligaments allow the bones to move in all directions. These ligaments meld together to form the joint capsule. A joint capsule is a watertight sack of tissue that surrounds the bones and holds them close tightly together; a joint capsule around ligaments is similar to a synovial sheath or bursa round tendons. The inside of the capsules are lined with synovial membrane, which secrete synovial fluid, a mucus-like fluid that provides lubrication and nutrition to the articulating surface of the bones. The articulating surfaces are covered with

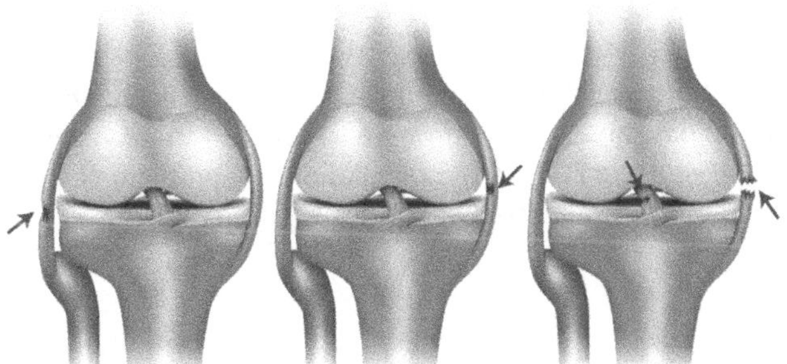

Figure 13.9 Knee ligaments in the knee are numerous and vary in direction to provide stability and support (Adapted with permission from Shutterstock)

cartilage. Cartilage becomes very slippery when lubricated with synovial fluid so that the adjacent surfaces slide against each other with very little resistance.

The biomechanical properties of ligament damage parallel those of tendons and were discussed previously. Ligaments are subject to creep, compression, deformation, and strain. Blood flow to ligaments is very slow, therefore, hastens recovery. Muscles have good blood flow, therefore, heal faster than tendons and ligaments.

Cartilage

Articulating surfaces of bones are covered with a shiny, smooth, white connective tissue called cartilage. Cartilage contributes to the synovial capsule of a joint and protects the underlying bone. The function of articular cartilage is to absorb shock and provide an extremely smooth surface to make motion easier. Articular cartilage can be up to one-quarter of an inch thick in the large, weight-bearing joints such as a hip. It is a bit thinner in joints such as the elbow, which does not support as great of a weight.

When movement occurs, joint cartilage is subject to two types of stress: gravitational pressure, particularly from weight-bearing joints of the legs and feet, and friction from the movement itself (picture of body and weights). Cartilage is well adapted to these stresses being strong, resilient, and smooth. Thus, it can absorb shock and allow sliding of bones relative to each other. Cartilage can become damaged by either blunt trauma or excessive wear. Rheumatoid arthritis and osteoarthritis (OA) are two common diseases involving damage to joint cartilage accompanied by inflammation, pain, and stiffness of the joint and surrounding muscles.

Joint cartilage does not contain blood vessels. It receives nutrients from the synovial fluid. Damage to the joint capsule can lead to cartilage deterioration. Cartilage has poor regenerative capacity and is, therefore, reluctant to heal. In addition, cartilage serves as a shock-absorbing function to protect underlying bone; once it is worn down, more force is transferred to the bone. This cycle highlights the importance of reducing an accumulation of microtraumas by limiting exposure to the physical workplace risk factors (Table 13.1).

Nerves

The nerves that control our voluntary and involuntary movement exit the spinal column in different regions. Injury of any form to a specific vertebrae area of the spine can affect other areas of the body. For example, a neck problem can cause pain or reduced strength in the arms and hands (Figure 13.10).

Muscles

The exact number of muscles that are contained in the human body is unknown. There are all sizes of muscles, ranging from the larger skeletal muscles of your leg, down to muscles that are literally “fibers” found attached to your spine. Muscles of the human body can be broken down into three categories: cardiac, skeletal, and smooth muscles. These three categories can be broken up into two subcategories: voluntary and involuntary muscles. In general, muscles have only two functions. Either they

TABLE 13.1 Stages of WMSD

Stage 1 – Usually shows aches and fatigue during the working hours, but with rest at night and days off work these aches seem to settle. This stage:

- Shows no drop in performance
- May persist for weeks or months
- Can be reversed

Stage 2 – Same symptoms occur early in the work shift and sleep does not settle the pain, in fact, sleep may be disturbed. This stage:

- Shows performance of the task is reduced
- Usually persists over months
- Can be reversed

Stage 3 – Symptoms persist while resting. Pain occurs while performing nonrepetitive movements. This stage:

- The person is unable to perform even light tasks
- May last for months or years
- Usually not reversible

contract or they relax. In the MS system, we are most concerned with the voluntary skeletal muscles.

Muscles are composed of thousands of tiny fibers all running in the same direction. They are red because they are filled with many blood vessels that supply them with oxygen and nutrients and carry away carbon dioxide and waste materials. Muscles can be injured in three ways. First, muscles fibers can be strained or irritated, causing temporary aching and swelling. Second, and more seriously, a small group of fibers can be torn apart. Third, a muscle may be subjected to a severe blow and crushed, breaking many of its blood vessels and causing blood to seep into a broader area.

Skeletal Muscles Of the two major muscle groups, voluntary and involuntary, the skeletal are voluntary muscles. Over 600 skeletal muscles have been discovered and named. Again, it is important to remember that our muscles either contract or relax. It is how we sit up; in general, the muscles of our thigh contract as the muscles of our posterior leg relax. Muscles can work in an antagonistic way (a muscle that opposes the action of another muscle, as by relaxing while the other one contracts, thereby producing smooth, coordinated movement) or agonistic way (working with another muscle or group of muscles to perform a certain action). So, why are skeletal muscles classified under the subcategory of "voluntary"? Voluntary muscle movements mean that we have control over their actions. For example, if they decide they want to flex their bicep, or decide to walk up the stairs, or throw a right hook, they are voluntarily

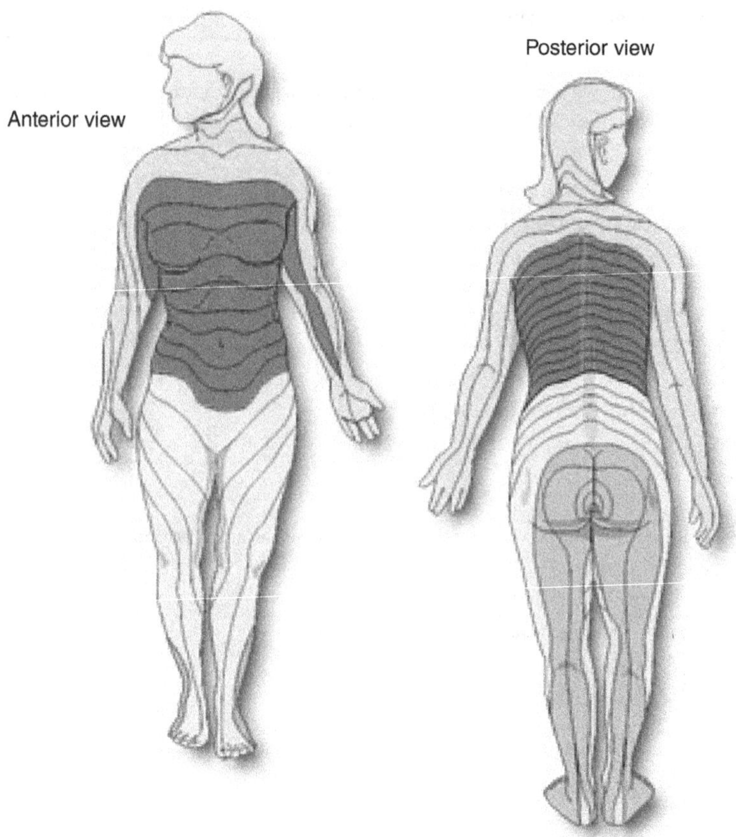

Figure 13.10 Nerve paths in the body (With permission Lippincott Williams & Wilkins, Baltimore, MD)

creating these actions through a series of signals you send and receive to and from the brain.

Skeletal muscles originate at a certain part of a bone, and insert at another part of the bone, either on the same bone or a different bone. When we contract and relax our skeletal muscles, we cause movement of our bones. It can be stated that the larger the muscle, the stronger it is; and generally, the larger the muscle, the greater range of motion. Vern Putz-Anderson (1998) summarizes of various WMSDs identified by profession. For example,

- Letter carriers are prone to shoulder problems and thoracic outlet syndrome from carrying heavy loads with a shoulder strap.
- Musicians are prone to carpal tunnel and epicondylitis from repetitive wrist movements and palmer-based pressure.
- Material handlers are prone to thoracic outlet and shoulder-tendinitis from carrying heavy loads.

- Workers who grind are at risk of tenosynovitis, thoracic outlet, and de Quervain's from repetitive wrist motions and prolonged flexed shoulder postures. Vibration and forceful ulnar deviations and repetitive forearm pronation.

INTRODUCTION TO THE SPINE

The spine is one of the most important, sophisticated, and complex structures in the body. The spine is affected by almost every movement made.

The following structures are critical to proper spine function. Injury to any of the components of the spine may provoke disease and pain. Some forms of back injuries can be caused by a single event such as a heavy awkward lift. Disc injury can also occur from wear and tear caused by repeated trauma over a period of months or years, for example, loading a delivery truck.

The Vertebral Column

The vertebral column comprises the posterior aspect of the trunk. A normal vertebral column is comprised of 33 irregular-shaped bones that are called vertebrae. The vertebral column, also called the spine, extends from the base of the neck to the pelvis. An inaccurate assumption is to think that the spine is a rigorous supporting rod. The vertebrae are connected in such a way that a movable "S"-curved structure is created. Its major functions are to distribute weight of the trunk evenly to the lower limbs while protecting the spinal cord and providing attachment points for the ribs and muscles of the neck and back as well as providing a pathway for nerves from the central nervous system of the brain to the distal regions of the body.

Spinal Regions

We can think of the vertebral column as a series of fixed segments (the vertebrae that are bones) having mobile connections (discs, tendons, and ligaments). Movement of individual vertebrae is compounded such that the entire structure has considerable mobility in three dimensions. The type and extent of mobility varies with different spinal regions, depending on size and shape of vertebrae, body, and other factors. Each vertebra is attached to its neighbor by three joints, the intervertebral disc (posterior) and two pair of facets. The bodies are joined by the fibrocartilaginous. Joints allow motion (articulation). The joints in the spine are commonly called facet joints (Figure 13.11).

Each vertebrae pair is limited in motion; however, each small movement over the length of the vertebral column allows a wide range of motion. Flexion (bending forward) and extension (bending backward) movements encompass a wider range of motion when compared to lateral flexion (bending sideways) and rotation of the spine.

The cervical spine beginning at the base of the skull has seven vertebrae with eight pairs of cervical nerves responsible for controlling the neck, arms, and upper body (reference Figure 13.11). This vertebral group is sturdy enabling it to support the weight of the head. The amount of stress placed on these vertebrae varies with movement and posture. This area of the spine is very movable.

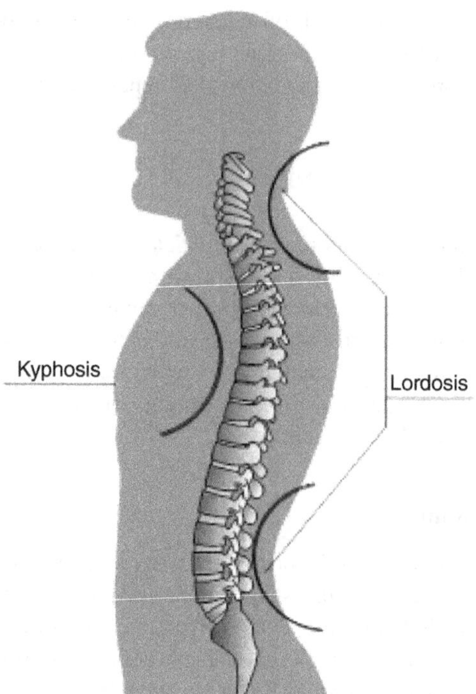

Figure 13.11 Curves and sections of the spine (Adapted with permission Lippincott Williams & Wilkins, Baltimore, MD)

Directly below the cervical region the thoracic portion of the spine starts. There are 12 vertebrae and 12 pairs of ribs, as well as nerve roots that control the midsection of the body. The ribs form the chest wall and protect many internal organs. This section of the spine is fairly immovable (try to move this area without moving your neck or lower back).

The third part of the spine is the lumbar vertebrae. There are normally five lumbar bones. However, some people have one less or one more. These vertebrae are the largest and strongest of the three regions (cervical, thoracic, and lumbar) because these carry the bulk of the body's weight. Five pairs of lumbar nerves manipulate the movement and sensory functions of the lower extremities. The lumbar region is the segment of the spine under the most pressure because it is on the bottom of the spine. When a person bends over, the lumbar region is asked to bend but still maintain the strength to hold up the spine.

A mass of five smaller bones, which are naturally fused together, is after the last lumbar vertebra. This bone mass is named the sacrum. Below the sacrum is the coccyx or tailbone made up of four little bones fused together. The sacrum and coccyx do not look like the other vertebrae. The pairs of nerve roots originating from this area are responsible for the action of the pelvic organs and buttock muscles. The sacrum and tailbone could be considered the basement and sub-basement of the building.

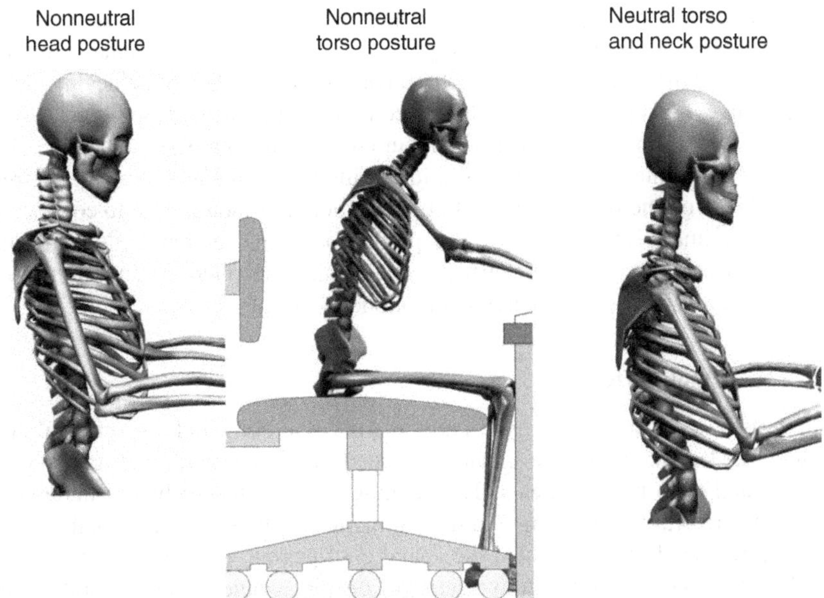

Figure 13.12 Neutral and awkward spinal postures (Adapted with permission from Ergonomic Image Gallery)

Maintaining the Curves

There are several characteristics curvatures of the vertebral column (reference Figures 13.11 and 13.12).

- Sacrum, convent toward the back
- Concave lumbar region (the term lordosis can refer either to an exaggeration of this curvature or to the normal condition
- Convex thoracic region (kyphosis)
- Concave cervical region.

Exact form of these curvatures varies between people.

Reducing the biomechanical loading on the spine is a key element in prevention against spinal injury. This can be done by:

- practicing proper postural alignment or a neutral spine when sitting, standing, and moving;
- including aerobic and flexibility exercises;
- getting enough rest.

The word posture comes from the Latin verb ponere, which means "to put or place." The general concept of human posture refers to "the carriage of the body as a whole, the attitude of the body, or the position of the limbs (the arms and legs)."

Webster's New World Medical Dictionary defines neutral posture as the stance that is attained "when the joints are not bent and the spine is aligned and not twisted. Neutral posture has given rise to the idea of achieving "ideal posture." Ideal posture indicates proper alignment of the body's segments such that the least amount of energy is required to maintain a desired position. The benefit of achieving this ideal position would be that the least amount of stress is placed on the body's tissues. In this position, a person is able to completely and optimally attain balance and proportion of his or her body mass and framework, based on his or her physical limitations. Good posture optimizes breathing and affects the circulation of bodily fluids.

Spinal Discs and Vertebra

Each vertebra consists of two main parts: the massive body and the vertebral arch (Figure 13.13). The arch in turn can be divided into many parts that are sites for ligament and tendon attachments. The opening between the body and the arch is called the vertebral foramen. As foramina of many vertebrae are lined up, they form the vertebral canal through which the spinal cord passes. The spaces between the pedicles of adjacent vertebra from a series of openings called intervertebral foramina. As spinal nerves branch off the spinal cord, they exit through these foramina (Figure 13.14).

Each vertebrae is attached to its neighbor by three joints, two facets and a disc. Each vertebrae has two sets of facet joints. Posteriorly, two facets of the top vertebra connect two superior facets of the bottom vertebra. The bodies of the vertebrae are joined by the fibrocartilaginous intervertebral. These discs are comprised of a nucleus pulposus (inside) and an annulus fibrosus (outer rim). The nucleus pulposus, or fluid-like substance, which occupies the center portion of the disc, acts as a cushion and a shock-absorbing apparatus during activities such as walking, running, and

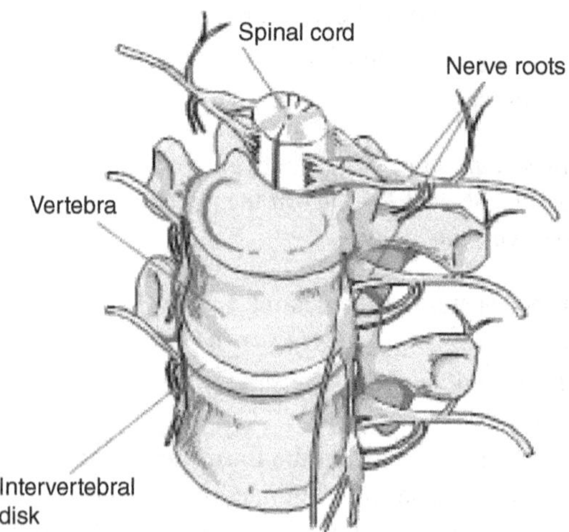

Figure 13.13 Parts of the vertebral body (Adapted with permission Lippincott Williams & Wilkins, Baltimore, MD)

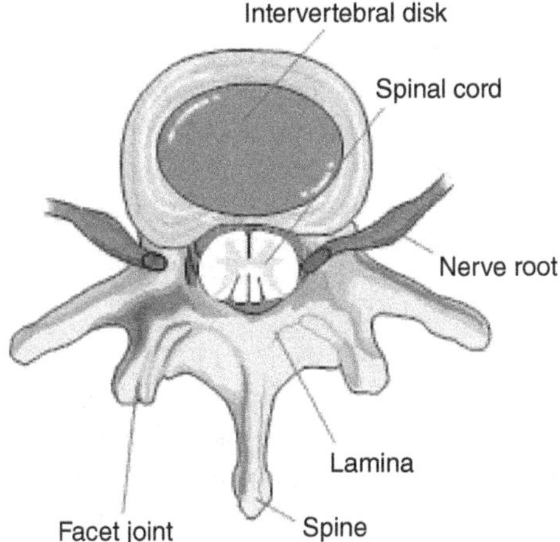

Figure 13.14 Parts of the spinal disc (Adapted with permission Lippincott Williams & Wilkins, Baltimore, MD)

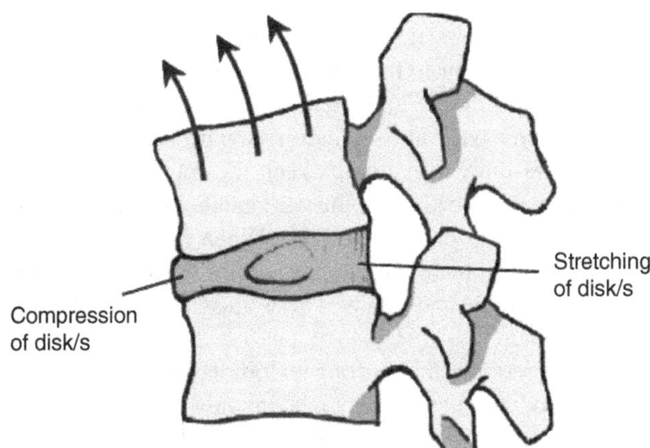

Figure 13.15 Disk movement during bending (Adapted with permission from Ergonomics Image Gallery)

jumping. The pulpy nucleus flattens and the annulus, or outer sheath that encompasses the nucleus pulposus, bulges when weight is applied, as occurs during standing and more so during lifting. During flexion and extension movements, the nucleus pulposus serves as a fulcrum. The annulus is simultaneously placed under compression on one side and tension on the other (Figure 13.15).

At birth, 80% of the disc is composed of water. In order for the disc to function properly, it must be well hydrated. The nucleus pulposus is the major carrier of the body's axial load and relies on its water-based contents to maintain strength and pliability.

Clinical evidence shows that occupational low-back injuries occur most often at the intervertebral disc between L5 and the sacrum S1. Damage to cartilaginous end place of the intervertebral disc may be one of the causes of these injuries. Cumulative damage to the cartilage end places hampers the flow of important nutrients to the disc and provides disc degeneration. As the disc degenerates, there is an increased risk of disc failure, subsequent alteration of the spinal geometry and eventually pain resulting from vertebral pressure on adjacent spinal nerves. Damage to the spinal discs can go undetected until significant degeneration occurs due to the lack of nerve fibers in the cartilage endplate.

The Core Muscles

The "core" is a group of muscles that surrounds the trunk. The main function of the core is to stabilize and protect the spine and pelvis when the body is in motion. Major muscles included are the pelvic floor muscles (PFM), transversus abdominis (TA), multifidus (MF), internal and external obliques, rectus abdominis, erector spinae (sacrospinalis) especially the longissimus thoracis, and the diaphragm. Minor core muscles include the latissimus dorsi, gluteus maximus, and trapezius (Figure 13.16).

There are four main muscle groups that make up the core: transversus abdominus, multifidus, PFM, and the diaphragm. TA is the deepest abdominal muscle that wraps around your abdomen like a corset and is connected to tissue surrounding the spine. When TA contracts, it is similar to the corset being tightened, therefore, assisting in increasing the pressure inside the abdomen that provides increased stability to the spine. MF is a deep lower back muscle that makes up the back part of the core. It is an important postural muscle that helps keep the spine erect. The PFMs are the bottom part of the core or the bathroom muscles and help to stabilize the pelvis. The diaphragm makes up the top part of the core. When all of these muscles contract simultaneously, they help to maintain the pressure in the abdomen, which then provides the stability to the spine and pelvis. These muscles can best contract when the spine is in a neutral posture.

A common misconception is that "strong abdominals protect the spine." In fact, as described above, the abdominal muscles make up only one part of the core. Furthermore, only the deep abdominal muscle, TA, is involved in protecting the spine. The famous "6-pack" or rectus abdominis muscle that many fitness fanatics train actually plays no role in protecting the spine.

Adequate core stability not only reduces strain on the spine but also helps maintain optimal postural alignment that will help reduce risk of injury. Core stability is also an important part of any Total Health, Safety, or Ergonomics program. A strong core means a strong foundation from which our limbs can move more safely, with more power and efficiency, and with less risk of injury.

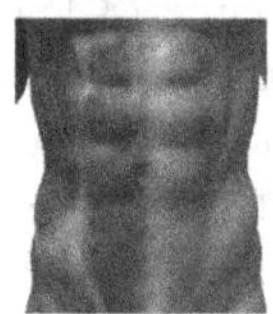

Figure 13.16 Frontal view of the core muscles (Adapted from Clip Art LLC)

Spinal WMSDS

Why do so many people experience back pain? Our core muscles are involved in every move we make.

- New use – starting up a new activity; muscles may not be used to the range of motion, force, or speed required. For example, the first time you snowboard, even if you are an avid skier.
- Misuse – cumulative effect of bad body use over a long period (e.g., poor postural alignment) or pushing the body too far too often.
- Overuse – repetitive use of one muscle group causing an imbalance. For example, carrying your child on the same hip.
- Disuse – lack of exercise may not cause a back problem but one can result when we attempt an activity requiring a certain degree of strength or flexibility. A certain percentage of people stop exercise or physical activity when they experience back pain.

Any of these causes can be found in an occupational setting.

Classifications of Spinal WMSDs The following list is an excerpt from ergonomics in back pain:

- Discogenic: disc hernia (most common at L 4-5 and L 5-S1)
- Neurological: nerve irritation, compression and/or tumors involving nerve roots
- Muscular/ligamentous tension: resulting from stress and nerve or ligament tension
- Trauma: acute injury or cumulative type
- Strain: small tears within the muscle/tendon
- Postural imbalance: creates uneven stresses on the musculoskeletal system
- Spasm: muscle contraction that produces an uncontrolled contraction
- Weakness: poor muscle tone
- Myofascitis: inflammation and tenderness of the muscle and the sheaths that envelop the muscle known as the fascia
- Structural: spondylolysis – a defect of the bony segment joining the articulations above and below a given segment
- Scoliosis: abnormal curve of the spine
- Compression fractures
- Dislocation degenerative disease annular tears
- Osteoarthritis: degenerative disorder that affects the facet joints and disk
- Stenosis – narrowing of a channel.

Neck disorders are commonly associated with prolonged exposure to static and awkward postures, typically as a consequence of visual requirements of a task. There is evidence that flexion beyond 30° leads to more rapid onset of fatigue, whereas low-back disorders are typically caused by repeated loading or high forces.

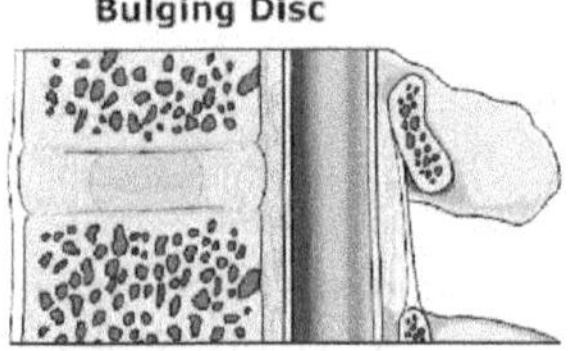

Normal, everyday pressure on the disc forces its outer surface to bulge slightly.

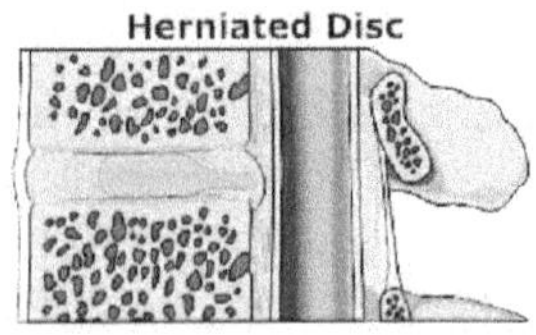

As a disc degenerates due to injury, disease or wear, the inner core extrudes back into the spinal canal, which is known as a disc herniation.

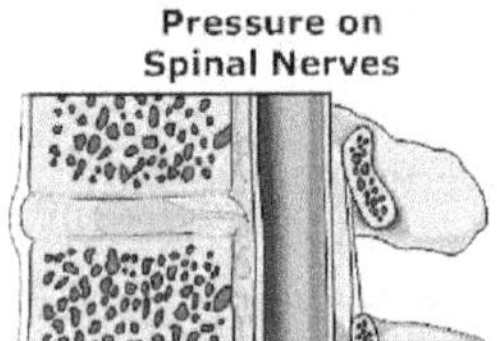

A herniated disc in the lower back can cause lower back pain and/or leg pain (sciatica) by putting pressure on the nerve root.

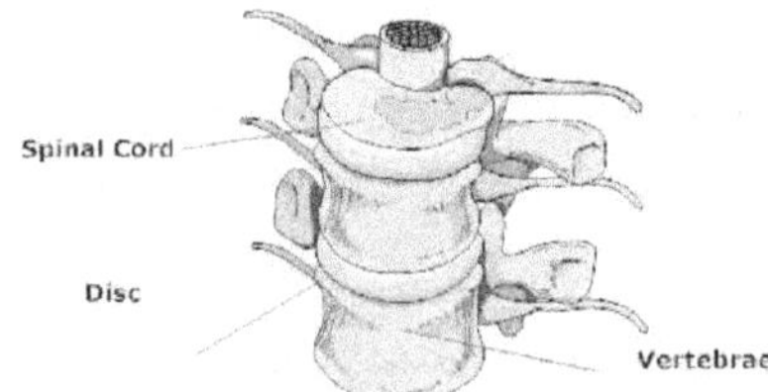

Discs are soft, rubbery pads between the vertebrae of the spinal column. They act as shock absorbers and allow the spine to flex. Discs are composed of a thick, outer ring of cartilage and inner, gel-like substance.

Figure 13.17 Types of disc disorders (Adapted with permission Lippincott Williams & Wilkins, Baltimore, MD)

Stages of Disc Degeneration The intervertebral disc changes over time from misuse, overuse, lack of use and new use. Early in our development, the spinal disc is spongy and firm. The nucleus portion in the center of the disc contains water and is plump separating the spinal vertebrae. This gives the disc its ability to absorb shock and protect the spine from heavy and repeated forces (Figure 13.17).

The first change that occurs is that the annulus around the nucleus weakens and begins to develop small cracks and tears. The body tries to heal the cracks with scar tissue, but scar tissue is not as strong as the tissue it replaces. The torn annulus can be a source of pain for two reasons. First, there are pain sensors in the outer rim of the annulus. They signal a painful response when the tear reaches the outer edge of the annulus. Second, like injuries to other tissues in the body, a tear in the annulus can cause pain due to inflammation.

Over time if the disc does not have a chance to heal, or if the **condition that caused the cracks remains unchanged**, the disc begins to lose water, causing it to lose some of its fullness and height. As a result, the vertebra begins to move closer together. Scar tissue does not allow recovery as readily as uninjured tissue.

As the disc continues to degenerate, the space between the vertebra shrinks (vertebra move closer together). This compresses the facet joints along the spinal column. As these joints are forced together, extra pressure builds on the articular cartilage on the surface of the facet joints. This extra pressure can damage the facet joints. Over time, this may lead to arthritis in the facet joints. These degenerative changes in the disc, facet joints, and ligaments cause the spinal segment to become loose and unstable. The extra movement causes even more wear and tear on the spine. As a result, more and larger tears occur in the annulus.

The nucleus may push through the torn annulus and into the spinal canal. This is called a herniated or ruptured disc. The disc material that squeezes out can press against the spinal nerves. The disc also emits enzymes and chemicals that produce inflammation. The combination of pressure on the nerves and inflammation caused by the chemicals released from the disc causes pain.

If the degeneration continues, bone spurs develop around the facet joints and around the disc. No one knows exactly why these bone spurs develop. Most doctors think that bone spurs are the body's attempt to stop the extra motion between the spinal segments. These bone spurs can cause problems by pressing on the nerves of the spine, where they pass through the neural foramina. This pressure around the irritated nerve roots can cause pain, numbness, and weakness in the low back, buttocks, and lower limbs and feet.

A collapsed spinal segment eventually becomes stiff and immobile. Thickened ligaments and facet joints, scarred and dried disc tissue, and protruding bone spurs prevent normal movement. Typically, a stiff joint does not cause as much pain as one that slides around too much. So this stage of degeneration may actually lead to pain relief for some people.

Back Pain Solutions

Prevention is the best health insurance. The first step to a healthy mind and body is the understanding about lifestyle choices that contribute to overall wellness. Correct posture, balanced diet, regular exercise, and stress reduction are all important factors. Stretching and strengthening can help maintain or regain range of motion. At the first sign of pain or discomfort that is not alleviated by rest, that wakes you up in the night or limits your after work or work activities, contact a healthcare professional.

A professional evaluation usually includes questions about symptoms, past health problems, family history of disease, work habits, daily activities, and sleeping positions. A physical examination is performed to analyze posture, spinal, and joint alignment. Muscles may be checked for trigger points in the areas of discomfort and radiographs (X-rays) and other tests may be recommended to aid in diagnosis. When the examination and evaluation are complete, the healthcare professional can discuss the condition and prescribe a treatment plan that will aid in recovery. Treatment options may include rest, core muscle stabilization, physical therapy, over-the-counter anti-inflammatory medicines, prescription medication, pain management therapy, steroidal injections, behavioral and lifestyle changes, and minor to major surgery. Other forms of recovery include nontraditional medicines such as acupuncture and yoga. Regardless of the treatment, if the cause of the disorder is not altered, the treatment may not be effective in the long term (Figure 13.18).

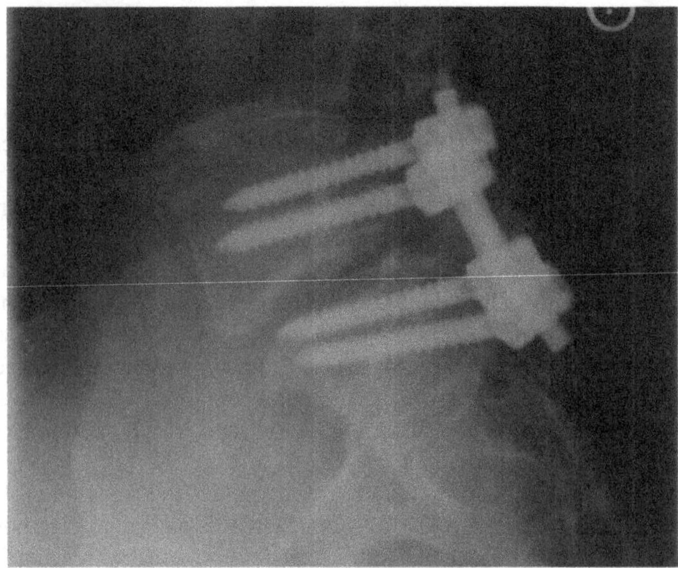

Figure 13.18 Posterior spinal between L5 and S1 to correct a structural failure between the segments

HAND, WRIST, ARM AND SHOULDER WMSDS

For a complete discussion of upper limb musculoskeletal disorders, refer "Cumulative Trauma Disorders – A Manual for Musculoskeletal Diseases of the Upper Limbs" edited by Vern Putz-Anderson and published by Taylor and Francis.

The structures of the upper extremities are particularly vulnerable to soft tissue injury. A main reason is that almost all work requires the constant and active use of the arms and hands and that long tendons attach the forearm muscles around the elbow to the fingers (Health, 1988).

Nerve Disorders

The nerves and blood vessels that run into the arm and hand exit the spinal column at C6, C7, and C8. The nerves travel between two muscles in the neck called scalene muscles, over the top of the rib cage, under the collar bone, through the arm pit, and down the arm into the hand. The area where the nerves and vessels leave the neck between the two scalene muscles and over the first rib is known as the thoracic outlet (reference Figure 13.10).

The radial nerve runs along the thumb-side edge of the forearm. It wraps around the end of the radius bone toward the back of the hand. It gives sensation to the back of the hand from the thumb to the third finger. It also supplies the back of the thumb and just beyond the main knuckle of the back surface of the ring and middle fingers.

The median nerve travels through a tunnel within the wrist called the carpal tunnel. This nerve gives sensation to the thumb, index finger, long finger, and half of the ring finger. It also sends a nerve branch to control the muscles of the thumb (thenar).

The ulnar nerve travels through a separate tunnel, called Guyon's canal. This tunnel is formed by two carpal bones – the pisiform and hamate – and the ligament that connects them. After passing through the canal, the ulnar nerve branches out to supply feeling to the little finger and half the ring finger. Branches of this nerve also supply the small muscles in the palm and the muscle that pulls the thumb toward the palm.

The nerves that travel to the hand are subject to problems. Constant bending and straightening of the wrist and fingers can lead to irritation or pressure on the nerves within their tunnels and cause problems such as pain, numbness, and weakness in the hand, fingers, and thumb.

Historically, the median nerve was called the musician's nerve due to problems associated with plucking motions (such as plucking a harp) it is responsible for pulling the hand back. The ulnar nerve was called the carpenters nerve due to problems associated with repeated hammering. The ulnar nerve runs along the outside of the elbow and can snap or slip with repetitive/forceful hammering type motions. We all know where our ulnar nerve is. It creates that pins and needles sensation when we strike our elbow.

Carpal tunnel syndrome is one of the most well-known nerve disorders because of its prevalence in the mid-1990s when computers with mice were becoming common place in work settings and homes.

Carpal Tunnel Syndrome The term carpal tunnel is a medical term for the space in the wrist where nerves and tendons pass from the forearm into the hand. Carpus is a word derived from the Greek word "karpos" which means "wrist."

The wrist is surrounded by a band of ligament tissue that normally functions as a support for the joint and creates a capsule. The tight space between this fibrous band and the wrist bone is called the carpal tunnel. Knowing the structure of the hand helps understand why keeping the wrist straight while performing tasks is important. Looking closely, you can see that the median nerve lies just under the transverse carpal ligament. The nerve can be easily compressed or strained because it is just below a soft structure and subject to direct pressure from contact stress and internal pressure from awkward postures. Figure 13.19 shows how the **median nerve** innervates the thumb and the first three fingers of the hand. The nerve transmits signals to and from the hand. If the median nerve is damaged, the sensation and strength of the hand is often compromised or lost.

Causes of Carpal Tunnel Syndrome Carpel Tunnel Syndrome (CTS) may develop when the median nerve is compressed or squeezed at the wrist. CTS is often the result of a combination of factors that increase pressure on the median nerve and tendons in the carpal tunnel, rather than a problem with the nerve itself. For example, damage to the tendon sheath can reduce blood flow to the nerve and cause its function to be impaired. Contributing factors include trauma or injury to the wrist that causes swelling, such as a sprain or fracture, overactivity of the pituitary gland, hypothyroidism, rheumatoid arthritis, mechanical problems in the wrist joint, work stress, repeated use of vibrating hand tools, fluid retention during pregnancy or menopause, or the development of a cyst or tumor in the canal.

There is no one cause of CTS but one or more of the following work activities may lead to CTS:

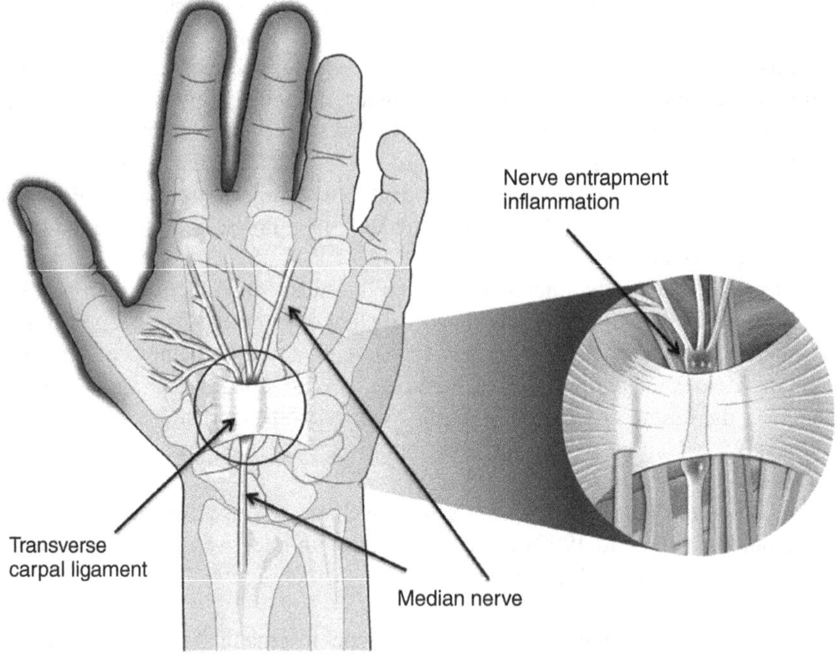

Figure 13.19 Carpal tunnel and median nerve (Adapted with permission from Shutterstock)

- Using a "pinch grip" with your fingers
- Extreme force involving the fingers, hand, and wrist
- Pressure on your hand, especially into the palm for long periods
- Repeated, rapid movements of the hand and wrist
- Tasks done while the hand or wrist is held in a bent or awkward position
- Low-frequency vibration to the hand
- Wearing gloves that do not fit
- Exposing the hands to cold temperatures for long periods
- Using tools that do not fit the hand requiring extreme pinch or oblique grips.

Jobs requiring highly repetitive and forceful motions are prime targets. The risk of developing CTS is not confined to people in a single industry or job; it is especially common in manufacturing, sewing, finishing, cleaning, and meat, poultry, or fish packing, and maintenance.

Symptoms of Carpal Tunnel Syndrome Symptoms usually start gradually, with frequent burning, tingling, or itching and numbness in the palm of the hand and the fingers, especially the thumb and the index and middle fingers. Some carpal tunnel sufferers say their fingers feel useless and swollen, even though little or no swelling is apparent. The symptoms often first appear in one or both hands during the night, since many people sleep with flexed wrists. A person with CTS may wake up feeling the need to "shake out" the hand or wrist. As symptoms worsen, tingling is felt during the day. Decreased nerve conduction will impact strength may make it difficult to form a fist, grasp small objects, or perform other manual tasks.

Tendinitis

Tendinitis (or tendonitis) is a tendon inflammation that is associated with a muscle/tendon, which is repeatedly tensed, moved, bent, vibrated, or in contact with a hard surface. Tendons themselves are cords of tough, fibrous connective tissue that attach muscles to bones and are found throughout the entire human body. Tendinitis is the inflammation and irritation of a tendon. If the normal smooth gliding motion of a tendon is impaired, the tendon will become inflamed and tendinitis will start to occur. Tendinitis occurs largely in the arms, hands, and wrist because of the length of the tendons operating the fingers, and the exposure to repetitive motions, but tendinitis can occur in any tendon group.

Symptoms of Tendinitis

- **When the tendon is under pressure, pain** is the first tendinitis symptom to develop. This pressure could come from lifting objects, playing sports, long-term hyperextension of the arm, repeated work with the hands or any type of manual job. In the first stages of tendinitis, pain usually only occurs when the tendons are under pressure. As the tendinitis develops, pain will start to occur throughout the day whether the tendon is under pressure or not. The pain will occur when you touch the tendon and move the joint.
- **Movement is restricted**, for example, if the tendinitis has developed in the bicep the individual may not be able to extend the arm fully.
- **Burning sensation** is felt mostly after exercise or manual labor and in the morning or late at night.
- **Affected area is swollen, red, warm, or lumpy.** The tendon sheaths may be visibly swollen from the accumulation of fluid and inflammation. This is a sign that tendinitis has advanced.

Recovery from Tendon and Ligament Damage Tendon and ligaments do not heal as quickly as muscles. Because of this, an injured worker may return to work after the initial pain or inflammation subsides but with a tendon/ligament that is not fully repaired, the repeated trauma from returning to unrestricted work too early can cause further damage.

Tendon/ligament repair is hastened due to oxygen supply. The body recovers through cell regeneration and cell energy is due to adenosine triphosphate (ATP). The more ATP cells make, the more energy cells have, and the faster they can reproduce to fill in the voided tissue. ATP is produced in the cell through processes such as glycolysis, the citric acid cycle, and the electron transport chain. Glycolysis and the citric acid cycle can only produce a few ATP (2 each), while the electron transport chain can produce many more (32–36). Without getting into physiology, the key to making the most energy possible is oxygen.

If one can introduce oxygen to a cell, cells can undergo what is called "aerobic respiration." Basically, this means that cells can perform all three processes (glycolysis, citric acid cycle, and electron transport chain). Without oxygen, cells can only perform two of those processes (glycolysis and the citric acid cycle). This is less than ideal for recovery. Without oxygen, the cells produce about 1/8 the energy that would have been produced with oxygen in the cell. Ligaments and tendons do not receive a lot of oxygen. Blood is how oxygen is transported throughout the body, but since the fibrous structures do not have much vasculature, they do not see very much oxygen. The skin, by comparison, has tiny blood vessels throughout that constantly provide oxygen to the cells. This is why a tendon/ligament injury will heal much slower than a superficial cut on the skin.

It is not uncommon for tendinitis to develop as a result of another tendon or joint injury. For example, shoulder tendinitis is often developed after a rotator cuff injury and knee tendinitis can be developed after having knee surgery. In these cases, tendinitis usually develops because the injury has not completely healed.

de Quervain's Tendinitis

de Quervain's is a condition brought on by irritation or inflammation of the wrist tendons at the base of the thumb. The inflammation causes the compartment (a tunnel or a sheath) around the tendon to swell and enlarge, making thumb and wrist movement painful. Making a fist and grasping or holding objects are common painful movements with de Quervain's tendinitis.

What Causes de Quervain's Tendinitis Repeatedly performing hand and thumb motions such as grasping, pinching, squeezing, or wringing may lead to inflammation. This inflammation can lead to swelling, which hampers the smooth gliding action of the tendons within the tunnel.

New mothers are especially prone to this type of tendinitis. Caring for an infant often creates awkward hand positioning and hormonal fluctuations associated with pregnancy and nursing further contribute to its occurrence. A wrist fracture can also predispose a patient to de Quervain's tendinitis, because of increased stresses across the tendons.

Tenosynovitis

This disorder occurs inside the tendon synovial sheath. Tenosynovitis develops when the inner (synovial) lining of the tendon sheath becomes injured or inflamed. It occurs most often in the hands, wrists, and elbows again due to the length of the tendons and many tunnels in which they pass though. The inner synovial lining is separate from the fibrous outer sheath covering the tendon, impacting nourishment and lubrication to the tendon. Irritation to the synovial lining can be caused by injury, overuse, repetitive strain, trauma, rheumatoid arthritis, gout, or infection. This condition is typically a precursor to CTS. Tenosynovitis can also occur when direct pressure is placed on a tendon and, yet the tendon is required to repeatedly move.

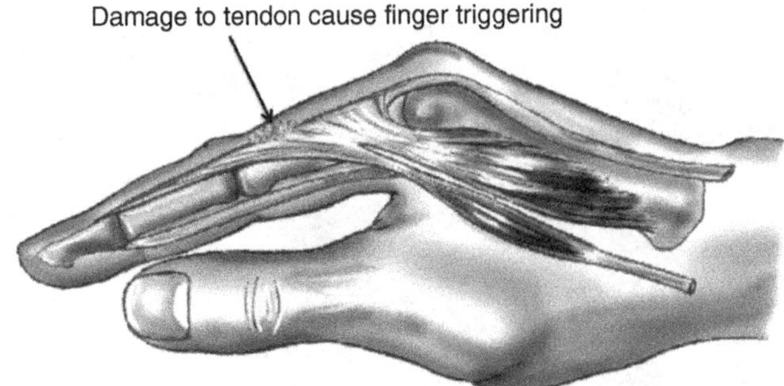
Damage to tendon cause finger triggering

Figure 13.20 Pully's and damager in the finger (Adapted with permission Lippincott Williams & Wilkins, Baltimore, MD)

Again even though we are discussing the upper limbs, these conditions can occur in the lower limbs as well.

Trigger Finger

"Trigger finger" or "trigger thumb" are the common names for stenosing tenosynovitis. Trigger finger involves the pulleys and tendons in the hands and fingers that work together to bend the fingers. The tendons work like long ropes connecting the muscles of the forearm with the bones of the fingers and thumb. In the finger, the pulleys are a series of rings that form a tunnel through which the tendons must glide, much like the guides on a fishing rod through which the line (or tendon) must pass. These pulleys hold the tendons close against the bone. The tendons and the tunnel have a slick lining that allows easy gliding of the tendon through the pulleys. Trigger finger occurs when the pulley at the base of the finger becomes too thick and constricting around the tendon, making it hard for the tendon to move freely through the pulley. Sometimes, the tendon develops a nodule (knot) or swelling of its lining. Because of the increased resistance to the gliding of the tendon through the pulley, you may feel pain, popping, or a catching feeling in the finger or thumb (triggering). When the tendon catches, it produces inflammation and more swelling. This causes a vicious cycle of triggering, inflammation, and swelling.

Signs and Symptoms Trigger finger may start with discomfort felt at the base of the finger or thumb, where they join the palm. This area is often tender to local pressure. A nodule may be found in this area and the finger or thumb becomes stuck or locked, and is hard to straighten or bend.

Figure 13.20 shows the area that becomes damaged from tools and, then the tendon does not slide through the opening as well. Once this condition occurs, the tendon is pretty much damaged for life. If our tendons did not have this arrangement, we could not curl our fingers (see Figure 13.21).

Nerve Disorders in the Elbow

The end of the humerus (arm bone) has two epicondyle, a lateral and a medial. Ligaments and tendons of the elbow joint are attached at the epicondyle and the nerves of

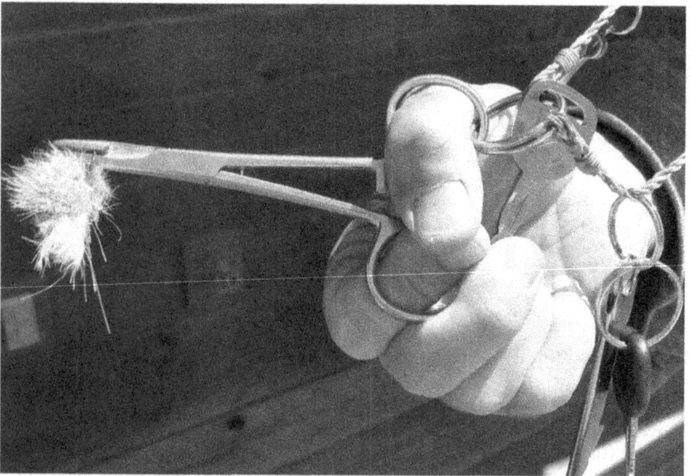

Figure 13.21 Activity that can cause trigger finger

the arm run over, beneath or between the elbows joint. Epicondylitis is inflammation at this attachment point of a tendon.

As you can see from Figure 13.22, all of the nerves that travel down the arm pass across the elbow. Three main nerves begin together at the shoulder: the radial nerve, the ulnar nerve, and the median nerve. These nerves carry signals from the brain to the muscles that move the arm. The nerves also carry signals back to the brain about sensations such as touch, pain, and temperature.

Some of the more common disorders of the elbow are problems of the nerves. Each nerve travels through its own tunnel as it crosses the elbow. Because the elbow must bend and rotate, the nerves must bend as well. Constant bending and straightening can lead to irritation as a nerve slides across a bone or pressure on the ligaments within their tunnels. For example, the cubital tunnel is a channel that allows the ulnar nerve to travel over the elbow and cubital tunnel syndrome is inflammation of the nerve within the tunnel or inflammation of the tunnel itself. Radial tunnel syndrome is caused by increased pressure on the radial nerve, which runs by the bones and muscles of the forearm and elbow.

Epicondylitis (Tennis-Elbow/Golfers Elbow)

There are several important tendons around the elbow. The biceps tendon attaches the large biceps muscle on the front of the arm to the radius. It allows the elbow to bend with force. One can feel this tendon crossing the front crease of the elbow when they tighten the biceps muscle.

The triceps tendon is the large tendon that connects the large triceps muscle on the back of the arm with the ulna. It allows the elbow to straighten with force, such as when you perform a push-up.

The muscles of the forearm cross the elbow and attach to the humerus. The outside, or lateral, bump just above the elbow is called the lateral epicondyle. Most of the

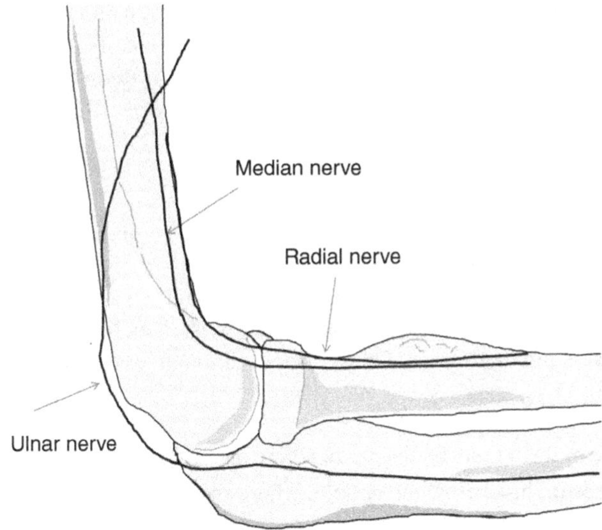

Figure 13.22 Median, radial, and ulnar never funneling through the elbow (Adapted with permission Lippincott Williams & Wilkins, Baltimore, MD; http://www.clipartof.com/download?txn=577e6995e66d8d9c7f0dc5d21e5f9aa2)

muscles that straighten the fingers and wrist all come together in one tendon to attach in this area. The inside, or medial, bump just above the elbow is called the medial epicondyle. Most of the muscles that bend the fingers and wrist all come together in one tendon to attach in this area. These two tendons are important to understand because they are a common location of tendonitis.

Lateral epicondylitis, commonly known as "tennis elbow," is an inflammation of the tendon fibers that attach the forearm extensor muscles to the outside of the elbow. These muscles lift the wrist and hand. Tennis elbow was first recognized by doctors more than 100 years ago. The name comes from a common cause; that is the back hand in tennis uses forceful wrist extension and rotation. Medial epicondylitis, known as "golfers elbow," is an inflammation of the tendon fibers that attach the forearm flexor muscles on the inside of the elbow. Medial epicondylitis or "golfers elbow" is a similar condition that occurs on the inside of the elbow. The name comes from the common cause, due to forceful rotations and flexion of the forearm.

Pain may be felt where these fibers attach to the bone on the outside/inside of the elbow or along the muscles in the forearm. Pain is usually more noticeable during or after stressful use of the arm. In severe cases, lifting and grasping even light things may be painful.

Causes of Epicondylitis Routine use of the arm or an injury to this area may stress or damage the muscle attachment and cause symptoms. Generally, people who develop this problem may be involved in activities with rotation of the forearm combined with forceful extension or flexion; for example, during painting. The condition is quite common in our late 30s and early 40s.

Signs and Symptoms of Epicondylitis The area of most pain is usually found near the bone on the outer or inner side of the elbow. This area is usually tender when touched and may be uncomfortable when gripping. In severe cases, almost any elbow movement can be uncomfortable.

Ganglion Cysts

Ganglion cysts, also known as Bible bumps, are more common in women, and 70% occur in people between the ages of 20 and 40. Ganglion cysts most commonly occur on the back of the hand (60–70%), at the wrist joint, and can also develop on the palm side of the wrist. When found on the back of the wrist, they become more prominent when the wrist is flexed forward. Other places, although less common, include these (Figure 13.26):

- The base of the fingers on the palm, where they appear as small pea-sized bumps
- The fingertip, just below the cuticle, where they are called mucous cysts
- The outside of the knee and ankle
- The top of the foot.

What Causes Ganglion Cysts The cause of ganglion cysts is not known. One theory suggests that trauma causes the tissue of the joint to break down forming small cysts, which then join into a larger, more obvious mass. The most likely theory involves a flaw in the joint capsule or tendon sheath that allows the joint tissue to bulge out.

Signs and Symptoms of Ganglion Cysts The ganglion cyst usually appears as a bump (mass) that changes size. It is usually soft, anywhere from 1 to 3 cm in diameter (about 0.4–1.2 in.) and does not move. The swelling may appear over time or appear suddenly, may get smaller in size, and may even go away, only to come back at another time. Most ganglion cysts cause some degree of pain, usually following acute or repetitive trauma, but up to 35% are without symptoms, except for appearance. The pain is usually nonstop, aching, and made worse by joint motion. When the cyst is connected to a tendon, one may feel a sense of weakness in the affected finger.

Thoracic Outlet Syndrome (TOS) – Neuro-Vascular Disorder Thoracic outlet syndrome is a combination of pain in the neck and shoulder, numbness and tingling of the fingers, and a weak grip. The thoracic outlet is the area between the rib cage and collarbone.

What Causes TOS TOS is a rare condition. Blood vessels and nerves coming from the spine or major blood vessels of the body pass through a narrow space near the shoulder and armpit on their way to the arms. As they pass by or through the collarbone (clavicle) and upper ribs, they may not have enough space. Pressure (compression) on these blood vessels or nerves can cause symptoms in the arms or hands. Problems with the nerves account for almost all cases of thoracic outlet syndrome.

Compression can be caused by an extra cervical rib (above the first rib) or an abnormal tight fibrous band connecting the spinal vertebra to the rib. People who suffer from TOS often have a history of injury to the area or overuse of the shoulder. TOS can be a repetitive stress injury. People with long necks and droopy shoulders may be more likely to develop this condition because of extra pressure on their nerves and blood vessels.

Signs and Symptoms of TOS Often symptoms are reproduced when the arm is positioned above the shoulder or extended. Individuals can have a wide range of symptoms from mild and intermittent, to severe and constant. Pains can extend to the fingers and hands, causing weakness. Symptoms of TOS may include the following:

- Pain, numbness, and tingling in the last three fingers and inner forearm
- Pain and tingling in the neck and shoulders (carrying something heavy may make the pain worse)
- Signs of poor circulation in the hand or forearm
- Weakness of the muscles in the hand.

Types of activities that can lead to TOS are those such as painting, sheet rocking, or overhead repairs on aircraft or vehicles.

LOWER LIMB

Lower limb WMSDs are currently a problem in many industries. "The epidemiology of these WMSDs has received until now modest awareness, despite this there is appreciable evidence that some activities (e.g., kneeling/squatting, climbing stairs or ladders, heavy lifting, walking/standing) are causal risk factors for their development." Other causes for acute lower limb WMSDs are related with slip and trip hazards (Laboratory, 2009). Despite the short awareness given to this type of WMSD they deserve significant concern, since they are often sources of high degrees of immobility and thereby can substantially degrade the quality of life (Laboratory, 2009). The most common lower limb WMSDs are (Laboratory, 2009) as follows:

- Hip and thigh conditions – osteoarthritis (most frequent), piriformis syndrome
- Trochanteritis, hamstring strains, sacroiliac joint pain
- Knee/lower leg – osteoarthritis, bursitis, beat knee/hyperkeratosis, meniscal lesions, patellofemoral pain syndrome, prepatellar tendonitis, shin splints, infrapatellar tendonitis, stress fractures
- Ankle/foot – achilles tendonitis, blisters, foot corns, hallux valgus (bunions), hammer toes, pes traverse planus, plantar fasciitis, sprained ankle, stress fractures, varicose veins, venous disorders.

Hip, Knees, Foot, and Ankle

Patellofemoral Pain Syndrome Patellofemoral syndrome is irritation to the underside of the patella where it meets the femur. Patellofemoral pain syndrome (PFPS) is a musculoskeletal disorder that has both work-related and nonwork-related origins. It is one of the least understood WMSDs and is often misdiagnosed.

The condition is generally caused by overuse and misuse of the knee joint, and risk increases with obesity or if the kneecap is not properly aligned. Women are more susceptible to PFPS, possibly due to the greater angle of the iliotibial band caused by the wider hips and shorter femur of the female anatomy. Kneecap alignment can be affected by unbalanced development of the quadriceps and iliotibial band muscle groups as well as tendon and ligament imbalances. These can be caused by performing repetitive work tasks that work or use the quadriceps and/or the iliotibial bands to a greater extent than the hamstrings. Forces put on the knee in sudden stopping and change of direction motions or change of direction motions while carrying weight, commonly seen in sports, also contribute to the condition. Changes in the cartilage under the kneecap, wearing down or becoming soft or rough can change the alignment between the kneecap and the tendons and bones of the upper and lower leg. The symptoms of the disorder include knee pain associated with extended sitting, squatting, jumping, and stair use (especially descending). The knee may at times "give out" and fail to support the body's weight and there may also be "popping" noises or a "grinding" feeling associated with movement of the joint.

Manual laborers experience the disorder due to overuse and misuse of the knee joint, namely, carrying loads and changing directions. Miners, construction workers, and carpenters, in general, report higher levels of all types of knee pain than people in other occupations. These types of professions involve frequent knee bending and heavy lifting and can also involve prolonged squatting and kneeling positions. People working at desk jobs may also experience the disorder (it is sometimes referred to as "theater-goers knee") because sitting for long periods of time with the knees hyperflexed causes the quadriceps muscles to pull on the kneecap, which can move it out of proper alignment over time. The hyperflexed knee seated position is an awkward posture. It involves placing the feet farther back than the knees nearly under the thighs, which puts the thighs at a downward angle from the torso and stretches the quadriceps, thus pulling the kneecap up toward the waist.

Treatment methods can sometimes cause PFPS to worsen, discouraging patients. Physical therapy can be effective, but relief is not always attributed to the therapy. Sometimes, it is the body's ability to heal that brings relief. NSAIDS can be a short-term relief but do nothing to get at the real cause of the pain. Figure 13.23 provides an explanation of the etiology of an ankle sprain.

Hip Bursitis Hip bursitis is a condition that causes pain over the outer part of the upper thigh. Bursitis occurs when the bursa, which is a small jelly-like sac that contains a small amount of fluid, becomes inflamed. There are a total of 160 bursae in the human body located around the major joints. Most bursae are found near the tendons around the large joints such as the shoulder joint, elbow joint, and hip and knee joint.

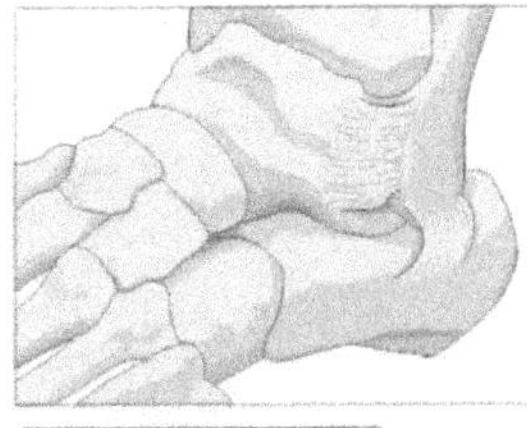

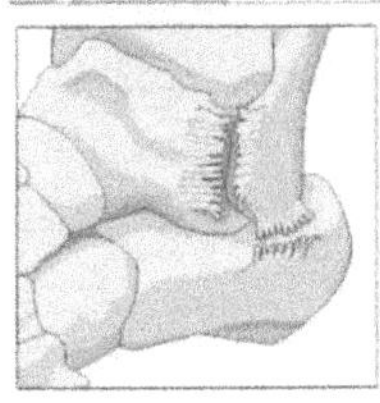

STRAINS

- A strain is caused by twisting or pulling a muscle or tendon.
- An acute strain is caused by trauma or an injury, or by improperly lifting heavy objects, or overstressing the muscles.
- A chronic strain is usually the result of overuse, prolonged, repetitive movement of the muscles and tendons.
- The usual signs and symptoms of a strain include pain, muscle spasm, and muscle weakness. Other symptoms and signs may include localized swelling, cramping, or inflammation, and possibly some loss of muscle function with a minor or moderate strain.

SPRAINS

- A sprain is an injury (a stretching or a tearing) to a ligament.
- Sprains often occur during sports or recreational activities.
- Sprains can occur in both the upper and lower parts of the body, but the most common site is the ankle.
- The usual signs and symptoms of a sprain include pain, swelling, bruising, and the
- loss of the ability to move and use the joint.

Figure 13.23 Ankle sprain (Adapted with permission by Lippincott Williams & Wilkins, Baltimore, MD; http://www.clipartof.com/download?txn=577e6995e66d8d9c7f0dc5d21e5f9aa2)

Bursae are fluid-filled sacs that cushion the joints of the body and allow joints, tendons, and muscles to glide over each other. They reduce friction and allow the human body to move without resistance. Bursae are generally healthy, but they can become inflamed and cause pain. When this occurs, it is known as bursitis. If the inflammation is not reduced, localized soft tissue trauma follows due to the increased stress between the tissues due to the added friction (Figure 13.24).

Bursitis is caused by the overuse of or direct trauma to the joint. Putting constant pressure on the hips can help contribute to this condition. This happens when workers are exposed to long durations of standing or sitting on hard surfaces. Awkward standing postures increase the risk to the worker, especially if the worker begins to favor one hip over the other. Receiving a direct, hard hit to a worker's hip can also higher the possibility of developing this WMSD. This can occur if an employee falls on their hip in the workplace. Repetitive motion of the hip joint is also a risk factor. An example of this is constantly climbing stairs or ladders. If a worker must climb a ladder while carrying a load, then there is a combination of risk factors; therefore, the risk of injury increases.

Knees

The knee is one of the most important joints of our body. It plays an essential role in movement related to carrying the body weight in horizontal (running and walking) and vertical (jumps) directions.

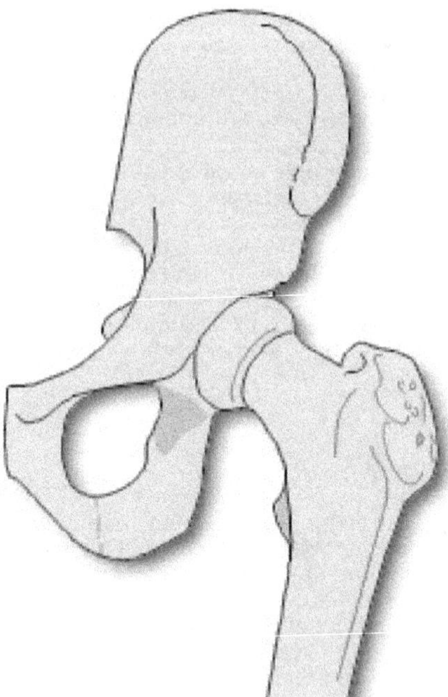

Figure 13.24 The hip joint (Adapted with permission Lippincott Williams & Wilkins, Baltimore, MD)

The knee joins the upper leg or the femur to the knee cap (patella), which is then joined to the lower leg or the tibia and fibula. Tendons connect the patella to the leg muscles, and ligaments stabilize the knee (Figure 13.25).

The anterior cruciate ligament (ACL) prevents the femur from sliding backward on the tibia (or the tibia sliding forward on the femur).

Two C-shaped pieces of cartilage called the medial and lateral menisci act as shock absorbers between the femur and tibia, and numerous bursae help the knee move smoothly.

Knee disorders, like other cumulative disorders of the body, build up over time through cumulative exposures. Knee disorders primarily consist of bursitis, meniscal lesions or tears, and osteoarthritis. Though kneeling and squatting are considered to be 2 of the primary risk factors correlated to these knee disorders, 12 other risk factors should also be contemplated. These 14 contributing risk factors include both occupational (extrinsic) and personal (intrinsic) variables that affect primarily the labor industries. Example industries include mining, construction, manufacturing, and custodial services where knee bending postural activities exist as a commonality. The risks are described in Table 13.2 (Reid, 2010).

ACL and MCL Injuries The knee is primarily stabilized by a pair of cruciate ligaments. The ACL stretches from the lateral condyle of femur to the anterior intercondylar area. The ACL is critically important because it prevents the tibia from being pushed too far forward relative to the femur. It is often torn during twisting

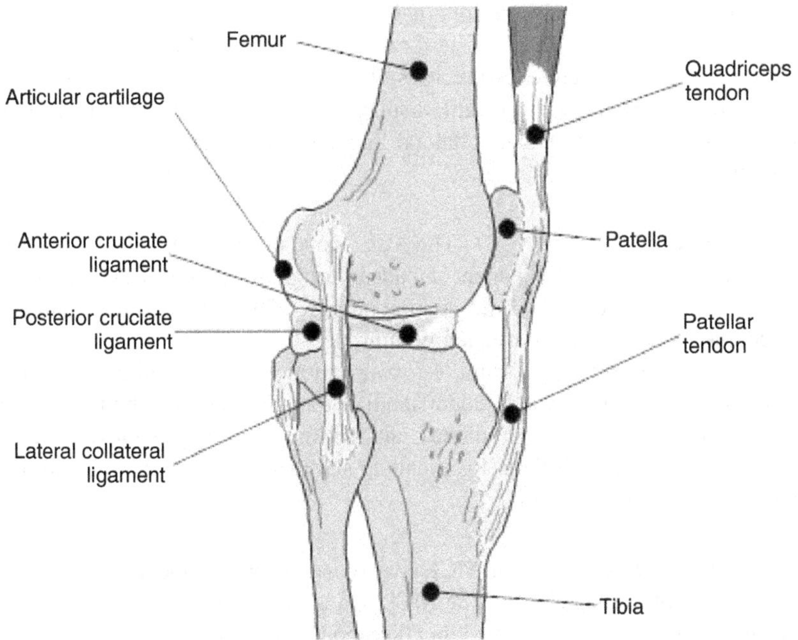

Figure 13.25 The complex knee (Adapted with permission Lippincott Williams & Wilkins, Baltimore, MD)

TABLE 13.2 Risk Factors for Knee Disorders

Risk	Extrinsic Risk	Intrinsic Risk	OA	Meniscal Disorders	Knee Bursitis
Kneeling	X		X	X	X
Squatting	X		X	X	
Crawling	X		X	X	
Stair/laddering climbing	X		X	X	
Lifting/carrying/moving	X		X	X	
Walking	X		X		
Standing up form a kneel/squat/crawl	X		X	X	
Chair sitting (while driving)	X			X	
BMI		X	X		
Past knee injury/surgery		X	X		
Age		X	X		
Using the knee as a hammer	X				X
Prolonged contact stress against the patella other than when kneeling	X				X
Physical intensive habits/hobbies that could affect the knee		X	X	X	

Source: Adapted from Reid (2010).

or bending of the knee. The posterior cruciate ligament (PCL) stretches from medial (middle) condyle of femur to the posterior intercondylar area. Injury to this ligament is uncommon but can occur as a direct result of forced trauma to the ligament. This ligament prevents posterior (backward) displacement of the tibia relative to the femur.

Tears of the ACL and medial collateral ligament (MCL) are two of the common knee injuries.

Anterior Cruciate Ligament (ACL) The ACL is one of four ligaments critical to the stability of the knee joint. A ligament is made of tough fibrous material and functions to control excessive motion by limiting joint mobility. Of the four major ligaments of the knee, the ACL is the most frequently injured.

The ACL is the primary restraint to forward motion of the shinbone (tibia). The anatomy of the knee is critical to understanding this relationship. The femur (thighbone) sits on top of the tibia (shinbone), and the knee joint allows movement at the junction of these bones. Without ligaments to stabilize the knee, the joint would be unstable and prone to dislocation.

Medial Collateral Ligament The MCL is also one of four ligaments that are critical to the stability of the knee joint. The MCL spans the distance from the end of the femur (thighbone) to the top of the tibia (shinbone) and is on the inside of the knee joint. The MCL resists widening of the inside of the joint or prevents "opening-up" of the knee.

Because the MCL resists widening of the inside of the knee joint, the MCL is usually injured when the outside of the knee joint is struck. This action causes the outside of the knee to buckle and the inside to widen. When the MCL is stretched too far, it is susceptible to tearing and injury. This is the injury seen by the action of "clipping" in a football game.

Ligament Injury Symptoms The most common symptom following a ligament injury is pain directly over the ligament. Swelling over the torn ligament may appear, and bruising, and generalized joint swelling is common 1–2 days after the injury. In more severe injuries, patients may complain that the knee is unstable, or feel as though their knee may "give out" or buckle.

Knee Osteoarthritis One of the most debilitating occupational knee disorders is knee osteoarthritis. This degenerative disease causes inflammation within the knee joint and is accompanied by cartilage rigidity and atrophy. The deformation and damage is exacerbated during movement and weight-bearing periods, which in turn affects additional joint structures within the knee. Inflammation, bone spurs, cartilage wear, narrowed joint space between the femur and tibia, and bone-to-bone contact are all signs of knee OA.

Osteoarthritis is the most common form of arthritis. It occurs when the protective cartilage on the ends of joint bones, most commonly in the hands, knees, hips, and/or spine, wears down over time. There is currently no cure for the disease; however, staying active, maintaining a healthy weight, and some medical treatments may slow progression of the disease and help improve pain and joint function.

Obesity is acknowledged by several studies as having an influence toward the development of knee OA.

Individuals that stand on their feet or work on hard surfaces seem to be more susceptible to both hip and knee OA. For example, janitors who are on their feet or bending frequently, road construction workers, carpet layers, and sheet metal workers all have an increased rate of OA. The constant standing on an unforgiving surface such as pavement or concrete seems to take a toll on their knees.

Knee Bursitis Although there are 11 bursae in the knee, according the Mayo Clinic Family Health Book, bursitis in the knee "most commonly occurs over the kneecap or on the inner side of the knee below the joint" (Litin, 2009).

Bursitis is the inflammation of a bursa sac accompanied by swelling (through fluid retention in the bursa sac) and thickening of the bursa walls. What is distinctive for knee bursitis is that its incidence is so common to certain industries that nicknames from those industries have been given. For example, in the coal mining industry, knee bursitis has come to be known as either "miner's knee" or "beat knee." In the carpet and floor laying industries, "carpet-layer's knee" is found in the literature. For house cleaning or custodial businesses, "housemaid's knee" is commonly spoken.

Two primary types of risk factors related to knee bursitis are using the knee as a hammer (sudden impact stress) and kneeling on or leaning against the knee (prolonged contact stress). Both types of stresses involve distributing forces through the knee to the knee bursae. Personal hygiene is considered to be intrinsic to individuals. Moore et al. (as cited in Reid, 2010) found that hair follicles on the skin of the knee that become infected can also possibly lead to knee bursitis. Proper hygiene by workers (such as cleaning and disinfecting knee pads as well as replacing worn ones) is a plausible mitigation strategy (Reid, 2010).

When bursa becomes inflamed due to trauma, symptoms can include swelling or tenderness around the knee when pressure is put on it. This can cause pain when moving or at rest and even cause mobility issues.

Foot and Ankle

Jobs that necessitate prolonged standing and walking activities are commonly associated with worker's complaints of foot and ankle pain. The foot and ankle are flexible structures of bones, joints, muscles, and soft tissues that let us stand upright and perform activities like walking, running, and jumping. The feet are divided into three sections:

- The forefoot contains the five toes (phalanges) and the five longer bones (metatarsals).
- The mid-foot is a pyramid-shaped collection of bones that form the arches of the feet. These include the three cuneiform bones, the cuboid bone, and the navicular bone.
- The hind-foot forms the heel and ankle. The talus bone supports the leg bones (tibia and fibula), forming the ankle. The calcaneus (heel bone) is the largest bone in the foot.

Muscles, tendons, and ligaments run along the surfaces of the feet, allowing the complex movements needed for motion and balance. The Achilles tendon connects the heel to the calf muscle and is essential for running, jumping, and standing on the toes. The Achilles tendon is the thickest and strongest tendon in the body. It is about 6 in. long.

The talus bone has a shiny joint surface covering that allows the ankle to glide effortlessly across the shiny undersurface of the **tibia**. When these two bones meet, they form the ankle joint. On the outside of the ankle, there is a smaller thin bone called the **fibula**. This bone helps prevent the ankle bone from shifting outward.

The stability of their ankle joint is dependent upon the ability of these bones to keep the central bone in place while the ankle moves back and forth. The joint is more stable when your foot is flat on the floor. The ankle is more rigidly held in place by the bony stabilizers of the fibula and malleolus because they are closer to the talus. However, when the toes are pointed, the ankle becomes unstable because the distance between the bony stabilizers of your ankle becomes larger. Thus, the ankle then relies more and more on the soft tissues including the ligaments to continue to provide stability (Figure 13.26). When an ankle twists, it is usually when the toes are pointed.

Injuries and Disorders of the Foot and Ankle Many foot and ankle disorders/injuries have common causes, which include the following:

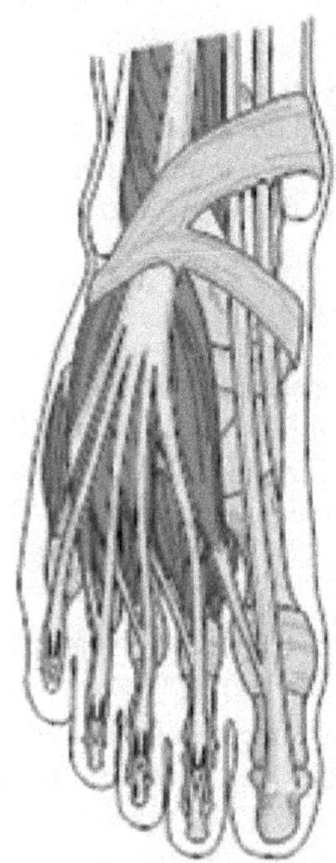

Figure 13.26 The structure of the foot is laden with crossing ligaments (Adapted with permission Lippincott Williams & Wilkins, Baltimore, MD)

- The feet roll inward too much when walking or standing (excessive pronation).
- The arch is high or flat.
- Walking, standing, or running for long periods of time, especially on hard surfaces.
- Wearing shoes with inadequate arch support.
- Being overweight.
- Placing high loads on the foot, for example, when jumping.

Plantar Fasciitis Plantar fasciitis is a common cause of heel pain. The plantar fascia is a flat ligament that connects the heel bone to the toes. Plantar fasciitis is caused by straining the ligament that supports the arch, which can lead to pain and swelling.

Bunion Bunion is an enlargement on the side of the foot near the base of the big toe (hallux). The enlargement is made up of a bursa (fluid-filled sac) under the skin. Bunions can be painful and can be aggravated by activity and wearing tight shoes.

Neuroma In the foot, a neuroma is a nerve that becomes irritated and swells up. If the nerve stays irritated, it can become thickened, which makes the nerve larger and causes more irritation. Pain from a neuroma is usually felt on the ball of the foot.

Corns Corns and callouses are areas of thick, hard skin. They usually develop due to rubbing or irritation over a boney prominence. The hard, thick skin is called a corn if it is on their toe and it is called a callous if it is somewhere else on the foot.

Toenail Fungus Fungi like a warm, moist, and dark environment (like inside a shoe). A fungal infection in your toenails may cause the nails to become discolored, thickened, crumbly, or loose. There are different causes and it is difficult to treat due to the hardness of the toenail.

Ingrown Toenail An ingrown toenail can occur for various reasons. The sides or corners of the toenail usually curve down and put pressure on the skin. Sometimes, the toenail pierces the skin and then continues to grow into the skin. This may cause redness, swelling, pain, and sometimes infection.

Hammer Toes A hammer toe is also sometimes referred to as a claw toe or mallet toe. It involves a deformity of the toe where there is an imbalance in the pull of the tendons. Either the tendon on top of the toe pulls harder or the tendon on the bottom of the toe pulls harder. This results in a curling up of the toe.

Plantar Warts Plantar warts are caused by a virus. Plantar means bottom of the foot, but warts can occur other places on the foot and toes as well. Plantar warts can be painful depending on where they are located. Sometimes, they are mistaken for calluses because layers of hard skin can build up on top of the wart.

Athlete's Foot Athlete's foot is a common skin condition that can affect everyone not just athletes. It is caused by a fungus. It may cause redness, itchiness, tiny bumps filled with fluid, or peeling skin. It is most commonly located between the toes or on the bottom of the feet.

Achilles Tendonitis Achilles tendonitis involves inflammation of the Achilles tendon. If the tendon stays inflamed long enough, it can lead to thickening of the tendon. Sometimes nodules or bumps can form in the tendon. Achilles tendonitis can become a long-term problem or can lead to rupture of the tendon.

SUMMARY

WMSDs are the unnecessary consequence of not fitting the task to the capabilities of the person. WMSDs currently account for more than one-third of all occupational injuries and illnesses, making them the largest job-related injury and illness problem in the United States today. WMSDs are injuries and disorders of the muscles, nerves, tendons, ligaments, joints, cartilage, and spinal discs that result from physically stressful activities and working conditions. WMSDs include carpal tunnel syndrome, sciatica, tendinitis, herniated spinal disc, and low-back strain.

WMSDs may result from the following:

- The physical demands of work, for example, including doing the same motion over and over again, maintaining the same position or posture while performing tasks, or sitting for a long time.
- The layout and condition of the workplace or workstation, for example, tasks that involve long reaches, working surfaces that are too high or too low, workstation edges or objects pressing into muscles or tendons, or equipment located in places that force the worker to assume awkward positions.
- Characteristics of objects handled, for example, the weight, size, center of gravity, and the equipment used to move heavy objects.
- Environmental conditions, for example, excessive exposure to cold temperatures while performing work tasks.

Ergonomics, as the science of fitting jobs to workers, seeks practical solutions that help prevent WMSDs in the workplace. The goal of ergonomics is to design office and industrial workstations, facilities, furniture, equipment, tools, and job tasks that are compatible with human dimensions, capabilities, and expectations with the ultimate goals of improved productivity, satisfaction, and decreased injuries and illnesses.

WMSDs are preventable if the warning signs and symptoms are acted upon and the physical workplace risk factors are changed (Figure 13.27).

KEY POINTS

WMSDs occur slowly over time and therefore can be reduced in frequency or severity with ergonomic improvements.

WMSDs can occur in any area of the body to any person.

Tip 1:
Warm up your muscles before stretching by walking or doing gentle movements.
Tip 2:
Don't bounce to intensify stretch, but instead slowly increase your stretch.
Tip 3:
Hold each position for about 20-30 seconds at stable resistance.
Tip 4:
Make sure to breathe slowly and rhythmically while stretching.
Tip 5:
Follow directions, for proper body positioning in important for an effective stretch.

Figure 13.27 Stretching for improved blood flow and flexibility (Adapted with permission Lippincott Williams & Wilkins, Baltimore, MD)

Increased blood flow helps the body heal.

REVIEW QUESTIONS

1. Is pain a WMSD?
2. What is one method for avoiding carpal tunnel syndrome?
3. Which soft tissue heals the fastest and why?

EXERCISE

1. Assign students a body region, for example, upper limb or spine. Have them research different WMSDs and discuss how working conditions can be casual. They can each provide a brief oral presentation to the rest of the class.

REFERENCES

Litin, S. (2009). *Mayo Clinic Family Health Book* 4th edn. Time Inc.

Reid, C. (2010). A Review of Occupational Knee Disorders. *Journal of Rehabilitation*, 489–501.

Vern Putz-Anderson. (1998) Cumulative Trauma Disorders., Taylor & Francis.

ADDITIONAL SOURCES

Ayub, A., Yale, S. H., Bibbo, C. Common Foot Disorders. *Clinical Medicine & Research*. 3(2), 2005, 116–119.

Center, U. o. (2013). Rehabilitation for Low Back Pain. Retrieved April 7, 2014, from University of Maryland Medical Center: http://umm.edu/programs/spine/health/guides/rehabilitation-for-low-back-pain.

Danis, C. G., Krebs, D. E., Gill-Body, K. M., Sahrmann, S. Relationship Between Standing Posture and Stability, *Journal of the American Physical Therapy Association*. 1998, 502–517.

Heintjes, E. (2004). Pharmacotherapy for Patellofemoral Pain Syndrome. *Cochrane Database System Reviews*, DC003470.

Ingraham, P. (2015). Save Yourself from Patellofemoral Pain Syndrome. Retrieved March 30, 2015, from Pain Science: https://www.painscience.com/tutorials/patellofemoral-pain-syndrome.php.

Institute, N. C. (2013). Pain Management Knee Pain. Retrieved January 15, 2015, from WedMD: http://www.webmd.com/pain-management/knee-pain/tc/patellofemoral-pain-syndrome-topic-overview.

Jindal, K. (2015). Knee Pain When Sitting at the Theatre. Retrieved February 6, 2015, from Ask the doctor: https://www.askthedoctor.com/knee-pain-when-sitting-at-the-theatre-15606.

Kendal, F. P., McCreary, E. K., Provance, P. G., Rodgers, M. M., Romani, W. A. (2005). *Muscles Testing and Function with Posture and Pain* 5th edn. Baltimore, MD: Lippincott Williams & Wilkins, pp. 49–65

Marieb, E. N. (2001). *Human Anatomy and Physiology*, San Francisco, CA: Person Education Inc., Publishing as Benjamin Cummings, pp. 215–216

National Library of Medicine. (2011). *National Library of Medicine – Medical Subjects Headings*. Retrieved April 18, 2013, from National Library of Medicine Musculoskeletal System: http://www.nlm.nih.gov/cgi/mesh/2011/MB_cgi?mode=&term=Musculoskeletal+System

Nunes, I. L. and Bush, P. M. Retrieved October 10, 2014, from: Work-Related Musculoskeletal Disorders Assessment and Prevention.

Okunribido, O. (2009). *Lower Limb MSD – Scoping Work to Help Inform Advice and Research Planning (RR706)*. Colegate, Norwich: Health and safety Executive.

Steilen, D. (2015). Knee Pain Patellofemoral Instability. Retrieved April 1, 2015, from Caring Medical Regenerative Medicine Clinics: http://www.caringmedical.com/prolotherapy-news/patellofemoral-instability-pain/.

The Merck Manual. (2007). Retrieved May 3, 2011, from Muscles: Biology of the Musculoskeletal System: http://www.merckmanuals.com/home/bone_joint_and_muscle_disorders/biology_of_the_musculoskeletal_system/introduction_to_the_biology_of_the_musculoskeletal_system.html.

Webster's New World Medical Dictionary. Definition of Neutral Posture.

14

HOW TO CONDUCT AN ERGONOMIC ASSESSMENT AND ERGONOMIC ASSESSMENT TOOLS

LEARNING OBJECTIVE

At the end of this module, you will be able to conduct an ergonomics assessment to include selecting and understanding the use of the applicable evaluation method for the task.

INTRODUCTION

An ergonomics assessment of a job or task leads one through the anticipation, recognition, evaluation, and prioritization and corrective action phases and directs their efforts toward the most effective improvements.

There are simple and complicated methods for the analysis of physical workplace risk factors but can be generalized into four categories:

- Checklist
- Interactive form based
- Observational
- Direct measurement.

No single assessment tool is perfect; however, some assessment tools can be of value to the ergonomist in the field (Marras & Karwowski, 2006). This chapter focuses on observational methods but urges the practitioner to consider using biomechanical modeling software as well as direct reading noise, temperature, and light meters where appropriate.

Occupational Ergonomics: A Practical Approach, First Edition.
Theresa Stack, Lee T. Ostrom and Cheryl A. Wilhelmsen.

BACKGROUND

Checklist

- Pros: easy to use, special equipment not required, reminders
- Cons: easily misused, observer bias, not flexible.

Interactive Form Based

- Pros: can be validated, potential respondent anonymity, sample large population quickly
- Cons: responder bias, costly to validate.

Observational

- Pros: easy to use, flexible, can be validated
- Cons: observer bias, observer presence can alter environment.

Direct Measurement

- Pros: can be validated, reliable
- Cons: equipment can be costly, test environment may represent true working conditions, and equipment presence can alter work methods.

PREPARING FOR THE SURVEY

Conducting an observational assessment usually requires the following 13 steps. It is very similar to conducting a job hazard analysis (JHA) or job safety analysis (JSA). The evaluator should become familiar with the technique they intend to employ and practice using the tool before entering the work site.

Step One: Select the Job or Task for Evaluation

- Review medical records or injury logs
- Query workers and supervisors on the following:
  - Which tasks are the most awkward or require the most uncomfortable postures?
  - Which tasks take the most effort?
  - Which tools or pieces of equipment are notoriously hard to work with?
  - Have there been changes in processes or procedures?
- Visually assess the jobs or tasks, looking for the ergonomics risk factors that occur for durations longer than 4 h/day and as frequently as three times per week.

Step Two: In Preparation of the Site Visit

The appropriate office should be notified, for example, industrial hygiene, safety, or occupational medicine. It can also be helpful to invite someone from the respective office to accompany them on the site visit. After the visit, these same personnel may be integral to implementing interventions. Select a time that is representative of the task or job they expect to evaluate and plan to spend most of the shift (4–6 h) in the work area.

Step Three: Become Knowledgeable About the Area You Plan to Enter

Obtain permission from the supervisor to enter the area and speak with employees so that everyone understands the purpose, process, and expected outcome. When determining the sample size, generally the rules of industrial hygiene apply and one tries to sample the square root of the number of operators that fall into a similar exposure group. Alternatively, find the most and least experienced operator, conduct the survey and compare results.

Research industries with similar jobs or tasks, looking for best practices or lessons learned. Become familiar with the process by reading standard operating procedures, taking an initial walk through, or discussions with management.

Step Four: Gather the Appropriate Tools for the Assessment

- Measuring tape
- Video or digital camera
- Audio recorder
- Notepad and writing utensil
- All personal protective equipment required for the area and task to be evaluated
- Appropriate checklist or assessment tool
- Light meter
- Force/weight measuring device
- Be sure to calibrate equipment pre- and postsampling. Note that some survey methods such as RULA/REBA and HAL are best employed with the evaluator calibrated.

Step Five: Gather Data

Begin by introducing yourself and explaining the purpose of the site visit. Explain what ergonomics is if the employees and supervisor are not knowledgeable. Explain that the purpose of the survey is to help the workers and this is not an inspection.

Remember, it can be difficult for workers to accept, at first, a suggested change if they have been performing the task the same way for years. Be understanding and make suggestions instead of demands. If an employee is reluctant to change, ask them to try it another way for a few days.

Obtain job or task information, which is primarily accomplished through the interview of the supervisor or designated escort.

- Job name, department, point of contact information
- Shift length

- Production standards
- Rotation schedule
- Number of employees (level of turnover/absenteeism)
- Basic population demographics (age/gender)
- Workplace injuries
- Required personal protective equipment
- Is there seasonal work or fluctuations in production?
- Production rates and quotas
- Is there job rotation?
- Number and frequency of rest breaks
- Process improvement suggestions
- Total exposure time at each subtask or workstation.

Gather user data by interviewing employees.

- Length of time working at job
- Ask the operator to walk them through the tasks.
- Any pain or discomfort associated with the job?
- Number and duration of breaks
- Involvement in health and safety committees
- What are the hardest tasks associated with the job?
- What tools or pieces of equipment are notoriously difficult to work with?
- Previously diagnosed work-related musculoskeletal disorders (WMSDs) or are you under a physicians care for a WMSD?
- Any suggestions for process improvement?
- If they had unlimited funds what would they change?

Do not interfere with the normal process or discuss personnel issues. Avoid leading questions about pain or discomfort. Typically, answers that are more accurate are gathered when workers are not interviewed in groups.

Perform the task analysis by first observing the entire work area and then dividing the major tasks into subtasks. Alternatively, use the information collected from the interviews on where to focus the assessment. Reference assessment tools for specific procedures related to the tool of choice. Each subtask is typically examined for the determination of exposure to ergonomics risk factors. For example:

- Job constraints
- Risk factors (including bending, stretching, twisting, static loads, and maintaining undesirable postures)
- Task associated with the job and frequency
- Measure
  - Part and tool weights
  - Grip forces and push/pull forces

  - Work heights
  - Reach distances
  - Carrying distances
  - Seat heights
  - Cart location
  - Product dimensions and weights
- What tools are being used? Reference hand tools
- Draw a diagram of the process flow
- Employ the proper survey tool
- Note organization of work surfaces and process flow
  - Is there wasted movement or material handling?

Video recordings or digital photographs are an important piece of documentation. Video recording is an opportunity to gather data, document and measure postures, cycle time, and work methods. Video tape also captures the difficulty of a task, is an audio record, can be played back in slow motion and can be used as a training tool. Some video and photographic considerations are as follows:

- Record the process from a variety of angles (front, back, right and left side, overhead)
- Record at least two of five angles
- Record both wide angle and close-up
- If feasible, record at least three cycles of a task
- Carry backup supplies:
  - Batteries, film, and memory cards

Upon conclusion of the data gathering:

- Thank everyone for their time and cooperation.
- Inform them of the expected outcome (e.g., report).
- Obtain contact information for follow-up questions.
- Brief appropriate personnel on the findings (in general) and when they can expect a copy of the outcome.
- If time permits, conduct a brainstorming session with employees, OSH or IH personnel, querying their suggestions for task/job improvements.

Step Six: Analyze the Data

Analyze the data to determine if the risk factors are present in a sufficient dose to increase the risk of injury. Incorporate the findings from all the methods employed during the data gathering effort. The assessment tools section contains methods to determine risk prevalence. If possible, brainstorm solutions with others during the report writing stage. An example risk assessment report can be found in the Exercises Appendix section. Report elements can include the following:

- History/background
  - Why was the survey requested?
  - Injury history.
    - Cost figures – actual paid and potential costs.
- Workplace Summary
  - Task summary
  - Demographics:
    - Number of workers
    - Age and gender mix
    - Reports of discomfort
- Survey findings
  - Summarize the methods used to identify the risk factors.
  - Summarize risk factor exposure.
  - Illustrate with photography or video stills.
  - Include the results of the Ergonomic Screening tool, Risk Factor Follow-on checklist; JR/PD or other evaluation method.
  - Discuss consequence of current design.
- Recommendations
  - Short- and long-term solutions.
  - Details of each recommendation and how it will address the identified problems.
  - Include diagrams of workstation or work flow suggestions.
  - Provide sources for tools/equipment changes.
  - Detail administrative controls and training recommendations.
  - Cost summary for short- and long-term recommendations.
  - Cost/benefit and payback projections.
  - Provide assumptions and relate to current workers' compensation, productivity, and/or mission's readiness issues.

Step Seven: Distribute Report

Assist with implementation of recommendations and follow-up.

AMERICAN CONFERENCE OF GOVERNMENTAL INDUSTRIAL HYGIENISTS THRESHOLD LIMIT VALUE (ACGIH TLV) FOR LIFTING

Introduction

The American Conference of Governmental Industrial Hygienists (ACGIH) has developed a threshold limit value (TLV) for lifting. A TLV is defined as workplace lifting conditions under which it is believed nearly all workers may be exposed repeatedly, day after day, without developing work-related lower back and shoulder disorders associated with repetitive lifting tasks.

The ACGIH TLV is a quick and easy tool to assess lifting tasks. The results can also direct the user to job redesign strategies. The information was adapted with permission.

Limitations

When using tables based on psychophysical data, some limitations apply. The tables are based on a realistic, and valid, assessment of what a worker feels he or she can tolerate. It is questionable whether a subject can anticipate how much can be tolerated over the long term without incurring an injury. Therefore, a psychophysical approach may underestimate the actual level of risk.

In addition, the tables assume good coupling between the load and the hands and between the feet and the floor, a two-handed handling, unrestricted posture, less than 30° of twisting, and the task cannot exceed 8 h/day or 360 lifts/h.

Applying the TLV

The tool consists of three charts; the chart used determines the TLV. The charts are a function of lifting duration (less than or greater than 2 h a day) and lifting frequency (how many average lifts). Then based on one of the four vertical zones (refer to Figure 14.1) and one of the three horizontal zones, a TLV for weight is given. In some cases, the chart indicates that there is no known safe limit for repetitive lifting under those conditions.

Step one: Determine the duration of the lifting exposure. Is the cumulative dose more than or less than 2 h/day?

Step two: Determine the frequency of lifting. How many lifts are performed in an hour or in a minute?

Step three: Locate the corresponding chart. Chart one is for infrequent lifting (Table 14.1), chart two for moderate lifting (Table 14.2), chart three for highly repetitive lifting (Table 14.3).

Typically, the average lifting scenario is chosen. For example, a worker unloads five delivery trucks a day; each truck has approximately 50 packages. It takes the worker 1 h to unload each truck. The duration of the exposure is 5 trucks × 1 h = 5 h. The frequency of lifting is 50 packages × 5 trucks = 250 packages in 5 h. 250 packages/5 h = 50 lift hours.

For this scenario, 5 h/day, 50 lifts/h, chart 3 is chosen.

Step four: Determine the horizontal zone at the start of the lift, reference Figure 14.1.

Step five: Determine the vertical zone at the start of the lift, reference Figure 14.1.

Step six: Locate the corresponding weight limit, this is the unadjusted TLV.

Step seven: Use your professional judgment to adjust (reduce) the TLV if any of the following conditions exist:

- High-frequency lifting >360 lifts/h (6 lifts/min)

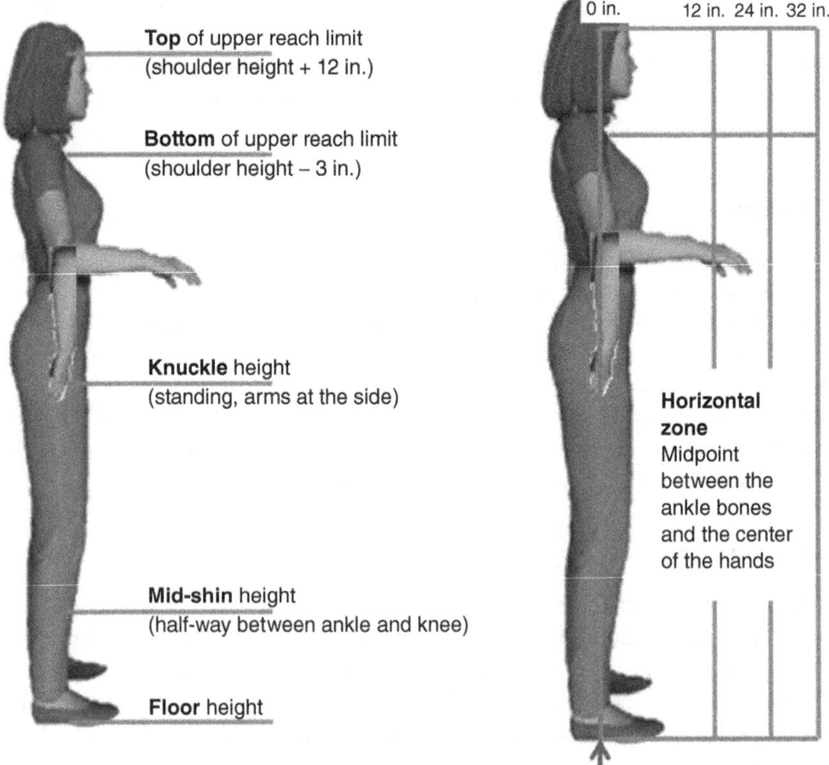

Figure 14.1 Vertical and horizontal lifting zone (Adapted with permission from Ergonomic Image Gallery)

- Lifting performed for longer than 8 h/day
- High asymmetry: lifting more than 30°
- One-handed lifting
- Constrained lower body posture, such as lifting while seated or kneeling
- Poor hand coupling: lack of handles, cutouts or other grasping points
- Unstable foot (e.g., inability to support the body with both feet while standing)
- During, or immediately after exposure to whole-body vibration at, or above the TLV for whole-body vibration.

Step eight: If the load is placed at the destination in a controlled manner, repeat steps four through seven for those parameters.

Step nine: If the recommended weight limit is exceeded, use the TLV tables (changing object location, handling frequency, or duration) to bring the task into a less hazardous environment. Refer the Manual Material Handling module and the Administrative Controls module for possible solutions.

TABLE 14.1 TLV for low frequency lifting tasks

≤2 h/day with ≤60 lifts/h or >2 h/day with ≤12 lift/h

Vertical Zone	Horizontal Zone[a]		
	Close<30 cm (12 in.)	Intermediate30–60 cm (12–24 in.)	Extended[b]>60–80 cm (24–32 in.)
Reach limit[c] or 30 cm (12 in.) above shoulder to 8 cm (3 in.) below shoulder	16 kg (35 lb)	7 kg (15 lb)	No safe limit for repetitive lifting[d]
Knuckle height[e] to below shoulder	32 kg (71 lb)	16 kg (35 lb)	9 kg (20 lb)
Middle shin to knuckle height[e]	18 kg (40 lb)	14 kg (30 lb)	7 kg (15 lb)
Floor to middle shin height	14 kg (31 lb)	No known safe limit for repetitive lifting[d]	No known safe limit for repetitive lifting[d]

Source: From ACGIH®, 2013 TLVs® and BEIs® Book. Copyright 2013. Reprinted with permission.

[a]Distance from midpoint between inner ankle bones and the load.

[b]Lifting tasks should not be started at a horizontal reach distance more than 80 cm (32 in.) to the midpoint between the inner ankle bones (Figure 14.1).

[c]Routine lifting tasks should not be conducted from starting heights greater than 30 cm (12 in.) above the shoulder or more than 180 cm above floor level (Figure 14.1).

[d]Routine lifting tasks should not be performed for shaded table entries marked "No known safe limit for repetitive lifting." While the available evidence does not permit identification of safe weight limits in the shaded regions, professional judgment may be used to determine if infrequent lifts of light weights may be safe.

[e]Anatomical landmark for knuckle height assumes the worker is standing erect with arms hanging at the sides.

LIBERTY MUTUAL MANUAL MATERIALS HANDLING GUIDELINES

Introduction

The complete Liberty Mutual Manual Materials Handling Guidelines manual is found at (https://libertymmhtables.libertymutual.com/CM_LMTablesWeb/taskSelection.do?action=initTaskSelection).

The goals of the tool are as follows:

- Providing an objective risk assessment of manual handling tasks such as lift, lower, push, pull, and carrying tasks
- Controlling costs associated with manual handling tasks.

Background

- Tables based on psychophysical research by Dr Snook and Dr Ciriello with the LM Research Institute for Safety (Snook & Ciriello, 1991)
- In the Liberty Mutual studies, the worker is given control of the weight of the object being handled and all other variables frequency, size, height, distance,

TABLE 14.2 TLV for moderate frequency lifting tasks

≤2 h/day with >60 and ≤360 lifts/h Or >2 h/day with >12 and ≤30 lifts/h

Vertical Zone	Horizontal Zone[a]		
	Close <30 cm (12 in.)	Intermediate 30–60 cm (12–24 in.)	Extended[b] >60–80 cm (24–32 in.)
Reach limit[c] or 30 cm (12 in.) above shoulder to 8 cm (3 in.) below shoulder	14 kg (31 lb)	5 kg (11 lb)	No safe limit for repetitive lifting[d]
Knuckle height[e] to below shoulder	27 kg (60 lb)	14 kg (31 lb)	7 kg (15 lb)
Middle shin to knuckle height[e]	16 kg (35 lb)	11 kg (24 lb)	5 kg (11 lb)
Floor to middle shin height	9 kg (20 lb)	No known safe limit for repetitive lifting[d]	No known safe limit for repetitive lifting[d]

Source: From ACGIH®, 2013 TLVs® and BEIs® Book. Copyright 2013. Reprinted with permission.
[a]Distance from midpoint between inner ankle bones and the load.
[b]Lifting tasks should not be started at a horizontal reach distance more than 80 cm (32 in.) to the midpoint between the inner ankle bones (Figure 14.1).
[c]Routine lifting tasks should not be conducted from starting heights greater than 30 cm (12 in.) above the shoulder or more than 180 cm above floor level (Figure 14.1).
[d]Routine lifting tasks should not be performed for shaded table entries marked "No known safe limit for repetitive lifting." While the available evidence does not permit identification of safe weight limits in the shaded regions, professional judgment may be used to determine if infrequent lifts of light weights may be safe.
[e]Anatomical landmark for knuckle height assumes the worker is standing erect with arms hanging at the sides.

etc. are controlled by the researcher. The worker monitors his/her feelings of exertion or fatigue and adjusts the weight of the object accordingly.

- The psychophysical methodology included measurements of O_2 consumption, heart rate, and anthropometrics.

Theory Behind Tables

- Designing manual material handling for greater than 75% of the female work population will offer protection from manual handling injuries.
- Tasks that have "population percentages of 10%" (only capable by less than 10% of the population) will most likely lead to injury.
- The tables can be used to perform "what-if scenarios."

Benefits of the Tables

- Simple to use.
- Recommended for moderate frequency tasks. Physiological models recommended for high-frequency tasks and biomechanical models for lifting tasks with lower frequency and heavier weights.

TABLE 14.3 TLV for high frequency and long duration lifting tasks

>2 h/day with >30 and ≤360 lifts/h

Vertical Zone	Horizontal Zone[a]		
	Close <30 cm (12 in.)	Intermediate 30–60 cm (12–24 in.)	Extended[b] >60–80 cm (24–32 in.)
Reach limit[c] or 30 cm (12 in.) above shoulder to 8 cm (3 in.) below shoulder	11 kg (24 lb)	No known safe limit for repetitive lifting[d]	No safe limit for repetitive lifting[d]
Knuckle height[e] to below shoulder	14 kg (31 lb)	9 kg (21 lb)	5 kg (11 lb)
Middle shin to knuckle height[e]	9 kg (20 lb)	7 kg (15 lb)	2 kg (4 lb)
Floor to middle shin height	No known safe limit for repetitive lifting[d]	No known safe limit for repetitive lifting[d]	No known safe limit for repetitive lifting[d]

Source: From ACGIH®, 2013 TLVs® and BEIs® Book. Copyright 2013. Reprinted with permission.
Source: Reproduced with permission from the American Conference of Governmental Industrial Hygienist Threshold Limit Value/Biological Exposure Indices Guide Book 2013.

[a]Distance from midpoint between inner ankle bones and the load.

[b]Lifting tasks should not be started at a horizontal reach distance more than 80 cm (32 in.) to the midpoint between the inner ankle bones (Figure 14.1).

[c]Routine lifting tasks should not be conducted from starting heights greater than 30 cm (12 in.) above the shoulder or more than 180 cm above floor level (Figure 14.1).

[d]Routine lifting tasks should not be performed for shaded table entries marked "No known safe limit for repetitive lifting." While the available evidence does not permit identification of safe weight limits in the shaded regions, professional judgment may be used to determine if infrequent lifts of light weights may be safe.

[e]Anatomical landmark for knuckle height assumes the worker is standing erect with arms hanging at the sides.

- Provides male and female population percentages able to perform the job without becoming unusually tired, weakened, overheated, or out of breath. Encompasses lifting, lowering, pushing, pulling, and carrying tasks.

Limitations to the Tables

There are a number of other considerations besides population percentage that the tables do not address that must be considered in task assessment including the following:

- Injuries
- Frequent bending
- Frequent twisting
- Frequent reaching(horizontal or increasing hand distance away from body and vertical hand distance above shoulder height)

- One-handed lifts
- Note: The containers used by subjects in the lab had handholds.
- Catching or throwing items.

Applying the Tables

Example 1 As mentioned, the tables are relatively self-explanatory. Given the following scenario, this example will illustrate the steps required to determine the population percentage for the task:

You observe the following task in a manufacturing plant lowering air compressor parts. The task has the following characteristics:

- Object weight: *45 lb*
- Hand distance: *10 in. from body*
- Initial hand height: *24 in. off the walking surface*
- Ending hand height: *6 in. off the walking surface*
    - Therefore, the lowering distance is (24–6 in.) = *18 in.*
    - Lifting frequency: *Every 30 s*

Step one: Find the correct male and female table that corresponds to the following:

a. Type of task (lifting, lowering, pushing, pulling, etc.)
b. Beginning/ending position of the hands and location
    - OR

Type of force applied (initial or sustained).

Given this information, select the appropriate table from the Table of Contents page:

Therefore, the correct table is "4M and 4F."

Step two: Find the appropriate male population percentages by entering the information into the websites calculator (Figures 14.2 and 14.3):

Step three: Find the appropriate female population percentage by entering the information into the website calculator (Figure 14.4).

The population percentage for this task is determined to be acceptable within the range of less than 10–57%.

As a general rule of thumb, designing manual tasks for greater than 75% of the female work population will offer the best protection from manual handling injuries. Studies have shown that two-thirds of low back claims from low percentage tasks (tasks capable of being performed by a small percentage of the population) can be prevented if the tasks are designed to accommodate at least 75% of the female work population (Snook et al., 1978). However, whenever manual handling and deep bending are combined, this significantly increases the risk of manual handling injuries. Therefore, additional protection for workers is recommended. When workers must bend noticeably during any lifting or lowering task, designing these manual tasks for

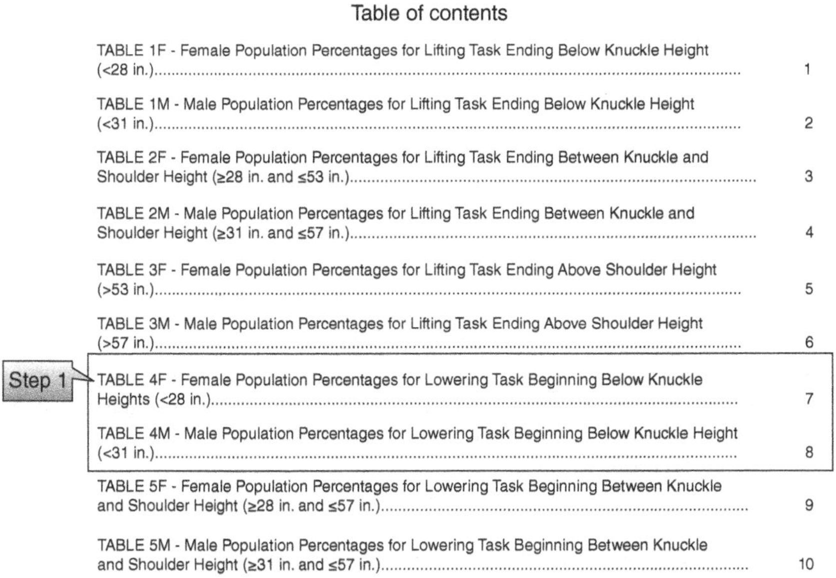

Table of contents

TABLE 1F - Female Population Percentages for Lifting Task Ending Below Knuckle Height (<28 in.)	1
TABLE 1M - Male Population Percentages for Lifting Task Ending Below Knuckle Height (<31 in.)	2
TABLE 2F - Female Population Percentages for Lifting Task Ending Between Knuckle and Shoulder Height (≥28 in. and ≤53 in.)	3
TABLE 2M - Male Population Percentages for Lifting Task Ending Between Knuckle and Shoulder Height (≥31 in. and ≤57 in.)	4
TABLE 3F - Female Population Percentages for Lifting Task Ending Above Shoulder Height (>53 in.)	5
TABLE 3M - Male Population Percentages for Lifting Task Ending Above Shoulder Height (>57 in.)	6
Step 1 → TABLE 4F - Female Population Percentages for Lowering Task Beginning Below Knuckle Heights (<28 in.)	7
TABLE 4M - Male Population Percentages for Lowering Task Beginning Below Knuckle Height (<31 in.)	8
TABLE 5F - Female Population Percentages for Lowering Task Beginning Between Knuckle and Shoulder Height (≥28 in. and ≤57 in.)	9
TABLE 5M - Male Population Percentages for Lowering Task Beginning Between Knuckle and Shoulder Height (≥31 in. and ≤57 in.)	10

Figure 14.2 Liberty mutual table guide for different populations and lifting heights (Adapted from https://libertymmhtables.libertymutual.com/CM_LMTablesWeb/taskSelection.do?action=initTaskSelection, Copyright 2005 Liberty Mutual Insurance, used with permission)

Male - Lifting Task Ending Below Knuckle Height (<31")

* Indicates Required Entry

* Object Weight (Pounds):	45
* Hand Distance:	10 inches
* Lifting Distance:	20 inches
* Frequency One Lift Every:	30 seconds
	Calculate
Population Percentage:	57%

Figure 14.3 Liberty mutual table guide for step two male population percentages (Adapted from https://libertymmhtables.libertymutual.com/CM_LMTablesWeb/taskSelection.do?action=initTaskSelection)

greater than 90% of the female work population will offer a more appropriate level of protection from manual handling injuries.

Finally, for certain jobs and industries, it is very difficult to design jobs that can be performed by 75% or greater of the female work population. These tables are very often used to perform what-if scenarios of various ergonomic interventions to help determine the most cost-effective and practical solution that offers the highest degree

Female - Lifting Task Ending Below Knuckle Height (<28")

* Indicates Required Entry

* Object Weight (Pounds): 45

* Hand Distance: 10 inches

* Lifting Distance: 20 inches

* Frequency One Lift Every: 30 seconds

Calculate

Population Percentage: Less Than 10%

Figure 14.4 Liberty mutual table guide for step three female population percentages Step four Interpret results (Adapted from https://libertymmhtables.libertymutual.com/CM_LMTablesWeb/taskSelection.do?action=initTaskSelection, Copyright 2005 Liberty Mutual Insurance, used with permission)

of control. There is no right answer or wrong solution. Whatever solution offers the most practical, cost-effective and highest degree of control possible is a good result.

PHYSICAL RISK FACTOR CHECKLIST

Introduction

The Physical Risk Factor Checklist is a tool used to identify and evaluate ergonomics stressors. The checklist was adapted from the Washington State Labor and Industries hazard and caution zone checklists and is used by the US Navy and can be found in OPNAVINST 5100.23 (series) Chapter 23 Ergonomics Program.

The Physical Risk Factor Checklist (Figure 14.5) examines individual body areas for risk based on posture, force, duration, and frequency.

Applying the Checklist

Start by identifying the ergonomics stressors, either through direct observation, or watching a videotape play back. Watching a video in slow motion enables greater accuracy.

This checklist is used for typical work activities that are a regular and foreseeable part of the job, occurring more than 1 day/week, and more frequently than 1 week/year.

The Physical Risk Factor Ergonomics Checklist is a tool used to identify physical stressors in the workplace. For each category determine whether the physical risk factors rate as a "caution" or "hazard" by placing a check (.) in the appropriate box. Make a notation if a category is not applicable.

If a hazard exists, it must be reduced below the hazard level or to the degree technologically and economically feasible. Ensure workers exposed to ergonomics stressors

			Caution		Hazard
	1. Working with the hand(s) above the head, or the elbow(s) above the shoulders	☐	More than 2 h total per day	☐	More than 4 h total per day
	2. Repeatedly raising the hands(s) above the head, or the elbow(s) above the shoulder(s) more than once per minute	☐	N/a	☐	More than 4 h total per day
	3. Working with the neck bent (without support and without the ability to vary posture)	☐	More than 30° for more than 2 h total per day	☐	More than 30° for more than 4 h per day More than 45° for more than 2 h total per day

Figure 14.5 Physical risk factor checklist that can be found in Appendix C (Adapted from OPNAVINST 5100.23 (G) Chapter 23, Appendix A)

at or above the "hazard" level have received general ergonomics training and provide a refresher of the ergonomics physical and contributing risk factors.

If the task rates a "caution," reevaluate at least yearly since changes in the work environment may create new ergonomics stressors.

Reduce to the lowest level feasible significant contributing physical risk factors and consider contributing personal risk factors in evaluation. Contributing risk factors contribute to but do not cause WMSDs. Physical contributing risk factors may include temperature extremes, inadequate recovery time, and stress on the job. Personal contributing risk factors may include but are not limited to age, pregnancy, obesity, thyroid disorder, arthritis, diabetes, or preexisting injuries such as wrist/knee/ankle strain or fracture, back strain, trigger finger, and carpal tunnel syndrome. Professional judgment should be used in instances where personal or physical contributing factors are present. The risk of developing a WMSD increases when ergonomics risk factors occur in combination. See the complete physical risk factor checklist in Appendix C of this manual.

AMERICAN CONFERENCE OF GOVERNMENTAL INDUSTRIAL HYGIENISTS THRESHOLD LIMIT VALUE FOR HAND ACTIVITY LEVEL (ACGIH HAL)

Introduction

ACGIH has developed an action limit (AL) for hand-arm activity levels (HAL) and a TLV. The TLV and AL are based on the peak hand force and the hand/arm activity level for a given task. Peak hand force can be measured quantitatively (i.e., strain gauge, biomechanical analysis), or qualitatively (observer or worker ratings, strength based on anthropometric percentiles). HAL can also be measured quantitatively (i.e., calculations based on the frequency of exertions and the work/recovery ratio), or qualitatively (worker or observer ratings). This chapter focuses on qualitatively. This material was adapted with permission.

Limitations and Benefits

The ACHIG TLV for HAL is designed to analyze monotasks. HAL is most valuable for jobs with repeatable steps that involve performing a similar set of motions or exertion repeatedly, for four or more hours per day. A few examples are working on an assembly line and using a keyboard and mouse.

Applying the Survey

HAL is a combination of average HAL and peak hand force (P_f) that indicate (from available data) when hand, wrist, and/or forearm WMSDs are as follows:

- Possible in many people
- Possible in some people
- Unlikely in most people.

Step one: The first step is observing at least three complete task cycles. A useful tool for collecting data can be found in Table 14.4. Disregard irregular or spurious actions and then determine the HAL using the observer ratings on a scale of 1–10 with anchors detailed in Figure 14.6. The scale is a modified Borg Scale. Compare the task with known examples as necessary. The rating is an average for the entire work cycle. Rate what you see. Ratings can then be adjusted for other factors, for example, increased or decreased production rates, and force. A data collection sheet can be used as a guide when gathering data.

TABLE 14.4 HAL and NPF Data Collection Sheet

Date:
Job:

	Left	Right
Hand-arm activity level (HAL)		
Refer to scale		
Normalized peak force (NPF)		
Refer to scale		
Ratio = NPF/(10 – HAL)		
Determine results	□ >TLV	□ >TLV
TLV = 0.78	□ AL to TLV	□ AL to TLV
AL = >0.56	□ <AL	□ <AL

Low		Medium		High	
0 Hands idle most of the time; no regular exertions	2 Consistent conspicuous long pauses; or very slow motions	4 Slow steady motion/ exertion; frequent brief pauses	6 Steady motion/ exertion infrequent pauses	8 Rapid steady motion/ exertion; infrequent pauses	10 Rapid steady motion or continuous exertion; difficulty keeping up

Figure 14.6 HAL-observed rating scale (From ACGIH®, 2013 TLVs® and BEIs® Book. Copyright 2013. Reprinted with permission)

TABLE 14.5 Estimation of Normalized Peak Force for Hand Activity Level

% MVC	Subjective Scale		Moore–Garg Observer Scale (Alternative Method)
	Score	*Verbal Anchor*	
0	0	Nothing at all	
5	0.5	Extremely weak (just noticeable)	Barely noticeable or relaxed effort
10	1	Very weak	
20	2	Weak (light)	Noticeable or defined effort
30	3	Moderate	
40	4		Obvious effort but unchanged facial expressions
50	5	Strong (heavy)	
60	6		Substantial effort with changed facial expression
70	7	Very strong	
80	8		
90	9		Uses shoulders or trunk for forces
100	10	Extremely strong (almost maximum)	

Source: Adapted from Borg's perceived exertion and pain scales (Borg, 1998).

Step two: The second step is to determine normalized peak hand forces. Disregard irregular or spurious actions and then estimate the normalized peak force (NPF) using observer ratings on a scale of 1–10 with anchors detailed in Table 14.5. Compare the task with known examples as necessary. The rating is an average for the entire work cycle. Rate what you see. Ratings can then be adjusted for other factors, for example, increased or decreased production rates, force.

Notes: The TLVs are guidance on the levels of exposure and conditions under which it is believed workers may be repeatedly exposed day after day, without adverse health effects. TLVs are not single numbers, but rather a combination of measured parameters and/or its effects on workers. Wide variations in individual susceptibility, exposure of an individual at, or even below, the TLV may result in annoyance, aggravation of a preexisting condition, or occasionally even physiological damage.

Step three: The third step is to locate the combination or intersection of HAL and NPF on the TLV graph. For example, an NPF of 4 and a HAL of 6 is considered an unacceptable exposure. See Figure 14.7.

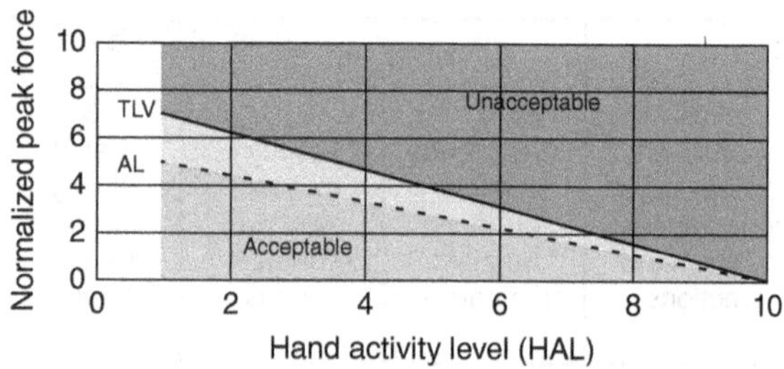

Figure 14.7 Normalized peak hand force and hand activity level threshold limit value determination (From ACGIH®, 2013 TLVs® and BEIs® Book. Copyright 2013. Reprinted with permission)

- Posture
- Sustained, nonneutral positions such as wrist flexion
- Extension, deviation or forearm rotation
- Contact stress
- Vibration
- Low temperature (<20 °C)
- Extended work shifts
- Preexisting injury.

RAPID UPPER LIMB ASSESSMENT (RULA)

Introduction

The Rapid Upper Limb Assessment (RULA) method has been developed by Dr Lynn McAtamney and Professor E. Nigel Corlett, ergonomists from the University of Nottingham in England (Dr McAtamney is now at Telstra, Australia). RULA is a postural targeting method for estimating the risks of work-related upper limb disorders. A RULA assessment gives a quick and systematic assessment of the postural risks to a worker. The analysis can be conducted before and after an intervention to demonstrate that the intervention has worked to lower the risk of injury.

The full RULA tables are shown in Appendix C– Survey Tools. Below each step is shown an illustration demonstrating how to score that step.

Applying the Assessment Tool

The following are the steps of a RULA analysis:

Step one: Locate upper arm position and score the position of the upper arm (Figure 14.8).

Step two: Locate the lower arm position and score the position of the lower arm (Figure 14.8).

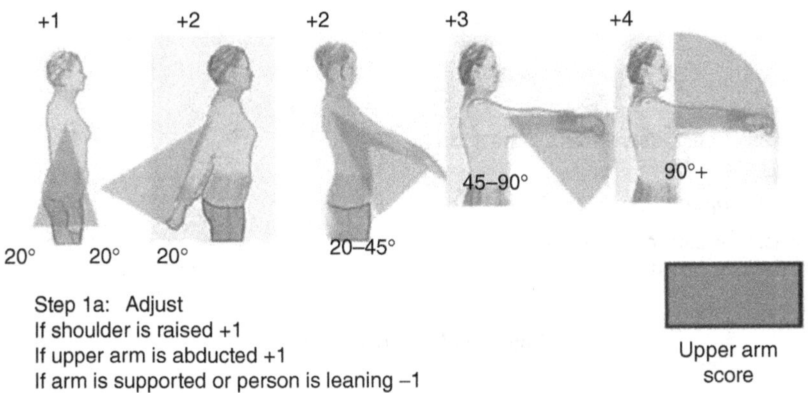

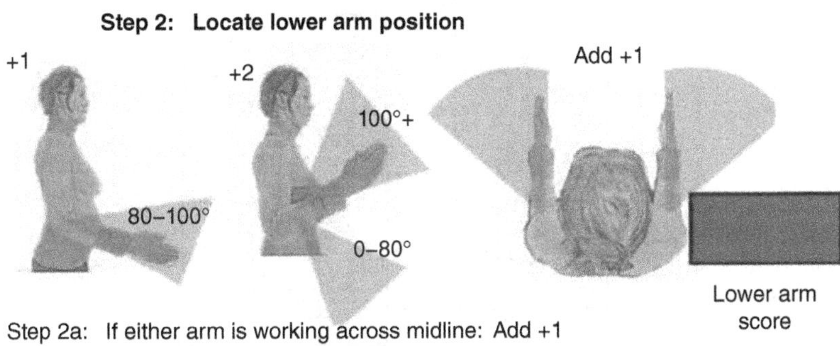

Figure 14.8 Upper and lower arms score (Created by Dr Ostrom)

Step three: Locate wrist position and score the wrist position (Figure 14.10).

Step four: Determine wrist twist, score accordingly (Figure 14.9).

Next we need to determine the score from **Table A** (Figure 14.10). This value is then fed into **Step five**.

Steps 6–8 are performed using data from the previous steps as shown in Figure 14.11.

Step 9: Locate and score the neck position and enter in the box (Figure 14.12).

Step 10: Locate the trunk position and then scoring the position in the box (Figure 14.12).

Step 11: Determine the score for supporting the legs and scoring (Figure 14.12).

We next need to determine the score from **Table B** to be used in **Step 12** (Figure 14.13).

Steps 13–15 are performed to determine the neck, trunk, and leg scores (Figure 14.14).

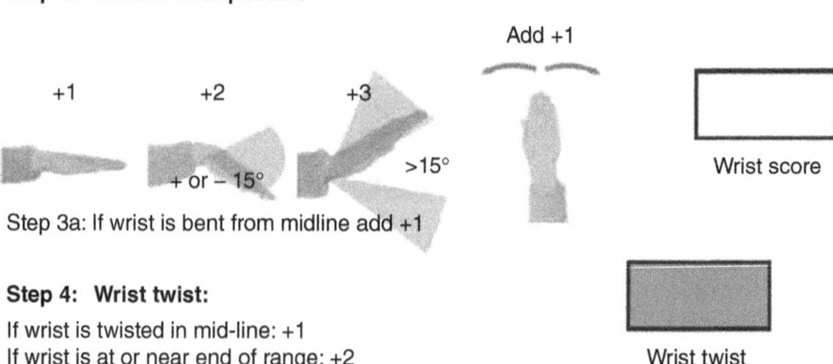

Figure 14.9 Wrist score

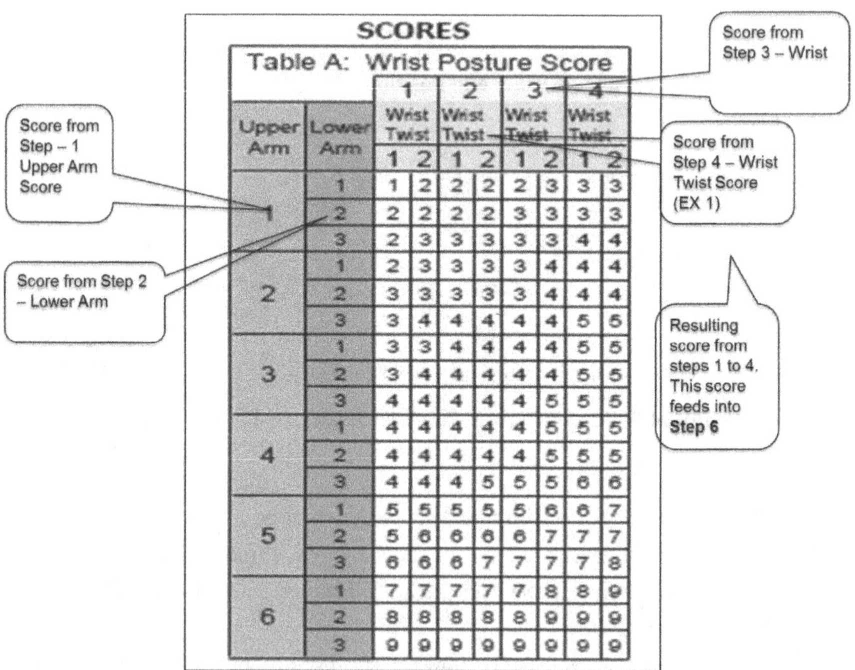

SCORES

Table A: Wrist Posture Score

Upper Arm	Lower Arm	1 Wrist Twist 1	1 Wrist Twist 2	2 Wrist Twist 1	2 Wrist Twist 2	3 Wrist Twist 1	3 Wrist Twist 2	4 Wrist Twist 1	4 Wrist Twist 2
1	1	1	2	2	2	2	3	3	3
	2	2	2	2	2	3	3	3	3
	3	2	3	3	3	3	3	4	4
2	1	2	3	3	3	3	4	4	4
	2	3	3	3	3	3	4	4	4
	3	3	4	4	4	4	4	5	5
3	1	3	3	4	4	4	4	5	5
	2	3	4	4	4	4	4	5	5
	3	4	4	4	4	4	5	5	5
4	1	4	4	4	4	4	5	5	5
	2	4	4	4	4	4	5	5	5
	3	4	4	4	5	5	5	6	6
5	1	5	5	5	5	5	6	6	7
	2	5	6	6	6	6	7	7	7
	3	6	6	6	7	7	7	7	8
6	1	7	7	7	7	7	8	8	9
	2	8	8	8	8	8	9	9	9
	3	9	9	9	9	9	9	9	9

Figure 14.10 Find the accurate value by finding the intersection of steps one through four on the matrix which is Table A

The final score is determined from RULA Table C (Figure 14.15).

The RULA action levels give you the urgency about the need to change how a person is working as a function of the degree of injury risk.

Action level 1 – RULA score 1–2 means that the person is working in the best posture with no risk of injury from their work posture.

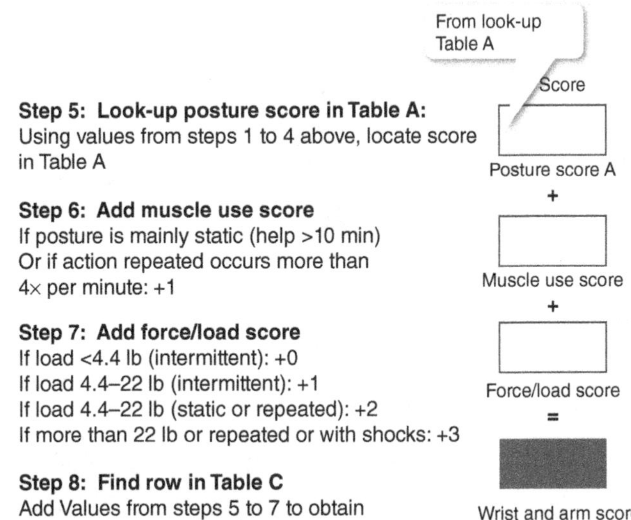

From look-up
Table A
Score
Step 5: Look-up posture score in Table A:
Using values from steps 1 to 4 above, locate score in Table A
Posture score A
+
Step 6: Add muscle use score
If posture is mainly static (help >10 min)
Or if action repeated occurs more than 4× per minute: +1
Muscle use score
+
Step 7: Add force/load score
If load <4.4 lb (intermittent): +0
If load 4.4–22 lb (intermittent): +1
If load 4.4–22 lb (static or repeated): +2
If more than 22 lb or repeated or with shocks: +3
Force/load score
=
Step 8: Find row in Table C
Add Values from steps 5 to 7 to obtain
Wrist and arm score

Figure 14.11 RULA force and muscle use values (Original graphic by Dr Ostrom)

Action level 2 – RULA score 3–4 means that the person is working in a posture that could present some risk of injury from their work posture, and this score most likely is the result of one part of the body being in a deviated and awkward position, so this should be investigated and corrected.

Action level 3 – RULA score 5–6 means that the person is working in a poor posture with a risk of injury from their work posture, and the reasons for this need to be investigated and changed in the near future to prevent an injury.

Action level 4 – RULA score 7–8 means that the person is working in the worst posture with an immediate risk of injury from their work posture, and the reasons for this need to be investigated and changed immediately to prevent an injury.

RAPID ENTIRE BODY ASSESSMENT (REBA)

Introduction

The Rapid Entire Body Assessment (REBA) method was developed by Dr Sue Hignett andDr Lynn McAtamney, ergonomists from University of Nottingham in England (Dr McAtamney is now at Telstra, Australia). REBA is a postural targeting method for estimating the risks of work-related entire body disorders. A REBA assessment gives a quick and systematic assessment of the complete body postural risks to a worker. The analysis can be conducted before and after an intervention to demonstrate that the intervention has worked to lower the risk of injury.

Appendix C contains the REBA tables.

The following are the steps of a REBA analysis. Below each step is shown an illustration demonstrating how to score that step:

B. neck, trunk, and leg analysis

Step 9: Locate Neck Position:

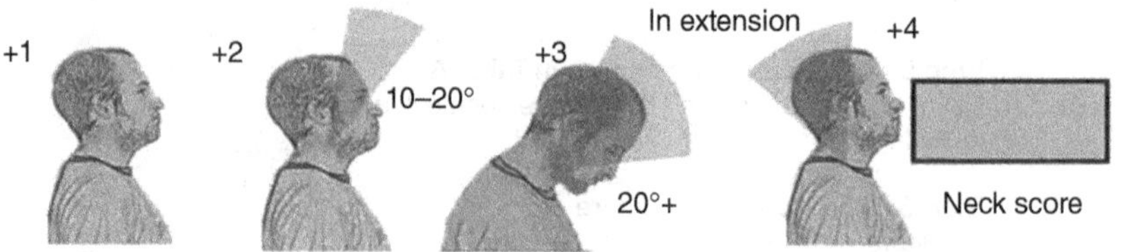

Step 9a: Adjust
If neck is twisted: +1
If neck is side bending: +1

Step 10: Locate trunk position

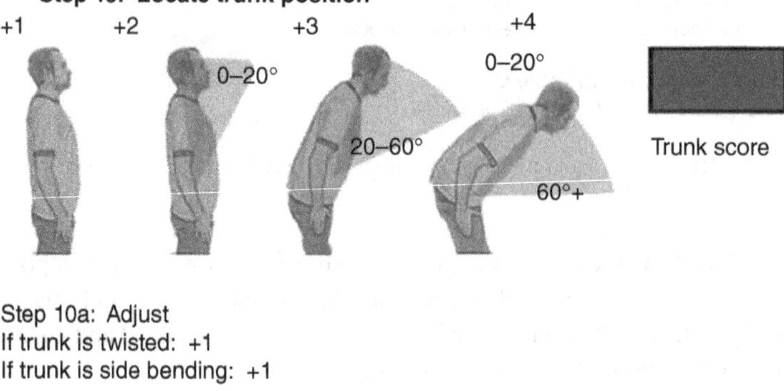

Step 10a: Adjust
If trunk is twisted: +1
If trunk is side bending: +1

Step 11: Legs
If legs and feet are supported: +1
If not: +2

Leg score

Figure 14.12 RULA neck, torso, and leg score

Step 1 is to determine the neck position and how to score (Figure 14.16).

Step 2 is to determine trunk position and how to score (Figure 14.16).

Step 3 is to determine score for the legs and enter (Figure 14.16).

Step 4 is to determine the posture score. To perform this you find the score on **Table A** using the values from Steps 1–3 above (Figure 14.17).

Steps 5 and 6 determine which row in **Table C** the final score will be located (Figure 14.18).

Step 7 determines the score for the upper arm posture and is placed in the box.

Step 8 determines the scoring for the lower arm (Figure 14.19).

Step 9 determines the score for the wrist (Figure 14.20).

Step 10 determines the posture score from **Table B** as shown in Figure 14.21.

Steps 11 and 12 determine which column in **Table C** the score is found.

Step 13 determines the activity score.

Step 14, final step, combines the scores from **Steps 6, 12, and 13.** This is the REBA Score. Figure 14.22.

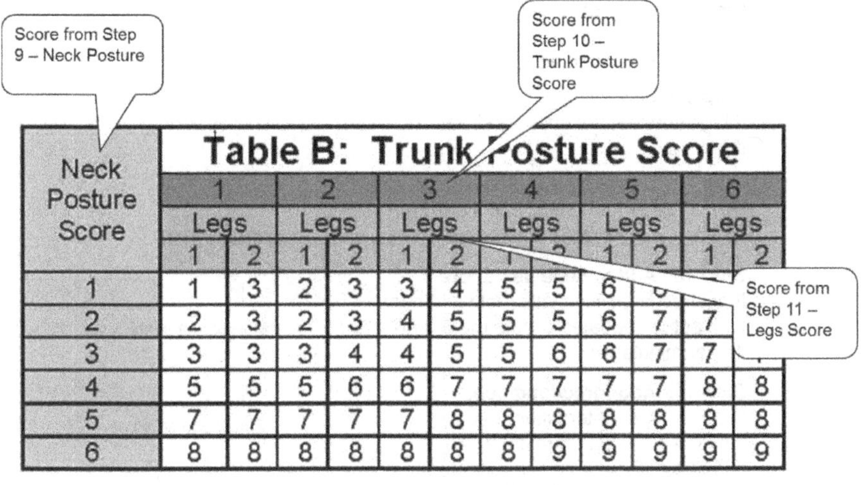

Neck Posture Score	Table B: Trunk Posture Score											
	1		2		3		4		5		6	
	Legs		Legs		Legs		Legs		Legs		Legs	
	1	2	1	2	1	2	1	[illegible]	1	2	1	2
1	1	3	2	3	3	4	5	5	6	[illegible]	[illegible]	[illegible]
2	2	3	2	3	4	5	5	5	6	7	7	[illegible]
3	3	3	3	4	4	5	5	6	6	7	7	[illegible]
4	5	5	5	6	6	7	7	7	7	7	8	8
5	7	7	7	7	7	8	8	8	8	8	8	8
6	8	8	8	8	8	8	8	9	9	9	9	9

Figure 14.13 RULA Table B trunk posture score

Table 14.6 shows the **Action Levels** based on the REBA score.

MIL-STD-1472F, MILITARY STANDARD, HUMAN ENGINEERING, DESIGN CRITERIA FOR MILITARY

Systems, Equipment, and Facilities (Section 5.9.11.3 Weight)

Introduction The Department of Defense Design Criteria Standard for Human Engineering 1472F (Defense, 1999) (MIL-STD 1472G) is intended to present human engineering design criteria to achieve mission success.

Limitations A limitation of the MIL-STD 1472 is its primary purpose is in the design and development of new systems and is intended to be used for infrequent lifting.

Applying the Standard

Step one: Observe the lifting task to obtain lifting frequency (lifts/min), cumulative lifting duration, maximum and average weight handled, average object size, lifting heights (ground to 3 ft or ground to 5 ft) and user population (male, female, or mixed). Where it is not possible to define the height to where the object is lifted, the shoulder height value can be used (5 ft).

Step two: The weight limits in Table 14.7, conditions A (lifting to 5 ft) and B (lifting to 3 ft), can be used as the maximum values in determining the weight of items lifted by one person with two hands. The weight limits can be double

Step 10a: Adjust
If trunk is twisted: +1
If trunk is side bending: +1

Step 11: Legs
If legs and feet are supported: +1
If not: +2

Leg score

Step 12: Look-up posture score in Table B:
Using values from steps 9 to 11 above
Locate score in Table B

Posture score B
+

Step 13: Add muscle use score
If posture is mainly static (held >10 min)
Or if action repeated occurs more than
4× per minute: +1

Muscle use score
+

Step 14: Add force/load score
If load <4.4 lb (intermittent): +0
If load 4.4–22 lb (intermittent): +a
If load 4.4–22 lb (static or repeated): +2
If more than 22 lb or repeated or with shocks: +3

Force/load score
=

Step 15: Find column in Table C
Add values from steps 12 to 14 to obtain
neck, trunk, and leg score, Find column in Table C.

Neck, Trunk, and leg score

Figure 14.14 Leg, posture, and muscle activity calculation

From Step 14
Wrist and Arm Score

From Step 8
Wrist and Arm Score

Table C	Neck, Trunk, and Leg Score						
Wrist and Arm Score	1	2	3	4	5	6	7+
1	1	2	3	3	4	5	5
2	2	2	3	4	4	5	5
3	3	3	3	4	4	5	6
4	3	3	3	4	5	6	6
5	4	4	4	5	6	7	7
6	4	4	5	6	6	7	7
7	5	5	6	6	7	7	7
8+	5	5	6	7	7	7	7

Figure 14.15 Final RULA score

if two people are performing the lift, with two hands and the load is evenly distributed.

- If one male is lifting, use the male-only population column.
- If one female is lifting, use the female and male population column.
- If two (or more) males are lifting, use the male-only population column.
- If two females (females and males) are lifting, use the male and female column.
- If the load is not evenly distributed, the weight limit applies to the heavier point.

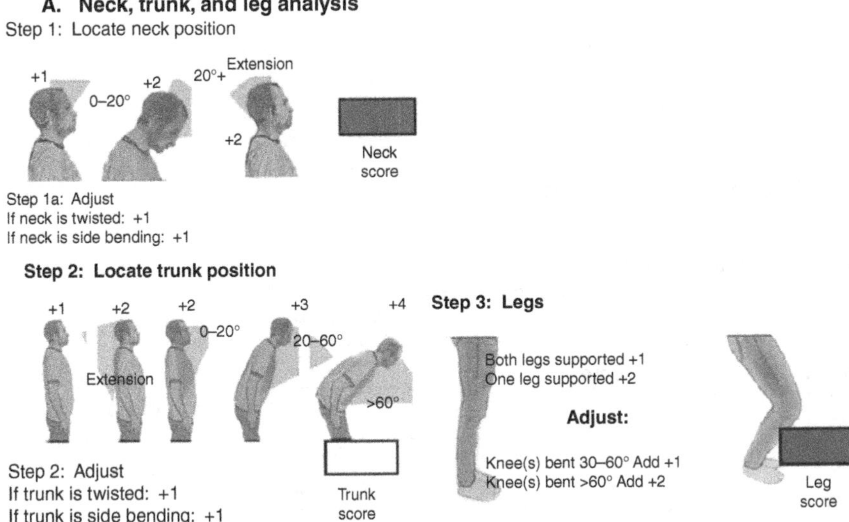

Figure 14.16 REBA neck, trunk, and leg score (Original graphic by Dr Ostrom)

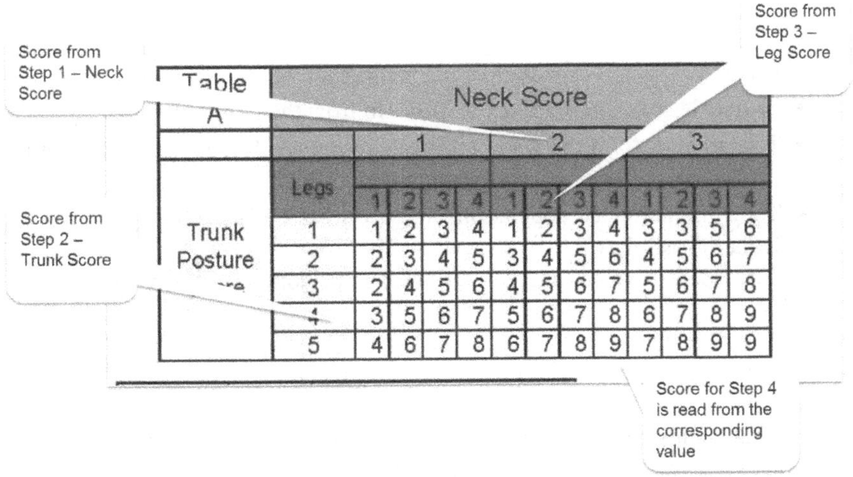

Table A		Neck Score											
		1				2				3			
	Legs	1	2	3	4	1	2	3	4	1	2	3	4
Trunk Posture Score	1	1	2	3	4	1	2	3	4	3	3	5	6
	2	2	3	4	5	3	4	5	6	4	5	6	7
	3	2	4	5	6	4	5	6	7	5	6	7	8
	4	3	5	6	7	5	6	7	8	6	7	8	9
	5	4	6	7	8	6	7	8	9	7	8	9	9

Figure 14.17 REBA Table A posture scoring

- Where three or more persons are lifting simultaneously, not more than 75% of the one-person value may be added for each additional lifter, provided that the object is sufficiently large that the lifters do not interfere.

For example:

- One male can safely lift 56 lb from the floor to 5 ft.
- One male can safely lift 87 lb from the floor to 3 ft.
- One female can safely lift 37 lb from the floor to 5 ft.

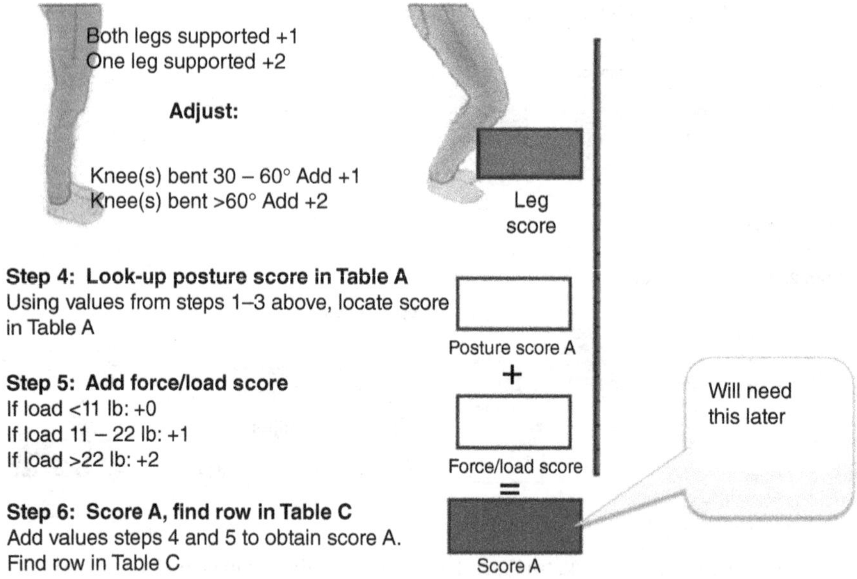

Figure 14.18 REBA posture and force score will be fed into REBA Table C at the end of the process (Original graphic by Dr Ostrom)

- One female can safely lift 44 lb from the floor to 3 ft.
- Two males can safely lift (56 × 2) 112 lb from the floor to 5 ft.
- Two females (or a male and female) can safely lift (37 × 2) 74 lb from the floor to 5 ft.
- Three lifters (mixed population) can safely lift 92.5 lb from the floor to 5 ft.

Calculation as follows:

The one-person value plus 75% of the one-person value for each additional lifter

$$37\,\text{lb} + (0.75 \times 37\,\text{lb} \times 2\,\text{persons}) = 92.5\ \text{lb maximum permissible weight}$$

Step three: adjust the maximum weight limit if the following conditions occur:

Lifting frequency: If the frequency of the lift exceeds 1 lift in 5 min or 20 lifts in 8 h, the permissible weight limit is reduced by (8.33 × LF) where LF is the lifting frequency in lifts for minute.

In the previous example, three people can safely lift 92.5 lb. If the lifting frequency is 6 lifts/min, then the maximum permissible weight is reduced by 50%. Calculation as follows:

$$8.33 \times \text{LF} = \text{reduction percent}$$
$$8.33 \times 6 = 50\%$$
$$0.5 \times 92.5\,\text{lb} = 46.25\,\text{lb maximum permissible weight}$$

B. Arm and wrist analysis

Step 7: Locate upper arm position:

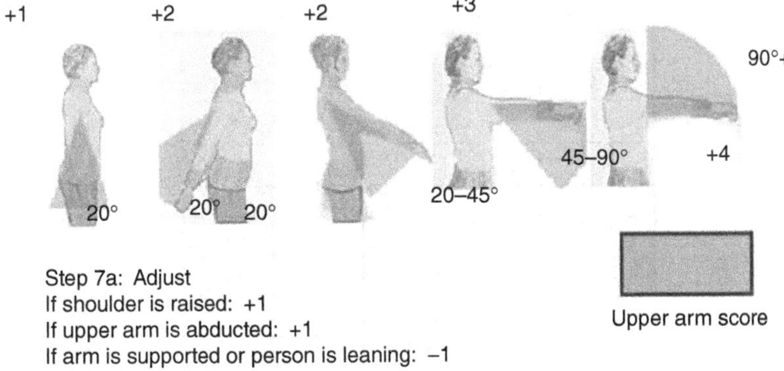

Step 8: Locate lower arm position:

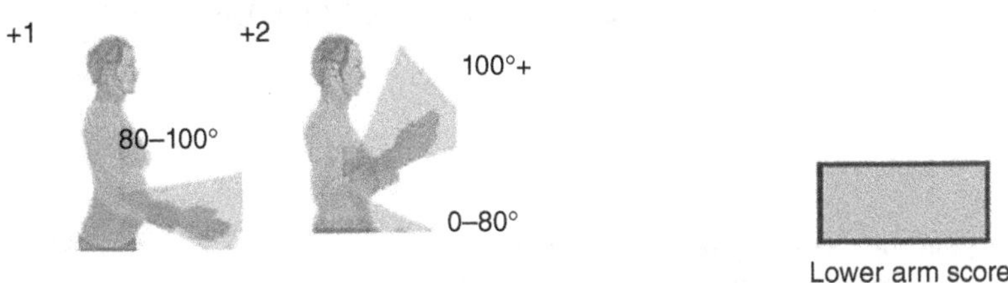

Figure 14.19 REBA upper and lower arm score

Step 9: locate wrist position:

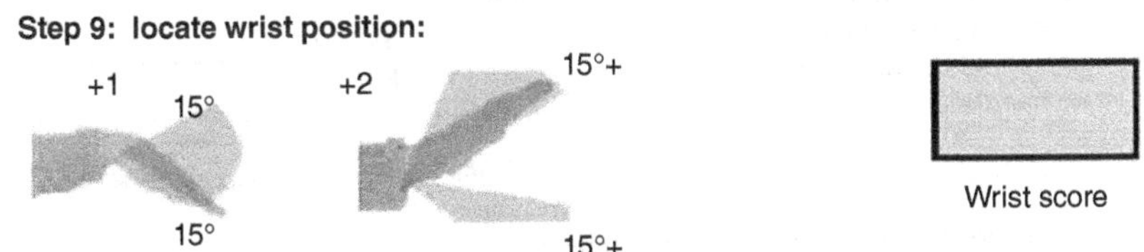

Figure 14.20 Reba posture scoring Table A for neck, trunk, and legs

Load size: The maximum permissible weight limits in Table 14.7 apply to an object of uniform mass that does not exceed 18 in. × 18 in. × 12 in. deep. This places the hand at half the depth or 6 in. from the body. If the depth exceeds 24 in., the permissible weight is reduced by 33%. If the depth exceeds 36 in., the permissible weight is reduced by 50% and by 66% if the depth exceeds 48 in.

Obstacles: If an object limits a lifter's approach, the maximum permissible limit is reduced by 33%.

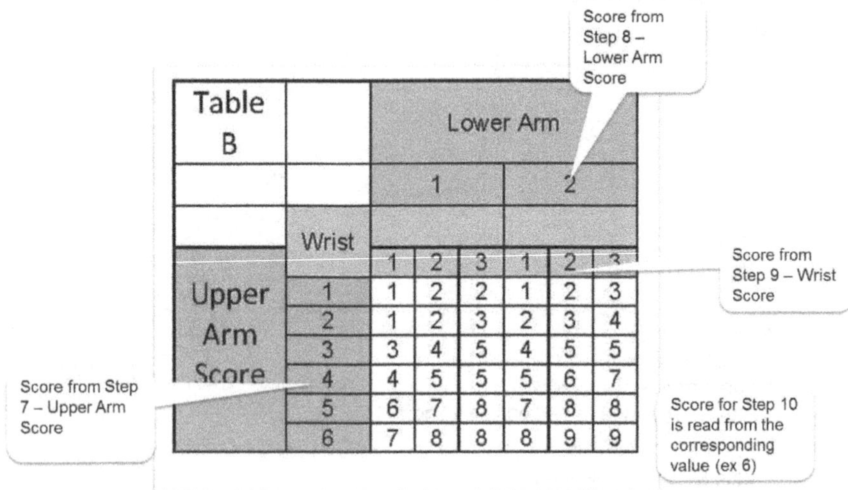

Table B		Lower Arm					
		1			2		
	Wrist	1	2	3	1	2	3
Upper Arm Score	1	1	2	2	1	2	3
	2	1	2	3	2	3	4
	3	3	4	5	4	5	5
	4	4	5	5	5	6	7
	5	6	7	8	7	8	8
	6	7	8	8	8	9	9

Figure 14.21 REBA posture scoring that will feed into REBA Table C at the end of the process

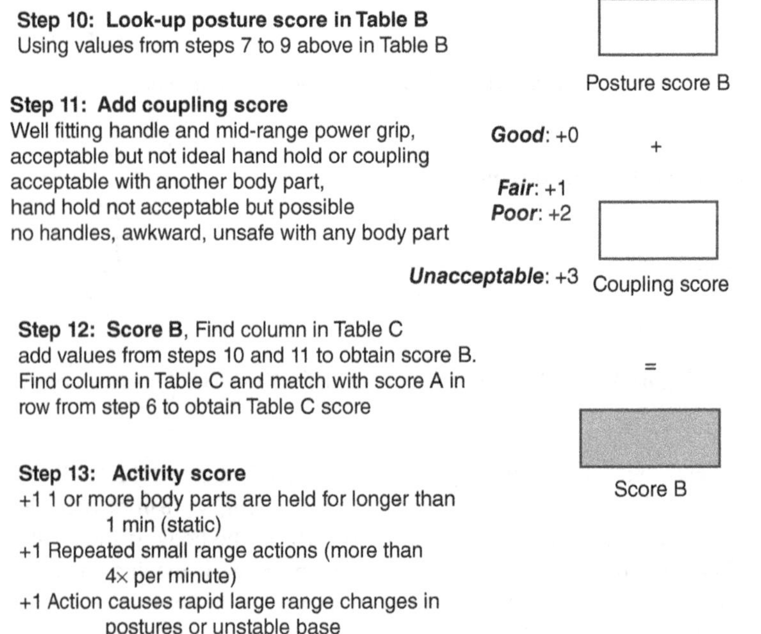

Figure 14.22 REBA Table C determines the final action-level score that is used to assess risk in the workplace

TABLE 14.6 REBA Action-Level Scores and Action Required

Action Level	REBA Score	Risk Level	Action Required
0	1	Negligible	None
1	2–3	Low	Maybe necessary
2	4–7	Medium	Necessary
3	8–10	High	Necessary soon
4	11–15	Very high	Necessary now

TABLE 14.7 Weight Limit Table, Values for Two Lifters in Parenthesis

Handling Function	Male and Female (lb)	Male Only (lb)
A. Lift an object from the floor and place on a surface not greater than 5 ft above the floor	37 (74)	56 (112)
B. Lift an object from the floor and place on a surface not greater than 3 ft above the floor	44 (88)	87 (174)

Source: Adapted from MIL-STD 1472G.

NATIONAL INSTITUTE FOR OCCUPATIONAL SAFETY AND HEALTH WORK PRACTICES LIFTING GUIDE

Introduction

In 1991, the National Institute for Occupational Safety and Health issued a revised work practices lifting guide (WPLG) for the design and evaluation of manual lifting tasks. The 1991 equation uses six factors that have been determined to influence lifting difficulty, combining the factors into one equation. Using the guide involves calculating values for the six factors in the equation for a particular lifting and lowering task, thereby generating a recommended weight limit (RWL) and lifting index (LI) for the task.

The equation incorporates a term called the lifting index, which is defined as a relative estimate of the level of physical stress associated with a particular manual lifting task. The estimate of the level of physical stress is defined by the relationship of the weight of the load lifted divided by the recommended weight limit.

- An LI below 1 indicates the task is safe for most healthy workers.
- An LI between 1 and 3 indicates that the object weight exceeded the RWL and should be addressed using either administrative or engineering controls. The task is safe for some but not all workers.
- An LI level greater than three indicates that the lifted weight exceeds the capacity to safely lift for most of the population, is likely to cause injury, and should be modified by implementation of engineering controls immediately.

Limitation

A limitation of the NIOSH WPLG is it assumes one person is performing a lift and does not take the stature of the individual into account. In addition, the RWL is within the strength capabilities of 75% of all women and 99% of all men.

NIOSH WPLG Assumptions

Application of the NIOSH WPLG assumes the following:

- Lifting task is two-handed and smooth.
- The hands are at the same height or level, and the load is evenly distributed between both hands.
- Manual handling activities other than lifting are minimal and do not require significant energy expenditure.
- Temperatures (66–79 °F) and humidity (35–50%) are within an acceptable range.
- One-handed lifts, lifting while seated or kneeling, lifting in a constrained or restricted work space, lifting unstable loads, wheelbarrows and shovels are not tasks designed to be covered by the WPLG.
- The shoe sole to floor surface coupling should provide for firm footing.
- Lifting and lowering assumes the same level of risk for low back injuries.
- Lifting outside of the ranges may increase the risk of injury.

Using the WPLG in situations that do not conform to these ideal assumptions will typically underestimate the hazard. The computed values of the lifting index are used by the OSH professional as a guide to estimate risk. The numbers by themselves do not identify a hazardous activity. The employer's incidence of injuries and lack of programs for training, work practice controls, and engineering controls related to lifting are elements used to determine the seriousness of the hazard.

Applying the Guide

The relevant task variables must be carefully measured and clearly recorded in a concise format. The Job Analysis Worksheet for either a single-task analysis or a multitask analysis provides a simple form for recording the task variables and the data needed to calculate the RWL and the LI values. The data needed for each task include the following:

The lifting equation for calculating the RWL is based on a multiplicative model of the six variables (Table 14.8).

$$\text{RWL} = \text{LC} \times \text{HM} \times \text{VM} \times \text{DM} \times \text{AM} \times \text{FM} \times \text{CM}$$

where if a formula results in a number less than one, use one. None of the values can be negative and therefore where appropriate the absolute value is used.

TABLE 14.8 NIOSH Work Practices Lifting Formula

Multiplier	Definition	Formula	Notes
LC	Load constant	51 lb	
HM	Horizontal multiplier	$10/H$	
VM	Vertical multiplier	$1 - (0.0075 \times \lvert V - 30 \rvert)$	
DM	Distance multiplier	$(0.82 \times (1.8/D))$ where D is $\lvert V_{start} - V_{stop} \rvert$	
AM	Asymmetric multiplier	$1 - (0.0032 \times A)$	A is between 0° and 130°
FM	Frequency multiplier	From look-up table	
CM	Coupling multiplier	From look-up table	

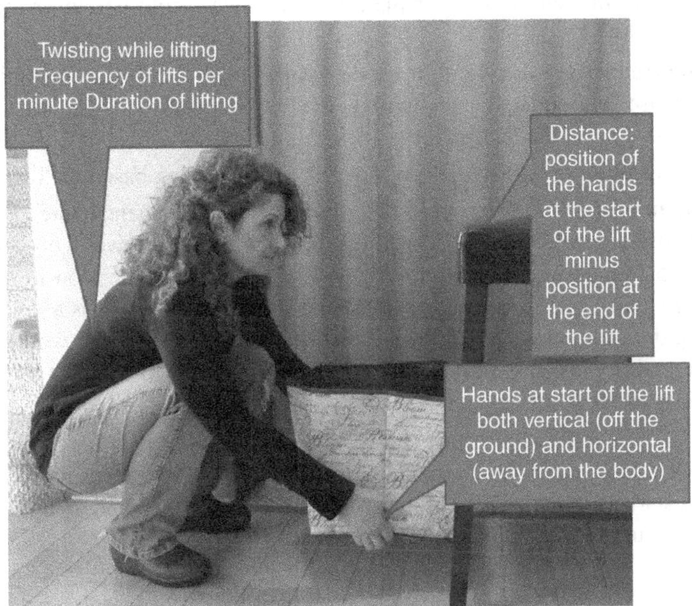

Figure 14.23 Anatomy of a lift

Anatomy of a Lift

Explained in Figure 14.23

Weight of the object lifted – Determine the load weight (L) of the object (if necessary, use a scale). If the weight of the load varies from lift to lift, record the average and maximum weights and calculate an LI for each.

Horizontal location of the hands with respect to the midpoint between the ankles (H) – The horizontal location of the hands at both the start (origin) and end (destination) of the lift is measured. The horizontal location is measured as the distance from the midpoint between the employee's ankles to a point projected on the floor directly below the midpoint of the hands grasping the object (the middle knuckle can be used to define the midpoint). The horizontal distance should be measured when the object is lifted (when the object leaves the surface).

Vertical location of the hands (*V*) – The vertical location is measured from the floor to the vertical midpoint between the two hands.

Travel distance (***D***) – The total vertical travel distance of the load during the lift is determined by subtracting the vertical location of the hands at the start of the lift from the vertical location of the hands at the end of the lift. This number is always positive.

Angle of asymmetry (*A*) – Determine the angle of asymmetry at the origin and destination of the lift. If the worker's hands are directly in front of the body and the worker is not twisting the angle would be 0. If the worker is twisting to where the hands are over the hips, the angle would be closer to 90°. See Figure 14.24.

Lifting frequency – First determine the total lifting time or duration of lifting and then use the frequency table to find the multiplier based on lifts per minute and the position of the hands, Table 14.8.

Coupling type – Classify the hand-to-container coupling based on Table 14.9.

Frequency of lift – Determine the average lifting frequency rate (*F*), in lifts per minute, periodically throughout the work session (average over at least an 15-min period). If the lifting frequency varies from session to session by more than 2 lifts/min, each work session should be analyzed as a separate task. The duration category, however, must be based on the overall work pattern of the entire work shift. The value is determined via a look-up table as seen in Table 14.10.

The RWL is a simple calculation once all the factors have been accounted for. The RWL is a starting point and is a recommendation based on the parameters of the lift. The RWL is used to calculate the lifting index.

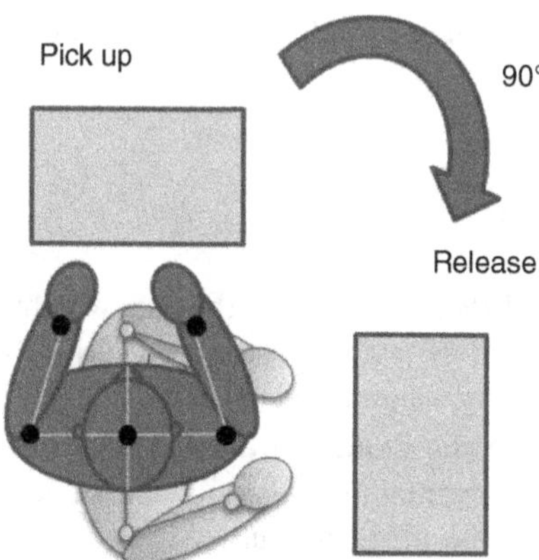

Figure 14.24 Asymmetry multiplier (Adapted from Ergonomics Plus Sweetster IN)

TABLE 14.9 Coupling Look-Up Multiplier Table

Good	Fair	Poor
CM = 1	$V < 30$ in. then CM = 0.95 $V \geq 30$ in. then CM = 1	CM = 0.9
For containers of optimal design, such as some boxes and crates, a "good" hand to object coupling would be defined as handles or handhold cutouts	For containers of optimal design a "fair" hand to object coupling would be defined as handles or handholds of less than optimal design (hand cannot form a 95° angle)	Container of less than optimal design or loose parts or irregular objects that are bully or hard to handle
For loose parts or irregular objects, which are not usually containerized such as casting and stock, a "good" hand to object coupling would be defined as comfortable grip in which the hand can be easily wrapped around the object		Lifting nonridge items

TABLE 14.10 Work Practices Lifting Guide Frequency Look-Up Table

Frequency (lifts/min)	≤1 h		>1 h but ≤2		>2 h but ≤8	
	$V < 30$ in.	$V > 30$ in.	$V < 30$ in.	$V > 30$ in.	$V < 30$ in.	$V > 30$ in.
0.2	1	1	0.95	0.95	0.85	0.85
0.5	0.97	0.97	0.92	0.92	0.81	0.81
1	0.94	0.94	0.88	0.88	0.75	0.75
2	0.91	0.91	0.84	0.84	0.65	0.65
3	0.88	0.88	0.79	0.79	0.55	0.55
4	0.84	0.84	0.72	0.72	0.45	0.45
5	0.8	0.8	0.6	0.6	0.35	0.37
6	0.75	0.75	0.50	0.50	0.27	0.57

LI = Actual Weight of the Load/RWL

Interpretation: increased risk of low back injury if the LI exceeds 1

<1 OK

=1 OK

>1 may have increased risk

>3 likely have increased risk.

Some believe that if workers are properly screened (based on the task requirements) and trained, they can safely work at lift indexes greater **than 1 but less than 3.**

Other Tools

The healthcare, hand tools, and manual material handling sections of this text include additional assessment tools as well as Appendix C.

SUMMARY

Workers exposure to physical workplace risk factors can be evaluated using many methods and no one method is correct in all situations. It is recommended that a variety of methods are used to improve the fit between the worker and the workplace.

While it is near impossible to eliminate a worker exposure to all the physical workplace risk factors, they can be reduced to lessen the probability of injury since it is the combination of risk factors that increase injury probability. For example, in a situation that has a high exposure to force in an awkward posture, changing the posture to neutral will reduce the force component as well. Small changes can make a large impact on a worker's overall exposure profile.

Control options can be identified and developed for each of the physical workplace risk factors found. A sound technique is to evaluate each solution against each other in a cost-benefit trade-off analysis.

Keep in mind the four basic approaches to controlling risk hazards proceeding from eliminating the hazard to using administrative controls. Full descriptions of the hazard control options are listed below in order of highest to lowest priority. As with other industrial hygiene and safety methods for controlling hazards, elimination, engineering, and substitution are preferred over administrative. In the context of ergonomics, few PPE solutions exist when it comes to hazard control.

A useful approach is to provide the risk reduction against each hazard control option and use it for a tired approach (Table 14.11).

KEY POINTS

- Small changes to a worker's exposure to the physical risk factors can result in a major impact on their safety, health, and overall well-being.
- Many tools exist to evaluate the physical workplace risk factors, but no tool is encompassing and therefore using multiple tools is a good approach.

REVIEW QUESTIONS

1. What are the four major parts of an assessment?
2. What tools should be brought to an assessment?
3. What is a limitation of the ACGIH TLV for lifting?
4. What is a limitation of the NIOSH WPLG?
5. What factors should be considered when evaluating a lifting task?

TABLE 14.11 Hierarchy of Control

Levels of Hazard Control
1. **Elimination** – A redesign or procedural change that eliminates exposure to an ergonomic risk hazard; for example, using a remotely operated soil compactor to eliminate vibration exposure
2. **Engineering controls** – A physical change to the workplace; for example, lowering the unload height of a conveyor
3. **Substitution** – An approach that uses tools/material/equipment with lower risk; for example, replacing an impact wrench with a lower vibration model
4. **Administrative** – This approach is used when none of the above can be used or are impractical to implement. Administrative controls are procedures and practices that limit exposure by control or manipulation of work schedule or the manner in which work is performed. Administrative controls reduce the exposure to ergonomic stressors and thus reduce the cumulative dose to any one worker. If you are unable to alter the job or workplace to reduce the physical stressors, administrative controls can be used to reduce the strain and stress on the work force. Administrative controls are most effective when used in combination with other control methods, for example, requiring two people to perform a lift

EXERCISE

1. Reference the Exercise appendix.

REFERENCES

Borg, G. (1998). Borg's Perceived Exertion and pain Scales. Champaign, IL: *Human Kinetics*.

Department of Defense. (1999). *MIL-STD-1472F, Department of Defense Design Criteria Standard: Human Engineering*. United States Government Printing Office.

Marras, W. S. & Karwowski, W. (2006). Interventions, Controls, and Applications in Occupational Ergonomics. *The Occupational Ergonomics Handbook* 2nd edn. CRC Press.

Snook, S. H., & Ciriello, V. M. (1991). The Design of Manual Handling Tasks: Revised Tables of Maximum Acceptable Weights and Forces. *Ergonomics*, 1197–1213.

Snook, S. H., Campanelli, R. A., & Hart, J. W. (1978a). A Study of Three Preventative Approaches to Low Back Injury. *Journal of Occupational Medicine*, 478–481.

ADDITIONAL SOURCE

Hignett, S. and McAtamney, L. (2000). Rapid Entire Body Assessment: REBA, *Applied Ergonomics*, 31, 201–205.

MacLeod, D. (1994). The Ergonomics Edge: Improving Safety, Quality, and Productivity. *Industrial Health and Safety*. Wiley.

McAtamney, L., & Corlett, E. N. (1993). RULA: A Survey Method for the Investigation of Work-Related Upper Limb Disorders. *Applied Ergonomics*, 91–99.

Opnavinst. (n.d.). Navy Occupational Safety and Health Instruction 5100.23G.2307. Retrieved February 2015, from Department of the Navy Issuances: https://acc.dau.mil/adl/en-US/377924/file/51114/ref%20r_ONI5100.23G_Navy%20SOH%20Manual.pdf.

Pheasant, S. (1991). *Ergonomics, Work and Health* 1st edn. Aspen.

Schumann, W. O., and Konig, H. (1954). The Observation of Atmospherics at the Lowest Frequencies. Retrieved from The Healers Journal: http://www.thehealersjournal.com/2012/05/21/the-schumann-resonance-earths-powerful-natural-vibration/#sthash.qrXSOhWM.dpuf.

Snook, S., Campanelli, R. & Hart, J. (1978b). A study of three preventative approaches to low back injury. *Journal of Occupational Medicine*, 20(7), 478–481.

The Ergonomics Group, Health and Environmental Laboratories, Eastman Kodak Company. (1989). *Ergonomic Design for People at Work, vol. 2.* Wiley.

15

ERGONOMICS IN THE HEALTHCARE INDUSTRY

LEARNING OBJECTIVE

Students will be able to identify causes of work-related musculoskeletal disorders (WMSDs) within the healthcare industry. They will be able to identify the key contributors to those WMSDs and solutions to reduce the potential WMSDs. They will also be able to identify the resource guides for patient care.

INTRODUCTION

The healthcare industry is huge and continues to grow at an astounding rate. Nurses and nursing aides are 2 of the top 15 professions that suffer from work-related musculoskeletal disorders (WMSDs) (Workplace Health Promotion, n.d.). In 2011, 50% of nurses and nurse aide's injuries and illnesses were contributed to WMSDs (Bureau of Labor Statistics, 2011). WMSDs associated with healthcare workers in the hospital healthcare industry and in the home healthcare industry were examined. Key contributors to WMSDs among healthcare workers are from repositioning patients in bed, turning patients, transferring patients from a bed to a gurney, pushing beds or gurneys far distances, transferring patients from a chair to a toilet, and transferring patients from a wheelchair to a vehicle. This chapter explores what causes and/or contributes to WMSDs among healthcare workers and preventive measures that can be implemented to reduce the potential for WMSDs in the healthcare industry.

Occupational Ergonomics: A Practical Approach, First Edition.
Theresa Stack, Lee T. Ostrom and Cheryl A. Wilhelmsen.

Healthcare workers are presented with many ergonomic challenges. In a healthcare environment, workers can work long hours and healthcare workers' duties involve being physically active for up to 16 h a day. They twist their bodies in awkward position throughout their day to deliver patient care. A trend shows nursing aides, orderlies, and attendants have the highest rates of WMSDs reported injuries through 2007–2011. In 2007, patient lifting injuries were reported for every 10,000 healthcare workers. Twenty-one employees suffered a lifting injury. WMSDs are one of the leading contributors for lost workday injury and illness cases.

ERGONOMIC RISK FACTORS/CONTRIBUTING FACTORS CAUSING HEALTHCARE INJURIES

Key factors contributing to WMSDs among healthcare workers are from repositioning patients in bed, manual lifting, turning patients, transferring patients from a bed to a gurney, pushing beds or gurneys far distances, transferring patients from a chair to a toilet, and transferring patients from a wheelchair to a car. Other risk factors that contribute to healthcare injuries are overexertion from quickly responding to patients in need, lifting more than 20 lifts/shift, lifting patients alone, lifting uncooperative patients, and moving bariatric patients. Also, there are issues with using certain pieces of medically related equipment such as pipettes and even the process of dealing with medical records. These topics are discussed in the following sections.

Patient Handling

In healthcare setting, healthcare workers routinely use manual lifting methods in life-threatening situations. Manual fore/aft lifts tend to put a high level of compression forces on the rear-facing worker, and the front-facing worker positions their body in an awkward position to transfer patients from one surface to another surface. The manual chicken or drag lift is used in healthcare setting to assist a patient after falling on the floor and repositioning the patient. The manual chicken or drag lift can cause lower back injuries, shoulder dislocation, and strain on the spinal cord. The manual cradle lift or basket lift is performed when healthcare providers put one hand under the patient's thigh, and one hand is placed on the back of the patient to ensure the patient positioning. The manual cradle lift or basket lift puts high levels of force on the spinal disc of the healthcare workers and repeatedly caused the lifters to bend their bodies in an awkward position. The manual three-person lift is used when transporting patients from a bed to a stretcher. A high level of stress is put on the spinal disc, neck, shoulders, and back of the worker when the worker bends forward at the waist to lift patients. The manual three-person lift puts excessive stress on the backs of the lifters because the weight of the patient is not equally distributed when the lifters reposition the patient. The manual belt lift is commonly used to lift patients up from falling on the ground. There is usually one person pulling on the transfer belt that puts excessive strain on the lifter's lower back and another person squats next to the patient assisting the standing worker by guiding the patient to a standing position, which can cause lower back strain. The manual blanket lifts are used in emergency situations when medical treatment is needed quickly for the patient. When using the manual blanket lifts method, three to five workers put excessive strain on their lower back, shoulders, and wrist. The manual two-person through arm lift or towel lift is used to reposition patients. The workers will position a towel under a patient's thigh to use as a sling to reposition the patient. Force is applied to the patient's thighs, arms, and shoulders. The manual one-person through arm lift is used to reposition a patient in bed. This lift puts strain on the workers' shoulder, back, and arms because the healthcare workers are unable to use their legs as leverage. The manual Australian shoulder lift is also used to reposition patients in bed. The healthcare workers position one hand under the patient's waist or thigh and the other hand on the patient's back for support. The healthcare worker's shoulder is placed under the patient's where the arm connects to the shoulder. All of these high-risk manual patients handling lifting method are the last resort method in safely handling patients. A safer alternative to manual patient handling would be to use ergonomic-friendly mechanical lifting equipment.

The Occupational Safety and Health Administration (OSHA) has produced a guidelines document on patient handling in nursing homes. It is entitled "Guidelines for Nursing Homes Ergonomics for the Prevention of Musculoskeletal Disorders" (OSHA). The following discussion on patient handling comes from that document.

Excerpts from Guidelines for Nursing Homes Ergonomics for the Prevention of Musculoskeletal Disorders

Assessing the potential for work to injure employees in nursing homes is complex because typical nursing home operations involve the repeated lifting and repositioning of the residents. Resident lifting and repositioning tasks can be variable, dynamic, and unpredictable in nature. In addition, factors such as resident dignity, safety, and medical contraindications should be taken into account. As a result, specific techniques are used for assessing resident lifting and repositioning tasks that are not appropriate for assessing the potential for injury associated with other nursing home activities. An analysis of any resident lifting and repositioning task involves an assessment of the needs and abilities of the resident involved. This assessment allows staff members to account for resident characteristics, while determining the safest methods for performing the task, within the context of a care plan that provides for appropriate care and services for the resident. Such assessments typically consider the resident's safety, dignity, and other rights, as well as the need to maintain or restore a resident's functional abilities.

The resident assessment should include examination of factors such as:

- the level of assistance the resident requires;
- the size and weight of the resident;
- the ability and willingness of the resident to understand and cooperate; and
- any medical conditions that may influence the choice of methods for lifting or repositioning.

These factors are critically important in determining appropriate methods for lifting and repositioning a resident. The size and weight of the resident will, in some situations, determine which equipment is needed and how many caregivers are required to provide assistance. The physical and mental abilities of the resident also play an important role in selecting appropriate solutions. For example, a resident who is able and willing to partially support their own weight may be able to move from his or her bed to a chair using a standing assist device, while a mechanical sling lift may be more appropriate for those residents who are unable to support their own weight. Other factors related to a resident's condition may need to be taken into account as well. For instance, a resident who has recently undergone hip replacement surgery may require specialized equipment for assistance in order to avoid placing stress on the affected area. A number of protocols have been developed for systematically examining resident needs and abilities and/or for recommending procedures and equipment to be used for performing lifting and repositioning tasks. The example presented here is as follows:

Patient Care Ergonomics Resource Guide

Safe Patient Handling and Movement is published by the Patient Safety Center of Inquiry, Veterans Health Administration, and the Department of Defense. This document provides flowcharts (Figures 15.1–15.6) that address relevant resident assessment factors and recommends solutions for resident lifting and repositioning problems. This material is one example of an assessment tool that has been used successfully. Employers can access this information from www.patientsafetycenter.com. Nursing home operators may find another tool or develop an assessment tool that works better in their facilities.

The nursing home operator should use an assessment tool that is appropriate for the conditions in an individual nursing home. The special needs of bariatric (excessively heavy) residents may require additional focus. Assistive devices must be capable of handling the heavier weight involved, and modification of work practices may be necessary. A number of individuals in nursing homes can contribute to resident assessment and the determination of appropriate methods for assisting in transfer or repositioning. Interdisciplinary teams such as staff nurses, certified nursing assistants, nursing supervisors, physical therapists, physicians, and the resident or his/her representative may all be involved. Of critical importance is the involvement of employees directly responsible for resident care and assistance, as the needs and abilities of residents may vary considerably over a short period of time, and the employees responsible for providing assistance are in the best position to be aware of and accommodate such changes.

The recommended solutions presented in the following pages are not intended to be an exhaustive list, nor does OSHA expect that all of them will be used in any given facility. The information represents a range of available options that a facility can consider using. Many of the solutions are simple, common sense modifications to equipment or procedures that do not require substantial time or resources to implement. Others may require more significant efforts. The integration of various solutions into the nursing home is a strategic decision that, if carefully planned and executed, will lead to long-term benefits. Equipment must meet applicable regulations regarding equipment design and use, such as the restraint regulations from the Centers for Medicare and Medicaid. In addition, administrators should follow any

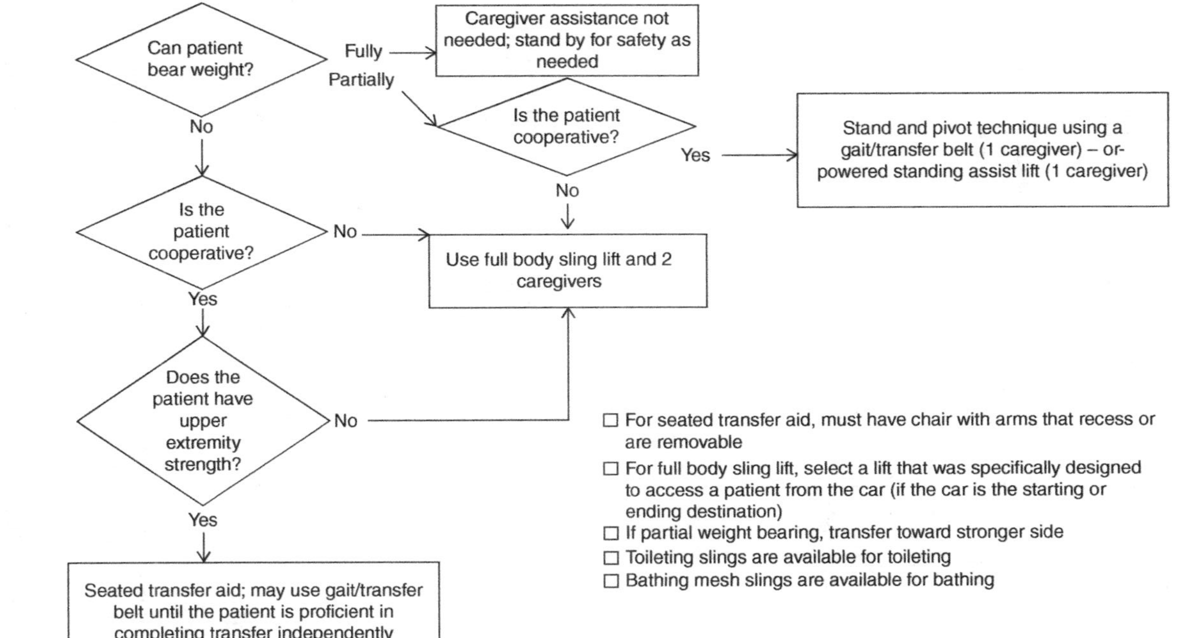

Figure 15.1 Transfer to and from: bed to chair, chair to toilet, chair to chair, or car to chair (The Patient Safety Center of Inquiry (Tampa, FL). Veterans Health Administration & Department of Defense. October 2001. Adapted from Guidelines for Nursing Homes Ergonomics for the Prevention of Musculoskeletal Disorders (OSHA, 2009))

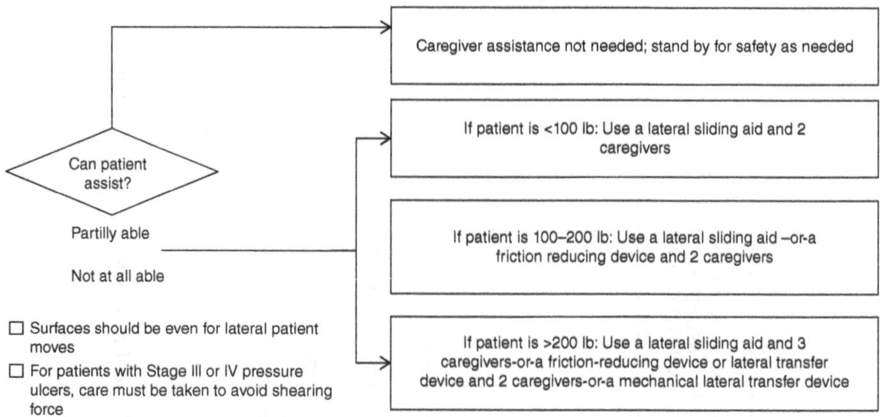

Figure 15.2 Lateral transfer to and from: bed to stretcher, trolley (The Patient Safety Center of Inquiry (Tampa, FL). Veterans Health Administration & Department of Defense. October 2001. Adapted from Guidelines for Nursing Homes Ergonomics for the Prevention of Musculoskeletal Disorders (OSHA, 2009))

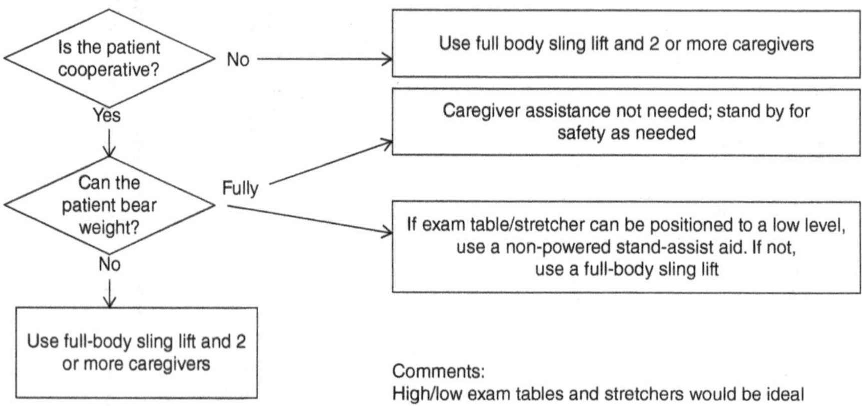

Figure 15.3 Transfer to and from: chair to stretcher (The Patient Safety Center of Inquiry (Tampa, FL). Veterans Health Administration & Department of Defense. October 2001. Adapted from Guidelines for Nursing Homes Ergonomics for the Prevention of Musculoskeletal Disorders (OSHA, 2009))

manufacturers' recommendations and review guidelines, such as the FDA Hospital Bed Safety Workgroup Guidelines, to help ensure patient safety (FDA, 2006).

Management should also be cognizant of several factors that might restrict the application of certain measures, such as residents' rehabilitation plans, the need for restoration of functional abilities, other medical contraindications, emergency conditions, and residents' dignity and rights.

The procurement of equipment and the selection of an equipment supplier are important considerations when implementing solutions. Employers should establish close working relationships with equipment suppliers. Such working

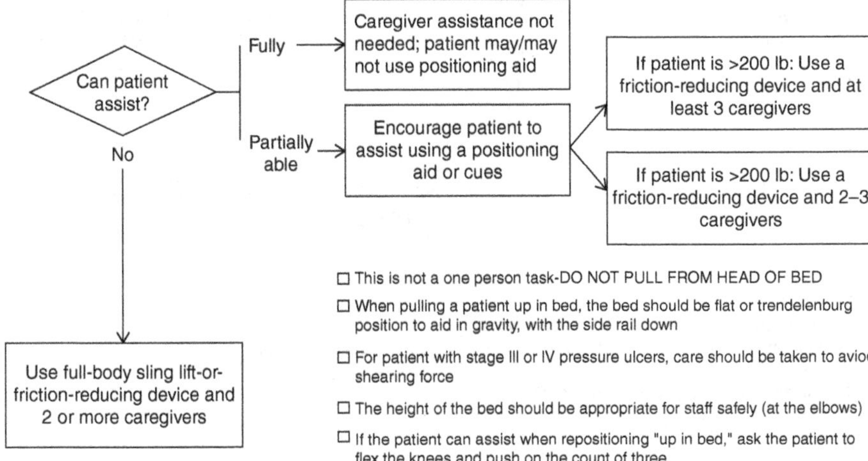

Figure 15.4 Reposition in bed: side-to-side, up in bed (The Patient Safety Center of Inquiry (Tampa, FL). Veterans Health Administration & Department of Defense. October 2001. Adapted from Guidelines for Nursing Homes Ergonomics for the Prevention of Musculoskeletal Disorders (OSHA, 2009))

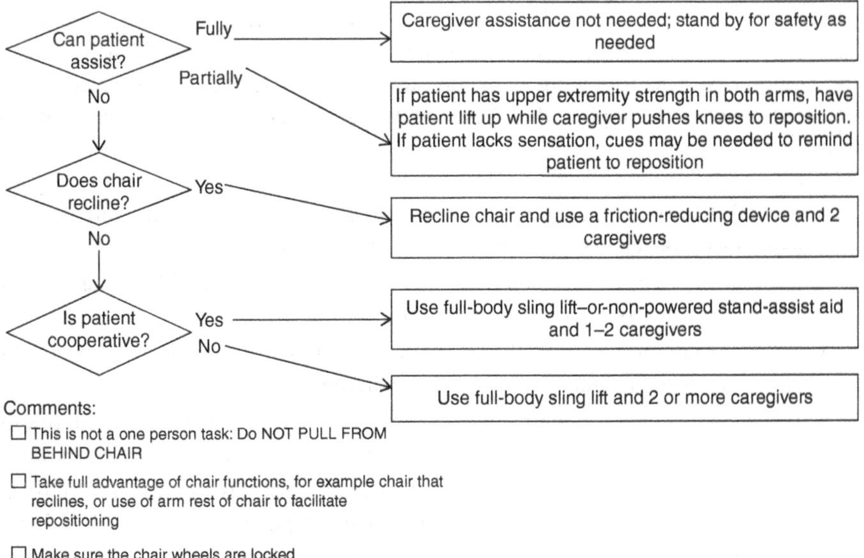

Figure 15.5 Reposition in chair: wheelchair and dependency chair (The Patient Safety Center of Inquiry (Tampa, FL). Veterans Health Administration & Department of Defense. October 2001. Adapted from Guidelines for Nursing Homes Ergonomics for the Prevention of Musculoskeletal Disorders (OSHA, 2009))

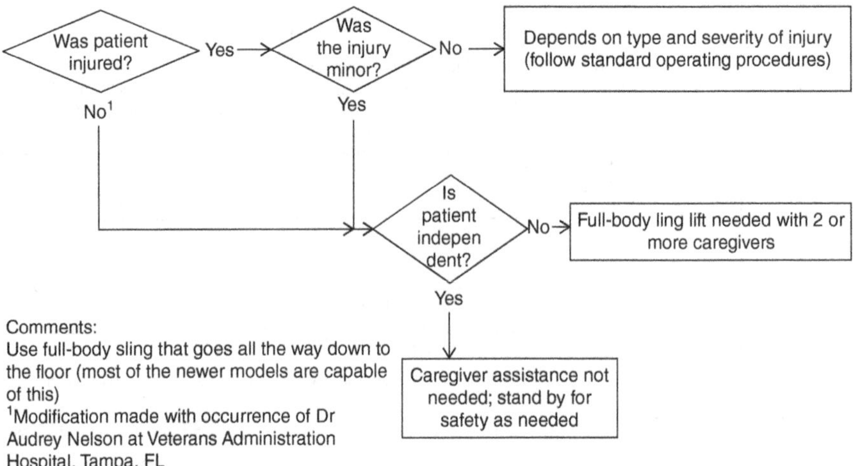

Figure 15.6 Transfer a patient up from the floor (The Patient Safety Center of Inquiry (Tampa, FL). Veterans Health Administration & Department of Defense. October 2001. Adapted from Guidelines for Nursing Homes Ergonomics for the Prevention of Musculoskeletal Disorders (OSHA, 2009))

relationships help with obtaining training for employees, modifying the equipment for special circumstances, and procuring parts and service when needed. Employers will want to pay particular attention to the effectiveness of the equipment, especially the injury and illness experience of other nursing homes that have used the equipment. The following questions are designed to aid in the selection of the equipment and supplier that best meets the needs of an individual nursing home.

- Availability of technical service – Is over-the-phone assistance, as well as onsite assistance, for repairs and service of the lift available?
- Availability of parts – Which parts will be in stock and available in a short time frame and how soon can they be shipped to your location?
- Storage requirements – Is the equipment too big for your facility? Can it be stored in close proximity to the area(s) where it is used?
- If needed, is a charging unit and backup battery included? What is the simplicity of the charging unit and space required for a battery charger if one is needed?
  - If the lift has a self-contained charging unit, what is the amount of space necessary for charging and what electrical receptacles are required? What is the minimum charging time of a battery?
- How high is the base of the lift and will it fit under the bed and various other pieces of furniture? How wide is the base of the lift or is it adjustable to a wider and lockable position?
- How many people are required to operate the lift for lifting of a typical 200-lb person?
- Does the lift activation device (pendant) have remote capabilities?

- How many sizes and types of slings are available? What type of sling is available for optimum infection control?
- Is the device versatile? Can it be a sit-to-stand lift, as well as a lift device? Can it be a sit-to-stand lift and an ambulation-assist device?
- What is the speed and noise level of the device? Will the lift go to floor level? How high will it go? Based on many factors, including the characteristics of the resident population and the layout of the facility, employers should determine the number and types of devices needed. Devices should be located so that they are easily accessible to workers. If resident lifting equipment is not accessible when it is needed, it is likely that other aspects of the ergonomics process will be ineffective.

If the facility can initially purchase only a portion of the equipment needed, it should be located in the areas where the needs are the greatest. Employers should also establish routine maintenance schedules to ensure that the equipment is in good working order. The following are examples of solutions for resident lifting and repositioning tasks.

Description: Powered sit-to-stand or standing assist devices.

When to Use: Use when transferring residents who are partially dependent, have some weight-bearing capacity, are cooperative, can sit up on the edge of the bed with or without assistance, and are able to bend hips, knees, and ankles. Transfers from bed to chair (wheel chair, Geri, or cardiac chair), or chair to bed, or for bathing and toileting. Can be used for repositioning where space or storage is limited.

Points to Remember: Look for a device that has a variety of sling sizes, lift height range, battery portability, handheld control, emergency shut-off, and manual override. Ensure device is rated for the resident weight. Electric-/battery-powered lifts are preferred to crank or pump type devices to allow smoother movement for the resident, and less physical exertion by the caregiver (Figure 15.7).

Description: Portable lift device (sling type); can be a universal/hammock sling or a band/leg sling.

When to Use: Lifting residents who are totally dependent, are partial- or non-weight bearing, are very heavy, or have other physical limitations. Transfers from bed to chair (wheel chair, Geri, or cardiac chair), chair or floor to bed, for bathing and toileting, or after a resident fall.

Points to Remember: More than one caregiver may be needed. Look for a device with a variety of slings, lift-height range, battery portability, handheld control, emergency shut-off, manual override, boom pressure-sensitive switch, that can easily move around equipment, and has a support base that goes under beds. Having multiple slings allows one of them to remain in place while resident is in bed or chair for only a short period, reducing the number of times the caregiver lifts and positions resident. Portable compact lifts may be useful where space or storage is limited. Ensure device is rated for the resident weight.

Figure 15.7 Powered sit-to-stand or standing assist devices (Adapted from Guidelines for Nursing Homes Ergonomics for the Prevention of Musculoskeletal Disorders (OSHA, 2009))

Electric-/battery-powered lifts are preferred to crank or pump type devices to allow a smoother movement for the resident, and less physical exertion by the caregiver. Enhances resident safety and comfort (Figure 15.8).

Description: Variable position Geri and cardiac chairs.

When to Use: Repositioning partial- or non-weight-bearing residents who are cooperative.

Points to Remember: More than one caregiver is needed, and use of a friction-reducing device is needed if resident cannot assist to reposition self in chair. Ensure use of good body mechanics by caregivers. Wheels on chair add versatility. Ensure that chair is easy to adjust, move, and steer. Lock wheels on chair before repositioning. Remove trays, footrests, and seat belts where appropriate. Ensure device is rated for the resident weight (Figure 15.9).

Description: Ambulation assist device.

When to Use: For residents who are weight bearing and cooperative and who need extra security and assistance when ambulating.

Points to Remember: Increases resident safety during ambulation and reduces risk of falls. The device supports residents as they walk and push it along during ambulation. Ensure height adjustment is correct for resident before ambulation.

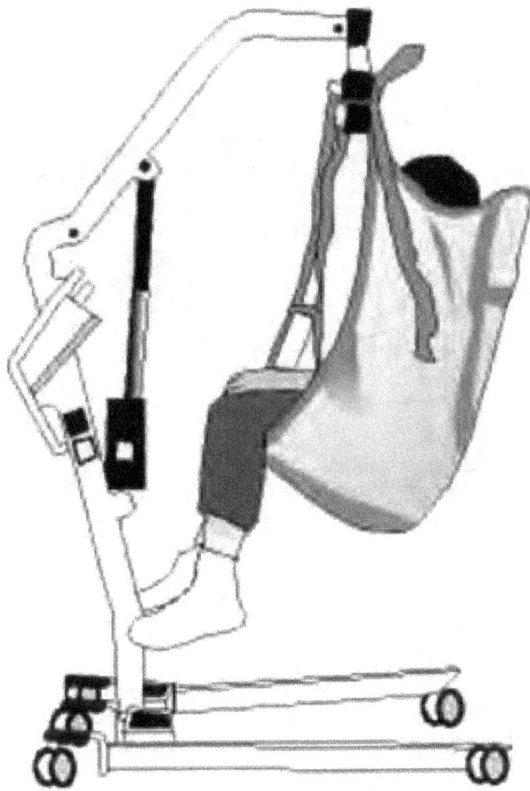

Figure 15.8 Portable lift device (sling type); can be a universal/hammock sling or a band/leg sling. (Adapted from Guidelines for Nursing Homes OSHA 3182-3R 2009)

Ensure device is in good working order before use and rated for the resident weight to be lifted. Apply brakes before positioning resident in or releasing resident from device (Figure 15.10).

Description: Ceiling-mounted lift device.

When to Use: Lifting residents who are totally dependent, are partial- or non-weight bearing, are very heavy, or have other physical limitations. Transfers from bed to chair (wheel chair, Geri, or cardiac chair), chair or floor to bed, for bathing and toileting, or after a resident falls. A horizontal frame system or litter attached to the ceiling-mounted device can be used when transferring residents who cannot be transferred safely between two horizontal surfaces, such as a bed to a stretcher or gurney while lying on their back, using other devices.

Points to Remember: More than one care giver may be needed. Some residents can use the device without assistance. May be quicker to use than portable device. Motors can be fixed or portable (lightweight). Device can be operated by handheld control attached to unit or by infrared remote control. Ensure

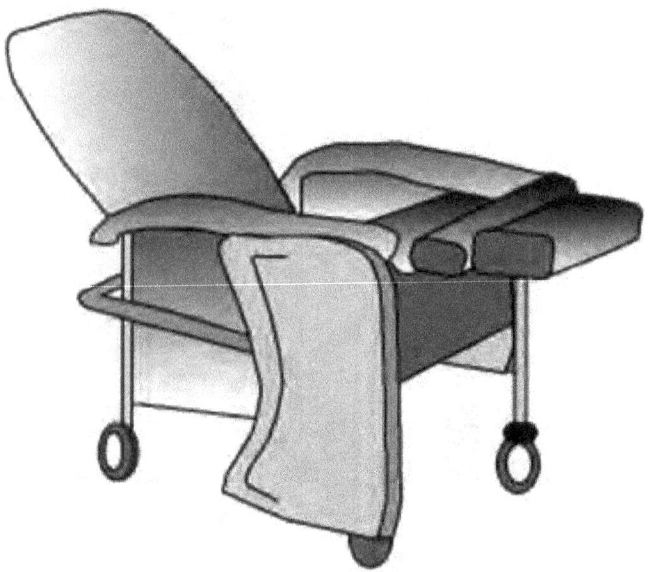

Figure 15.9 Variable position Geri and cardiac chairs. (Adapted from Guidelines for Nursing Homes OSHA 3182-3R 2009)

device is rated for the resident weight. It increases residents' safety and comfort during transfer (Figure 15.11).

Description: Devices to reduce friction force when transferring a resident such as a draw sheet or transfer cot with handles to be used in combination slippery sheets, low friction mattress covers, or slide boards; boards or mats with vinyl coverings and rollers; gurneys with transfer devices; and air-assist lateral sliding aid or flexible mattress inflated by portable air supply.

When to Use: Transferring a partial- or non-weight-bearing resident between two horizontal surfaces such as a bed to a stretcher or gurney while lying on their back or when repositioning resident in bed.

Points to Remember: More than one caregiver is needed to perform this type of transfer or repositioning. Additional assistance may be needed depending upon resident status, for example, for heavier or noncooperative residents. Some devices may not be suitable for bariatric residents. When using a draw sheet combination use a good hand-hold by rolling up draw sheets or use other friction-reducing devices with handles such as slippery sheets. Narrower slippery sheets with webbing handles positioned on the long edge of the sheet may be easier to use than wider sheets. When using boards or mats with vinyl coverings and rollers use a gentle push and pull motion to move resident to new surface. Look for a combination of devices that will increase resident's comfort and minimize risk of skin trauma. Ensure transfer surfaces are at same level and at a height that allows caregivers to work at waist level to avoid extended reaches and bending of the back. Count down and synchronize the transfer motion between caregivers (Figure 15.12).

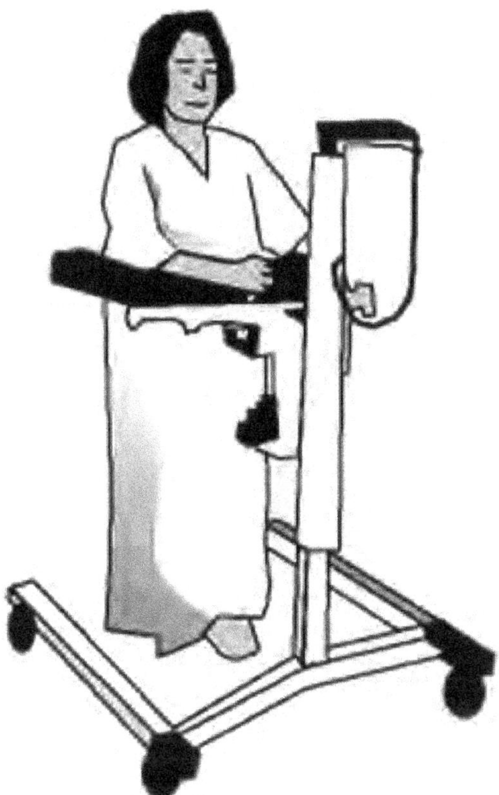

Figure 15.10 Ambulation assist device (Adapted from Guidelines for Nursing Homes OSHA 3182-3R 2009)

Description: Convertible wheelchair, Geri, or cardiac chair to bed; beds that convert to chairs.

When to Use: For lateral transfer of residents who are partial- or non-weight bearing. Eliminates the need to perform lift transfer in and out of wheelchairs. Can also be used to assist residents who are partially weight bearing from a sit-to-stand position. Beds that convert to chairs can aid repositioning residents who are totally dependent, non-weight bearing, very heavy, or have other physical limitations.

Points to Remember: More than one caregiver is needed to perform a lateral transfer. Additional assistance for lateral transfer may be needed depending on the resident's status, for example, for heavier or noncooperative residents. Additional friction-reducing devices may be required to reposition a resident. Heavy-duty beds are available for bariatric residents. Device should have easy-to-use controls located within easy reach of the caregiver, sufficient foot clearance, and wide range of adjustment. Motorized height-adjustable devices are preferred to those adjusted by crank mechanism to minimize physical exertion. Always ensure device is in good working order before use. Ensure

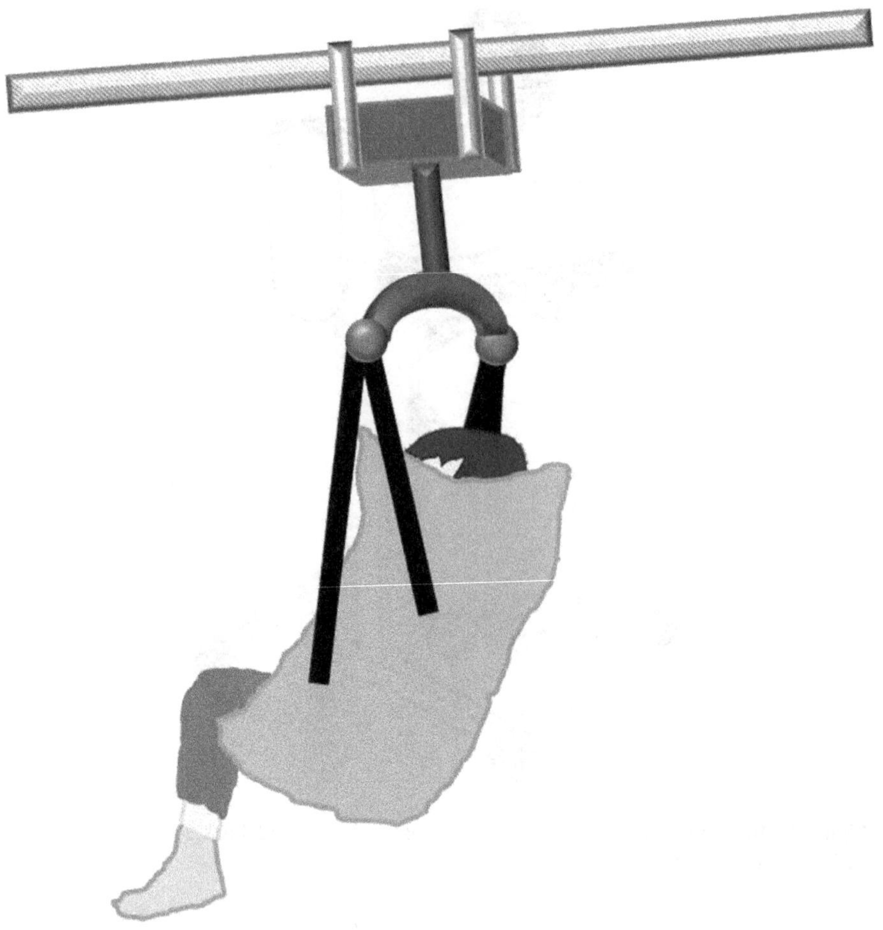

Figure 15.11 Ceiling-mounted lift device (Adapted from Guidelines for Nursing Homes OSHA 3182-3R 2009)

wheels on equipment are locked. Ensure transfer surfaces are at same level and at a height that allows caregivers to work at waist level to avoid extended reaches and bending of the back (Figure 15.13).

Description: Transfer boards – wood or plastic (some with movable seat).

When to Use: Transferring (sliding) residents who have good sitting balance and are cooperative from one level surface to another, for example, bed to wheelchair, wheelchair to car seat or toilet. Can also be used by residents who require limited assistance but need additional safety and support.

Points to Remember: Movable seats increase resident comfort and reduce incidence of tissue damage during transfer. More than one caregiver is needed to perform a lateral transfer. Ensure clothing is present between the resident's skin and the transfer device. The seat may be cushioned with a small towel for comfort, may be uncomfortable for larger residents. It's usually used in

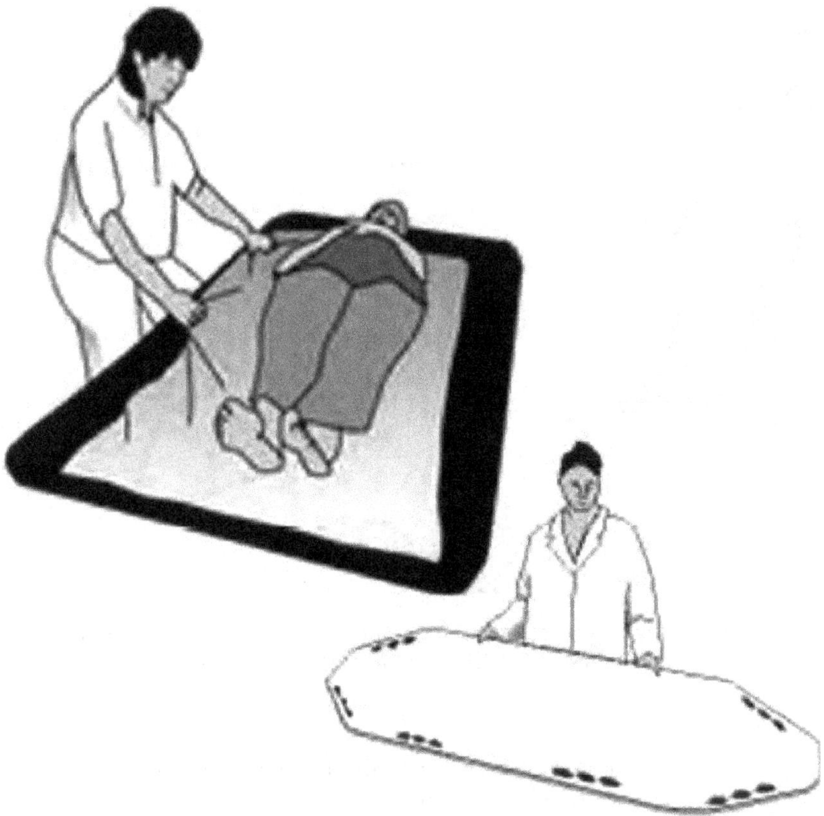

Figure 15.12 Devices to reduce friction force when transferring a resident (Adapted from Guidelines for Nursing Homes OSHA 3182-3R 2009)

conjunction with gait belts for safety depending on resident status. Ensure boards have tapered ends, rounded edges, and appropriate weight capacity. Ensure wheels on bed or chair are locked and transfer surfaces are at same level. Remove lower bedrails from bed and remove arms and footrests from chairs as appropriate (Figure 15.14).

Description: Lift cushions and lift chairs.

When to Use: Transferring residents who are weight bearing and cooperative but need assistance when standing and ambulating. Can be used for independent residents who need an extra boost to stand.

Points to Remember: Lift cushions use a lever that activates a spring action to assist residents to rise up. Lift cushions may not be appropriate for heavier residents. Lift chairs are operated via a handheld control that tilts forward slowly, raising the resident. Residents need to have physical and cognitive capacity to be able to operate lever or controls. Always ensure device is in good working order before use and is rated for the resident weight to be lifted, can aid resident independence (Figure 15.15).

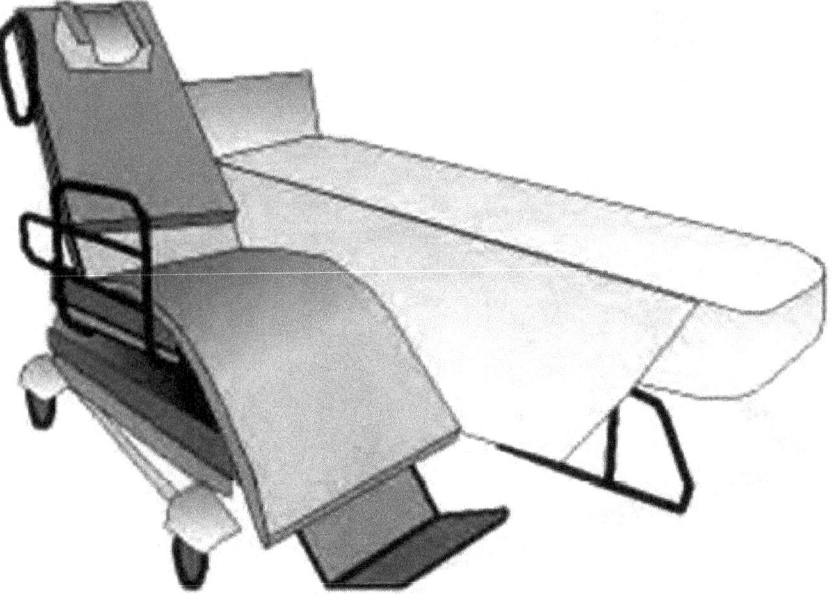

Figure 15.13 Convertible wheelchair, Geri, or cardiac chair to bed; beds that convert to chairs (Adapted from Guidelines for Nursing Homes OSHA 3182-3R 2009)

Description: Stand-assist devices can be fixed to bed or chair or be freestanding. There is a variety of such devices on the market.

When to Use: Transferring residents who are weight-bearing and cooperative and can pull themselves up from sitting to standing position. Can be used for independent residents who need extra support to stand.

Points to Remember: Check that device is stable before use and is rated for resident weight to be supported. Ensure frame is firmly attached to bed, or if it relies on mattress support that mattress is heavy enough to hold the frame, can aid resident independence.

Description: Scales with ramp to accommodate wheelchairs; portable-powered lift devices with built-in scales; beds with built-in scales.

When to Use: To reduce the need for additional transfer of partial- or non-weight bearing or totally dependent residents to weighing device.

Points to Remember: Some wheelchair scales can accommodate larger wheelchairs. Built-in bed scales may increase weight of the bed and prevent it from lowering to appropriate work heights (Figure 15.16).

Description: Gait belts/transfer belts with handles.

When to Use: Transferring residents who are partially dependent, have some weight-bearing capacity, and are cooperative. Transfers such as bed to chair, chair to chair, or chair to car; when repositioning residents in chairs; supporting

Figure 15.14 Transfer boards (Adapted from Guidelines for Nursing Homes OSHA 3182-3R 2009)

residents during ambulation; and in some cases, when guiding and controlling falls or assisting a resident after a fall.

Points to Remember: More than one caregiver may be needed. Belts with padded handles are easier to grip and increase security and control. Always transfer to resident's strongest side. Use good body mechanics and a rocking and pulling motion rather than lifting when using a belt. Belts may not be suitable for ambulation of heavy residents or residents with recent abdominal or back surgery, abdominal aneurysm, and so on. Should not be used for lifting residents. Ensure belt is securely fastened and cannot be easily undone by the resident during transfer. Ensure a layer of clothing is between residents' skin and the belt to avoid abrasion. Keep resident as close as possible to caregiver during transfer. Lower bedrails, remove arms and foot rests from chairs, and other items that may obstruct the transfer. For use after a fall, always assess the resident for injury prior to movement. If resident can regain standing position with minimal assistance, use gait or transfer belts with handles to aid resident. Keep back straight, bend legs, and stay as close to resident as possible. If resident cannot stand with minimal assistance, use a powered portable or ceiling-mounted lift device to move resident (Figure 15.17).

Description: Electric-powered height-adjustable bed.

When to Use: For all activities involving resident care, transfer, repositioning in bed, and so on, to reduce caregiver bending when interacting with resident.

Points to Remember: Device should have easy-to-use controls located within easy reach of the caregiver to promote use of the electric adjustment, sufficient foot clearance, and wide range of adjustment. Adjustments must be completed

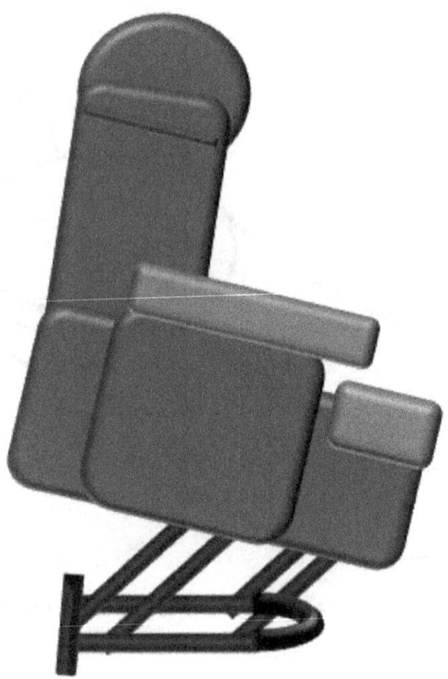

Figure 15.15 Lift cushions and lift chairs (Adapted from Guidelines for Nursing Homes OSHA 3182-3R 2009)

in 20 s or less to ensure staff use. For residents that may be at risk of falling from bed, some beds that lower closer to the floor may be needed. Heavy-duty beds are available for bariatric residents. Beds raised and lowered with an electric motor are preferred over crank-adjust beds to allow a smoother movement for the resident and less physical exertion to the caregiver (Figure 15.18).

Description: Trapeze bar; hand blocks and push up bars attached to the bed frame.

When to Use: Reposition residents that have the ability to assist the caregiver during the activity, that is, residents with upper body strength and use of extremities, who are cooperative and can follow instructions.

Points to Remember: Residents use trapeze bar by grasping bar suspended from an overhead frame to raise themselves up and reposition themselves in a bed. Heavy-duty trapeze frames are available for bariatric residents. If a caregiver is assisting, ensure that bed wheels are locked, bedrails are lowered, and bed is adjusted to caregiver's waist height. Blocks also enable residents to raise themselves up and reposition themselves in bed. Bars attached to the bed frame serve the same purpose. This may not be suitable for heavier residents and can aid resident independence (Figure 15.19).

Description: Pelvic lift devices (hip lifters).

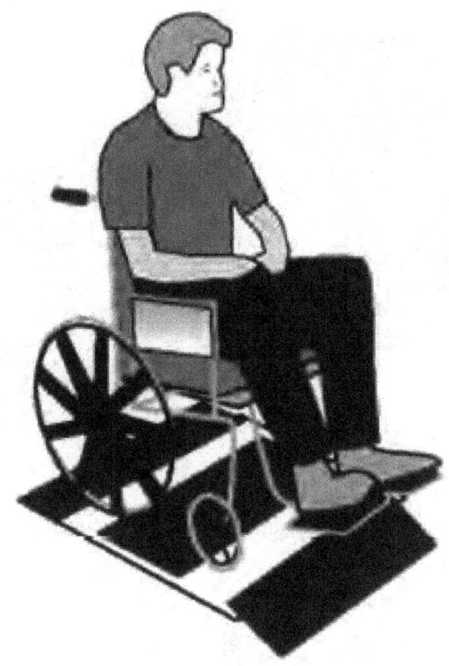

Figure 15.16 Scales with ramp (Adapted from Guidelines for Nursing Homes OSHA 3182-3R 2009)

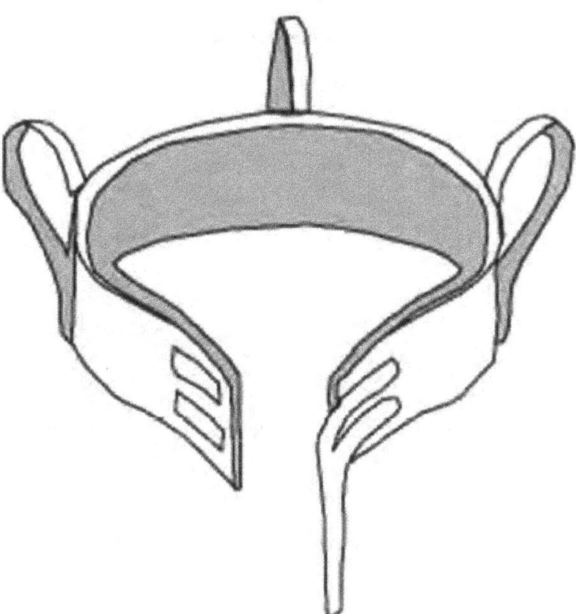

Figure 15.17 Gait belts/transfer belts with handles (Adapted from Guidelines for Nursing Homes OSHA 3182-3R 2009)

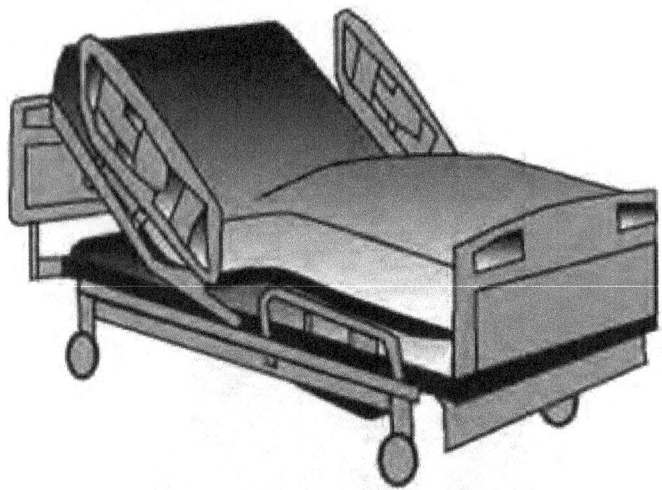

Figure 15.18 Electric-powered height-adjustable bed (Adapted from Guidelines for Nursing Homes OSHA 3182-3R 2009)

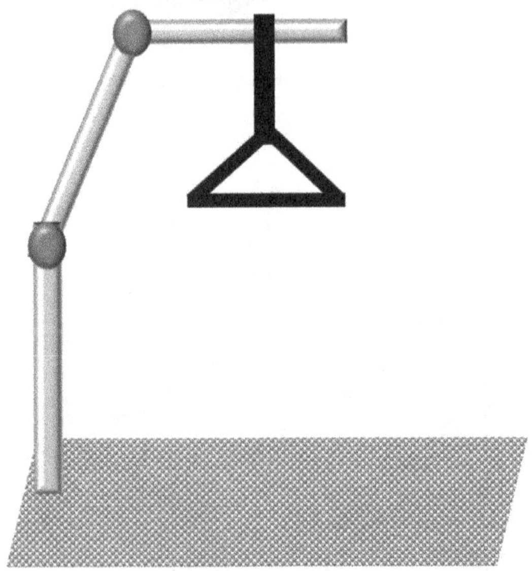

Figure 15.19 Trapeze bar (Adapted from Guidelines for Nursing Homes OSHA 3182-3R 2009)

When to Use: To assist residents who are cooperative and can sit up to a position on a special bedpan.

Points to Remember: Convenience of device may reduce need for resident lifting during toileting. Device is positioned under the pelvis. The part of the device located under the pelvis gets inflated so the pelvis is raised and a special bedpan put underneath. The head of the bed is raised slightly during this procedure.

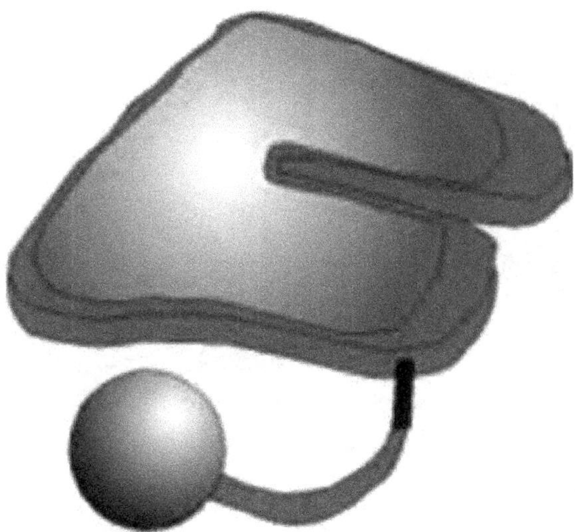

Figure 15.20 Pelvic lift devices (Adapted from Guidelines for Nursing Homes OSHA 3182-3R 2009)

Use correct body mechanics, lower bedrails, and adjust bed to caregivers' waist height to reduce bending (Figure 15.20).

Description: Height-adjustable bathtub and easy-entry bathtubs.

When to Use: Bathing residents who sit directly in the bathtub, or to assist ambulatory residents climb more easily into a low tub, or easy-access tub. Bathing residents in portable-powered or ceiling-mounted lift device using appropriate bathing sling.

Points to Remember: Reduces awkward postures for caregivers and those who clean the tub after use. The tub can be raised to eliminate bending and reaching for the caregiver. Use correct body mechanics, and adjust the tub to the caregiver's waist height when performing hygiene activities. Increases resident safety and comfort (Figure 15.21).

Description: Shower and toileting chairs.

When to Use: Showering and toileting residents who are partially dependent, have some weight bearing capacity, can sit up unaided, and are able to bend hips, knees, and ankles.

Points to Remember: Ensure that wheels move easily and smoothly; chair is high enough to fit over toilet; chair has removable arms, adjustable footrests, safety belts, and is heavy enough to be stable; and that the seat is comfortable, accommodates larger residents, and has a removable commode bucket for toileting. Ensure that brakes lock and hold effectively and that weight capacity is sufficient (Figure 15.22).

Description: Toilet seat risers.

Figure 15.21 Height-adjustable bathtub (Adapted from Guidelines for Nursing Homes OSHA 3182-3R 2009)

When to Use: For toileting partially weight-bearing residents who can sit up unaided, use upper extremities (have upper body strength), are able to bend hips, knees, and ankles, and are cooperative. Independent residents can also use these devices.

Points to Remember: Risers decrease the distance and amount of effort required to lower and raise residents. Grab bars and height-adjustable legs add safety and versatility to the device. Ensure device is stable and can accommodate resident's weight and size (Figure 15.23).

Description: Bath boards and transfer benches.

When to Use: Bathing residents who are partially weight bearing, have good sitting balance, can use upper extremities (have upper body strength), are cooperative, and can follow instructions. Independent residents can also use these devices.

Points to Remember: To reduce friction and possible skin tears, use clothing or material between the resident's skin and the board. Can be used with a gait or transfer belt and/or grab bars to aid transfer. Back support and vinyl padded seats add to bathing comfort. Look for devices that allow for water drainage and have height-adjustable legs, may not be suitable for heavy residents. If wheelchair is used, ensure wheels are locked, the transfer surfaces are at the

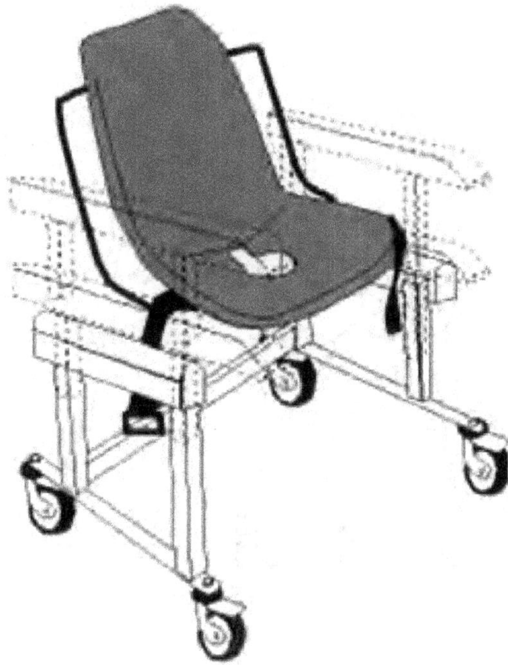

Figure 15.22 Shower and toileting chairs (Adapted from Guidelines for Nursing Homes OSHA 3182-3R 2009)

same level, and device is securely in place and rated for weight to be transferred. Remove arms and foot rests from chairs as appropriate and ensure that floor is dry (Figure 15.24).

Description: Grab bars and stand assists; can be fixed or mobile.
Long-handled or extended shower heads or brushes can be used for personal hygiene.

When to Use: Bars and assists help when toileting, bathing, and/or showering residents who need extra support and security. Residents must be partially weight bearing, able to use upper extremities (have upper body strength), and be cooperative. Long-handled devices reduce the amount of bending, reaching, and twisting required by the caregiver when washing feet, legs, and trunk of residents. Independent residents who have difficulty reaching lower extremities can also use these devices.

Points to Remember: Movable grab bars on toilets minimize workplace congestion. Ensure bars are securely fastened to wall before use (Figure 15.25).

Description: Height-adjustable shower gurney or lift bath cart with waterproof top.

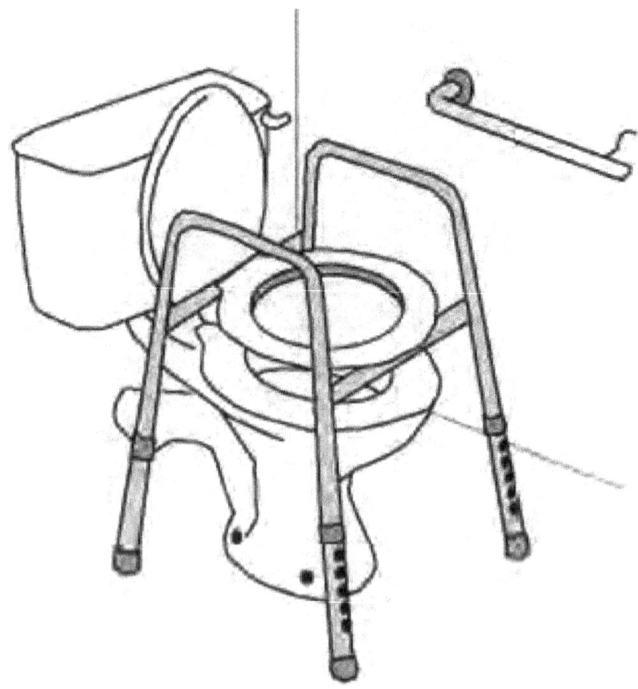

Figure 15.23 Toilet seat risers (Adapted from Guidelines for Nursing Homes OSHA 3182-3R 2009)

When to Use: For bathing non-weight-bearing residents who are unable to sit up. Transfer resident to cart with lift or lateral transfer boards or other friction-reducing devices.

Points to Remember: The cart can be raised to eliminate bending and reaching to the caregiver. Foot and head supports are available for resident comfort, may not be suitable for bariatric residents. Look for carts that are power-driven to reduce force required to move and position device (Figure 15.26).

Description: Built-in or fixed bath lifts.

When to Use: Bathing residents who are partially weight bearing, have good sitting balance, can use upper extremities (have upper body strength), are cooperative, and can follow instructions, useful in small bathrooms where space is limited.

Points to Remember: Ensure that seat rises so resident's feet clear tub, easily rotates, and lowers resident into water, may not be suitable for heavy residents. Always ensure lifting device is in good working order before use and rated for the resident weight. Choose device with lift mechanism that does not require excessive effort by caregiver when rising and lowering device (Figure 15.27).

Description: Use of carts.

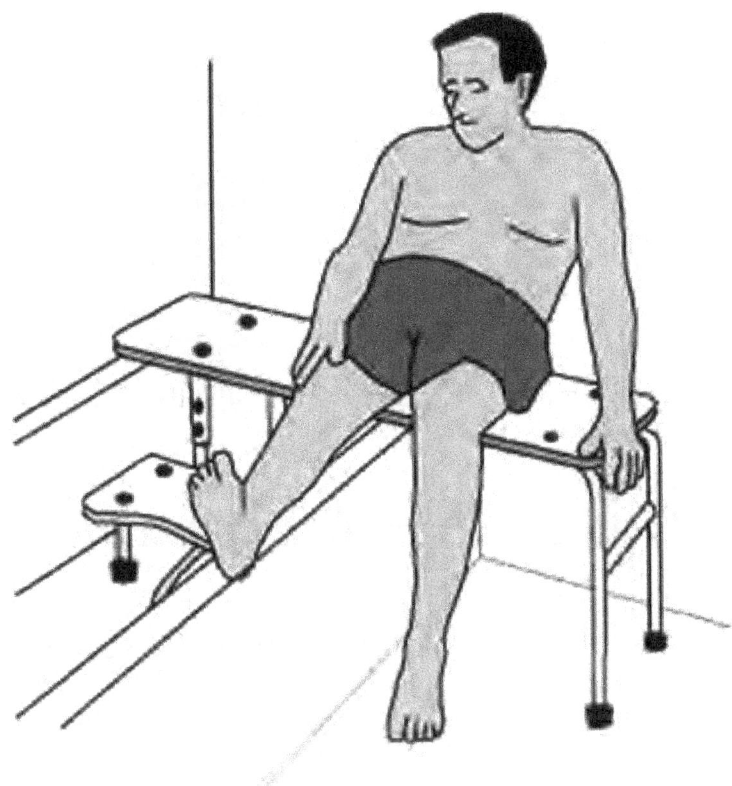

Figure 15.24 Bath boards and transfer benches (Adapted from Guidelines for Nursing Homes OSHA 3182-3R 2009)

When to Use: When moving food trays, cleaning supplies, equipment, maintenance tools, and dispensing medications.

Points to Remember: Speeds process for accessing and storing items. Placement of items on the cart should keep the most frequently used and heavy items within easy reach between hip and shoulder height. Carts should have full-bearing wheels of a material designed for the floor surface in your facility. Cart handles that are vertical, with some horizontal adjustability, will allow all employees to push at elbow height and shoulder width. Carts should have wheel locks. Handles that can swing out of the way may be useful for saving space or reducing reach. Heavy carts should have brakes. Balance loads and keep loads under cart weight restrictions. Ensure stack height does not block vision. Low-profile medication carts with easy-open side drawers are recommended to accommodate hand height of shorter nurses (Figure 15.28).

Description: Work methods and tools to transport equipment.

When to Use: When transporting assistive devices and other equipment.

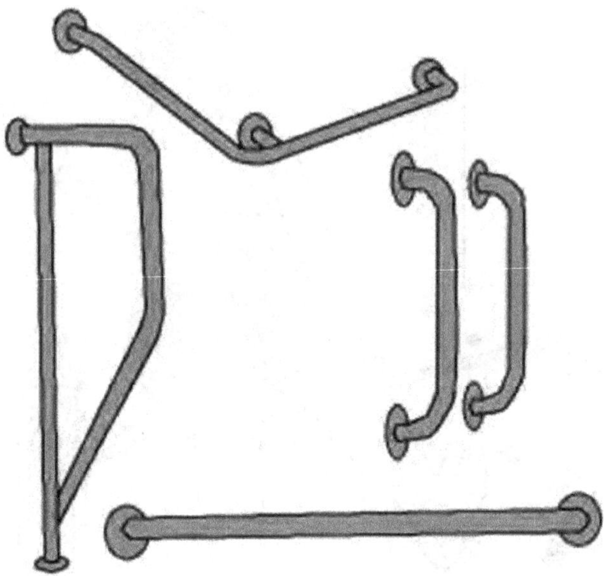

Figure 15.25 Grab bars and stand assists (Adapted from Guidelines for Nursing Homes OSHA 3182-3R 2009)

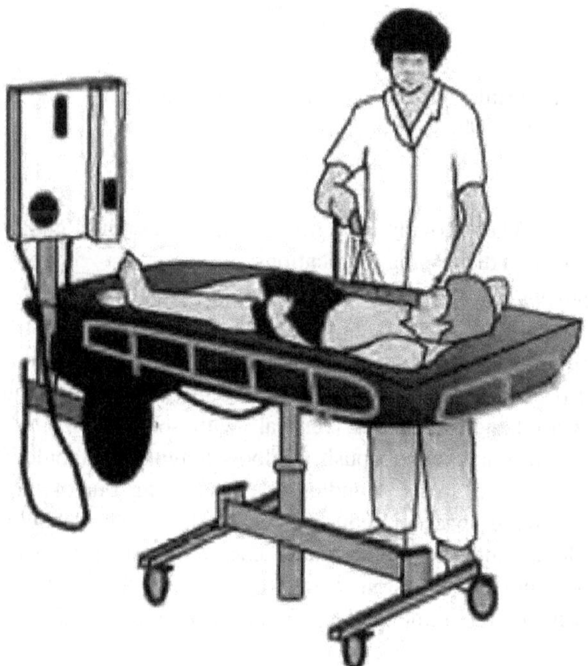

Figure 15.26 Height-adjustable shower gurney (Adapted from Guidelines for Nursing Homes OSHA 3182-3R 2009)

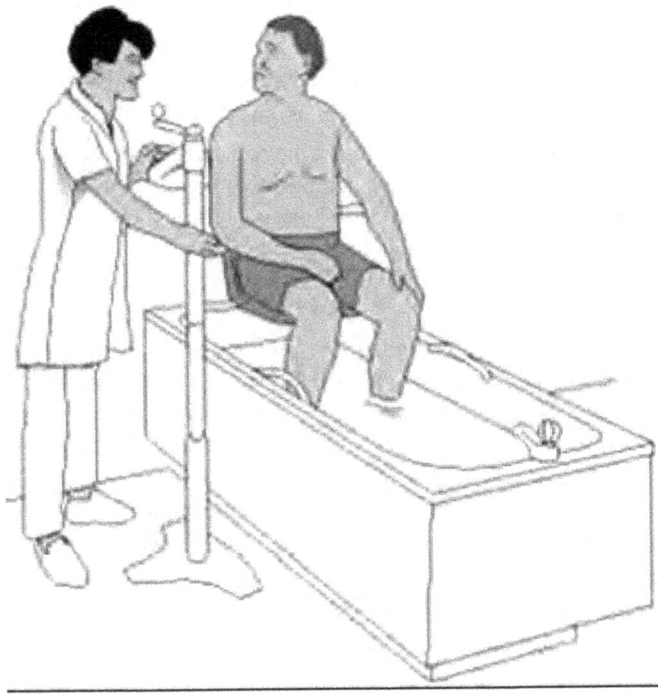

Figure 15.27 Built-in or fixed bath lifts (Adapted from Guidelines for Nursing Homes OSHA 3182-3R 2009)

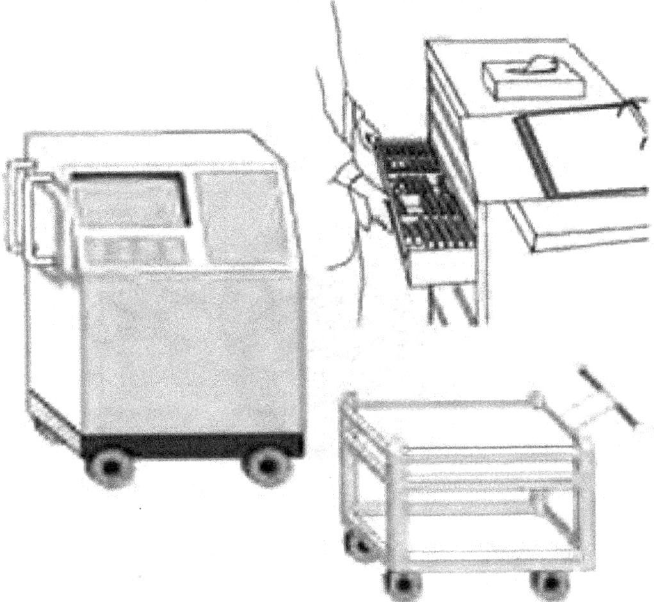

Figure 15.28 Carts (Adapted from Guidelines for Nursing Homes OSHA 3182-3R 2009)

Points to Remember:

- Oxygen tanks: Use small cylinders with handles to reduce weight and allow for easier gripping. Secure oxygen tanks to transport device.
- Medication pumps: Use stands on wheels.
- Transporting equipment: Push equipment, rather than pull, when possible. Keep arms close to the body, and push with whole body and not just arms. Remove unnecessary objects to minimize weight. Avoid obstacles that could cause abrupt stops. Place equipment on a rolling device if possible. Take defective equipment out of service. Perform routine maintenance on all equipment.
- Ensure that when moving and transporting residents, additional equipment such as oxygen tanks and IV/medication poles are attached to wheelchairs or gurneys or moved by another caregiver to avoid awkwardly pushing with one hand and holding freestanding equipment with the other hand (Figure 15.29).

Description: Filling and emptying liquids from containers.

When to Use: In housekeeping areas when filling and emptying buckets with floor drain arrangements.

Points to Remember: Reduces risk of spills, slips, speeds process, and reduces waste. The faucet and floor drain is used in housekeeping. Ensure that casters don't get stuck in floor grate. Use hose to fill bucket. Use buckets with casters to move mop bucket around. Ensure casters are maintained and roll easily (Figure 15.30).

Description: Filling and emptying liquids from containers.

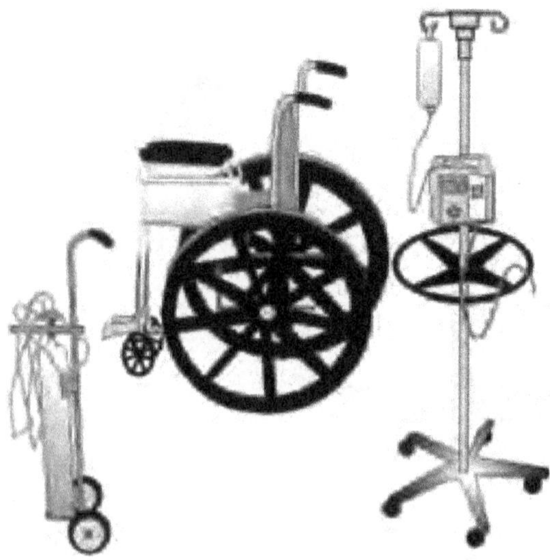

Figure 15.29 Transport equipment (Adapted from Guidelines for Nursing Homes OSHA 3182-3R 2009)

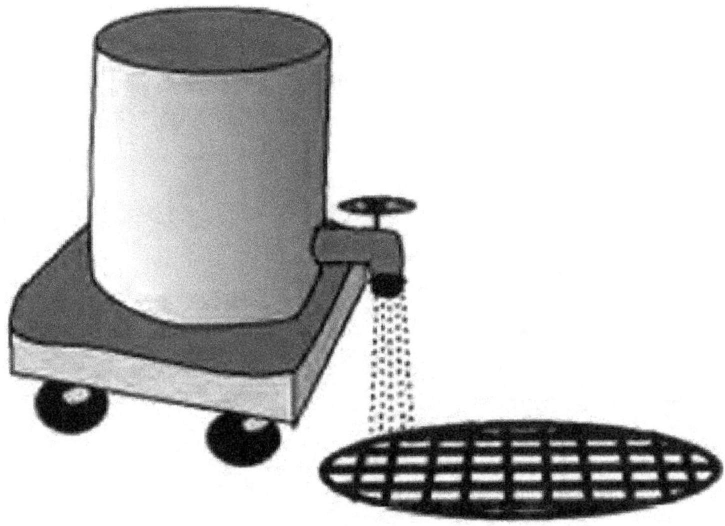

Figure 15.30 Liquid transfer equipment (Adapted from Guidelines for Nursing Homes OSHA 3182-3R 2009)

When to Use: In dietary when pouring soups or other liquid foods that are heavy.

Points to Remember: Reduces risk of spills and burns, speeds process, and reduces waste. Use an elevated faucet or hose to fill large pots. Avoid lifting heavy pots filled with liquids. Use ladle to empty liquids, soups, and so on, from pots. Small sauce pans can also be used to dip liquids from pots. If the worker stands for more than 2 h/day, shock-absorbing floors or insoles will minimize back and leg strain. With hot liquids, ensure a splash guard is included (Figure 15.31).

Description: Select and use properly designed tools.

When to Use: When selecting frequently used tools for the kitchen, housekeeping, laundry, and maintenance areas.

Points to Remember: Enhances tool safety, speeds process, and reduces waste. Handles should fit the grip size of the user. Use bent-handled tools to avoid bending wrists. Use appropriate tool weight. Select tools that have minimal vibration or vibration damping devices. Implement a regular maintenance program for tools to keep blades sharp and edges and handles intact. Always wear the appropriate personal protective equipment.

Description: Spring-loaded carts that automatically bring linen within easy reach.

When to Use: Moving or storing linen.

Points to Remember: Speeds process for handling linen, and reduces wear on linen due to excessive pulling. Select a spring tension that is appropriate for the weight of the load. Carts should have wheel locks and height-appropriate

Figure 15.31 Liquids (Adapted from Guidelines for Nursing Homes OSHA 3182-3R 2009)

handles that can swing out of the way. Heavy carts should have brakes (Figure 15.32).

Description: Equipment and practices for handling bags.

When to Use: When handling laundry, trash, and other bags.

Points to Remember: Reduces risk of items being dropped, and speeds process for removing and disposing of items. Receptacles that hold bags of laundry or trash should have side openings that keep the bags within easy reach and allow employees to slide the bag off the cart without lifting. Provide handles to decrease the strain of handling. Chutes and dumpsters should be positioned to minimize lifting. It is best to lower the dumpster or chute rather than lift materials to higher levels. Provide automatic opening or hardware to keep doors open to minimize twisting and awkward handling (Figure 15.33).

Description: Tools used to modify a deep sink for cleaning small objects.

When to Use: Cleaning small objects in a deep sink.

Points to Remember: Place an object such as a plastic basin in the bottom of the sink to raise the work surface. An alternative is to use a smaller porous container to hold small objects for soaking, transfer to an adjacent countertop for aggressive cleaning, and then transfer back to the sink for final rinsing. Store inserts and containers in a convenient location to encourage consistent use. This technique is not suitable in kitchens/food preparation.

Description: Front-loading washers and dryers.

Figure 15.32 Spring-loaded carts (Adapted from Guidelines for Nursing Homes OSHA 3182-3R 2009)

When to Use: When loading or unloading laundry from washers, dryers, and other laundry equipment.

Points to Remember: Speeds process for retrieving and placing items, and minimizes wear-and-tear on linen. Washers with tumbling cycles separate clothes, making removal easier. For deep tubs, a rake with long or extendable handle can be used to pull linen closer to the door opening. Raise machines so that opening is between hip and elbow height of employees. If using top-loading washers, work practices that reduce risk include handling small loads of laundry, handling only a few items at a time, and bracing your body against the front of the machine when lifting. If items are knotted in the machine, brace with one hand while using the other to gently pull the items free. Ensure that items go into a cart rather than picking up baskets of soiled linen or wet laundry (Figure 15.34).

Description: Work methods and tools to clean resident rooms with water and chemical products.

When to Use: When cleaning with water and chemical products and using spray bottles.

Points to Remember:

- Cleaning implement: use alternate leading hand, avoid tight static grip, and use padded nonslip handles.

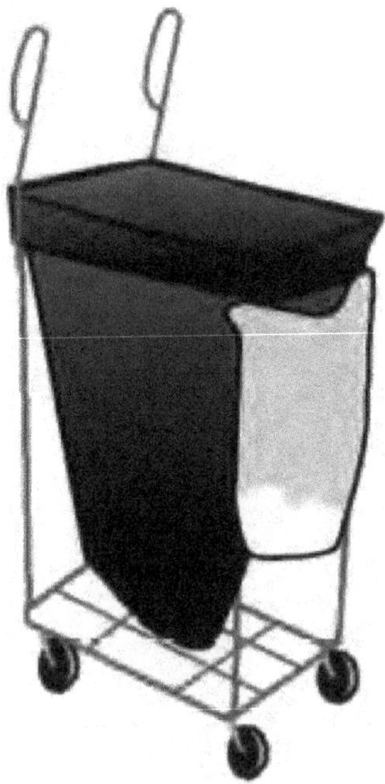

Figure 15.33 Laundry carts (Adapted from Guidelines for Nursing Homes OSHA 3182-3R 2009)

- Spray bottles: Use trigger handles long enough for the index and middle fingers. Avoid using the ring and little fingers.
- For all cleaning: Use chemical cleaners and abrasive sponges to minimize scrubbing force. Use kneepads when kneeling. Avoid bending and twisting. Use extension handles, step stools, or ladders for overhead needs. Use carts to transport supplies or carry only small quantities and weights of supplies. Ventilation of rooms may be necessary when chemicals are used.
- Avoid lifting heavy buckets, for example, lifting a large, full bucket from a sink. Use a hose or similar device to fill buckets with water. Use wheels on buckets that roll easily and have functional brakes. Ensure that casters are maintained. Use rubber-soled shoes in wet areas to prevent slipping.
- Cleaning wheelchairs: Cleaning workstation should be at appropriate height.

Description: Work methods and tools to vacuum and buff floors.

When to Use: Vacuuming and buffing floors.

Points to Remember:

- Both vacuum cleaners and buffers should have lightweight construction, adjustable handles, triggers (buffer) long enough to accommodate at least the index and middle fingers, and easy to reach controls. Technique is important for both devices, including use of appropriate grips, avoiding

Figure 15.34 Laundry machines (Adapted from Guidelines for Nursing Homes OSHA 3182-3R 2009)

tight grips, and for vacuuming, by alternating grip. The use of telescoping and extension handles, hoses, and tools can reduce reaching for low areas, high areas, and far away areas. Maintain and service the equipment and change vacuum bags when 0.5–0.75 full.

- Vacuums and other powered devices are preferred over manual equipment for moderate-to-long duration use. Heavy canisters or other large, heavy equipment should have brakes.

Training is critical for employers and employees to safely use the suggestions identified in these guidelines. Of course, training should be provided in a manner and language that all employees can understand. The following describes areas of training for nursing home employees, their supervisors, and program managers who are responsible for planning and managing the nursing home's ergonomics efforts. OSHA recommends refresher training be provided as needed to reinforce initial training and to address new developments in the workplace.

Nursing assistants and other workers at risk of injury. Employees should be trained before they lift or reposition residents, or perform other work that may involve risk of injury. Ergonomics training can be included with other safety and health training, or incorporated into general instructions provided to employees. Training is usually most effective when it includes case studies or demonstrations based on the nursing home's polices, and allows enough time to answer any questions that may arise. Training should ensure that these workers understand the following:

Training Recommendations

- Policies and procedures that should be followed to avoid injury, including proper work practices and use of equipment
- How to recognize musculoskeletal disorders (MSDs) and their early indications
- The advantages of addressing early indications of MSDs before serious injury has developed
- The nursing home's procedures for reporting work-related injuries and illnesses as required by OSHA's injury and illness recording and reporting regulation.

Training for Charge Nurses and Supervisors

Charge nurses and supervisors should reinforce the safety program of the facility, oversee reporting guidelines, and help ensure the implementation of resident and task-specific ergonomics recommendations, for example, using a mechanical lift. Because charge nurses and supervisors are likely to receive reports of injuries, and are usually responsible for implementing the nursing home's work practices, they may need more detailed training than nursing assistants on:

- methods for ensuring use of proper work practices;
- how to respond to injury reports; and
- how to help other workers implement solutions.

Training for Designated Program Managers

Staff members who are responsible for planning and managing ergonomics efforts need training so they can identify ergonomics concerns and select appropriate solutions. These staff members should receive information and training that will allow them to

- identify potential problems related to physical activities in the workplace through observation, use of checklists, injury data analysis, or other analytical tools;
- address problems by selecting proper equipment and work practices;
- help other workers implement solutions; and
- evaluate the effectiveness of ergonomics efforts.

USE OF MEDICAL EQUIPMENT

There is a wide range of equipment that is used in the medical field. Items include, but are not limited to scalpels, pipettes, needles, instrumentation/monitors, pumps of various types, IV bags, and so on. This section discusses the ergonomic risk factors associated with the use of pipettes.

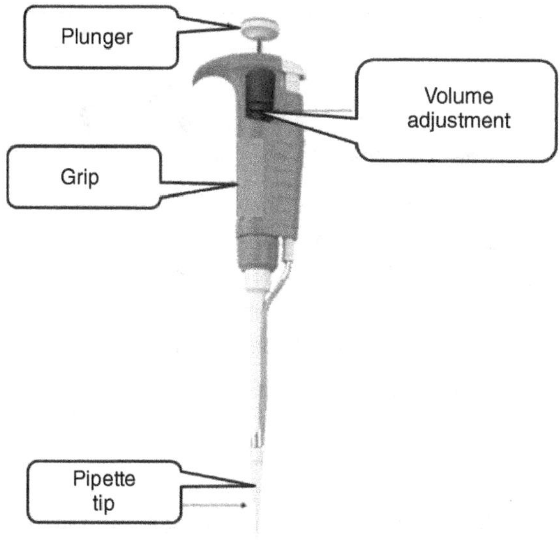

Figure 15.35 Typical mechanical pipette

Pipettes

A typical mechanical pipette, as shown in Figure 15.35, works in the following way. The user holds the pipette in a dagger grip (like an ice pick or similar tool would be held; also called reverse grip) with the index finger under a curved piece, called a finger hook, to keep the hand from sliding up the pipette. While gripping tightly, the user applies force using the hand, wrist, and arm, downward onto a pipette tip to firmly attach it to the end of the pipette. The user's thumb then extends to sit on the top of the pipette where the spring-loaded plunger is located. The thumb exerts pressure on the plunger, then releases, creating a vacuum, in order to aspirate liquid. Applying pressure to the plunger again will eject the liquid from the tip. Most mechanical pipettes have two "stops" on the plunger, one that discharges the liquid and one that is responsible for blowout, driving out any other residual liquid kept in the tip of the pipette. This, along with another trigger used to release the pipette tip from the pipette after use, again uses pressure from the thumb to work. To choose the quantity to dispense, the user twists the plunger, or for fine volume adjustments, the thumb or index finger spins a small dial to increase or decrease volume.

Ergonomic Stressors Related to Pipetting

Thumb/Force When pipetting, the thumb is used almost exclusively to draw out and deliver liquids as well as eject pipette tips. These tasks may be repeated hundreds of times daily and the force needed can vary, but all are significant to the total stress on the thumb. Figure 15.36 shows a representative force displacement curve documenting the strength needed to use a pipette plunger, from one repetition of aspiration to delivery (Asundi et al., 2005). Step 1 is pressing the plunger, using lateral pinch

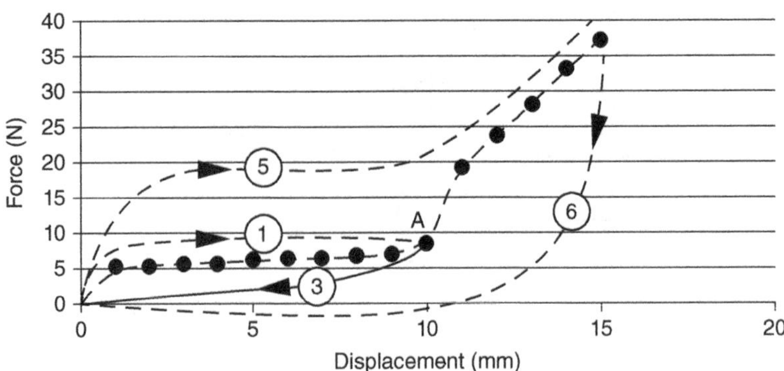

Figure 15.36 Force displacement curve for pipette use cycle (Adapted from Asundi et al. (2005))

strength, to the first stop before releasing to draw out the liquid in step 3. Notice that the greatest amount of force is needed, from approximately 20–35 N (2–3.6 kg) in step 5, in order to complete the blowout of the residual liquid. Step 6 is when the plunger is released. Although these measurements appear high, they are not as high as the amount needed for pipette tip ejection. The force needed to discharge the tip from the end of the pipette can easily reach up to 8 kg of force (Fox, 1999). It is also important to emphasize that the biomechanical force within the joint of the thumb can be over nine times higher the force experienced at the tip of the thumb (Erickson & Woodward, 2001).

All these stressors can lead to damage. One of the most common injuries for a laboratory worker using a pipette is de Quervain's Tenosynovitis. Especially with pipettes that have a longer plunger length and high plunger and tip ejection forces, this thumb injury can cause pain and swelling near the base of the thumb as well as a "sticking" sensation upon movement. If untreated, it can cause pain to travel up the thumb and, in extreme cases, into the forearm (Mayo Clinic, 2012). Other concerns are the wearing of the joint at the base of the thumb. Repetitive use can injure the carpometacarpal joint and lead to arthritis.

Hand and Wrist Movement/Repetitive Motion Pipette grip is an important aspect to pipetting. Hand grip is normally increased when seating a new pipette tip or ejecting one, as well as during the blow out phase. Laboratory workers often pipette multiple sample aliquots while firmly holding the pipette in a static position. This can lead to decreased blood flow in the hand causing fatigue. Nerve injuries such as carpal tunnel syndrome can also occur from holding the pipette too tightly, causing the tendons of the hand to swell, compressing the median nerve in the carpal tunnel area. This can be intensified by pressure on the nerve upon bending the wrist since usually the worker's wrist is not in a neutral position while pipetting. Typical wrist position is often tilted to one side in order to aspirate and deliver aliquots of liquids. According to a study of female laboratory technicians, those who used pipettes more than 300 h/year had a risk of hand ailments five times higher than those with less exposure (Bjorksten et al., 1994).

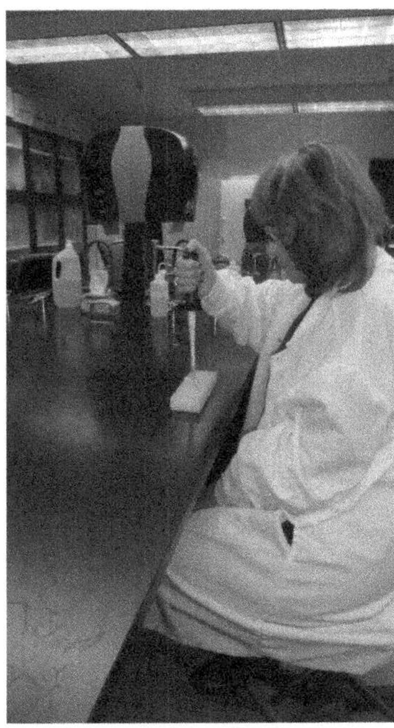

Figure 15.37 Posture of laboratory worker (Adapted photo by Ostrom)

Workstation/Awkward Postures The task of pipetting normally entails nonneutral static postures of the head, neck, and shoulder (Lichty et al., 2011). The head might be tilted forward and down, shoulders hunched, and arm extended at shoulder height in order to effectively complete work (Figure 15.37). Body positioning such as this can compress the blood vessels, as well as restrict the flow of blood in the neck and shoulders causing weakness. This posture can also lead to nerve injuries due to the compression of the head and neck muscles as well as nerves against bones in the upper body. In addition to the awkward posture from using the pipette, normal bench top work can have many ergonomic stressors associated with it. Most pipette work requires workers to sit or stand for long periods of time in a static posture that can lead to WMSDs.

Workstations vary depending on the work performed in a laboratory. Laboratory workers often utilize fume hoods or glove boxes to safely perform work. Unfortunately, while these workstations might mitigate the chemical or biological hazards of the work, they can possibly create ergonomic complications. Often when pipetting in a hood or glove box, reach becomes an issue. Having to fully extend the elbow and wrist to access necessary bottles or other items can lead to strain on the muscles and tendons responsible for straightening the wrist and fingers causing tennis elbow, or lateral epicondylitis (Erickson & Woodward, 2001). A worker may also have to pipette aliquots of liquid in containers while working around a hood sash or other obstruction putting strain on the shoulder and arm (Figure 15.38).

Elbow positioning is another ergonomic issue. In bench top applications, the worker naturally will place their elbow in only a few ways. One is on the bench

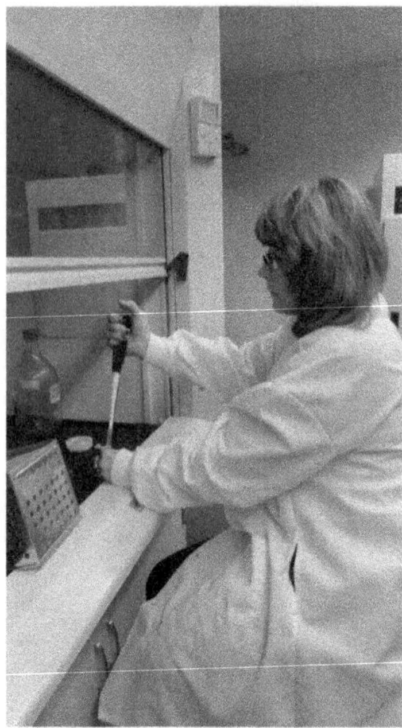

Figure 15.38 Pipetting in a hood (Adapted photo by Ostrom)

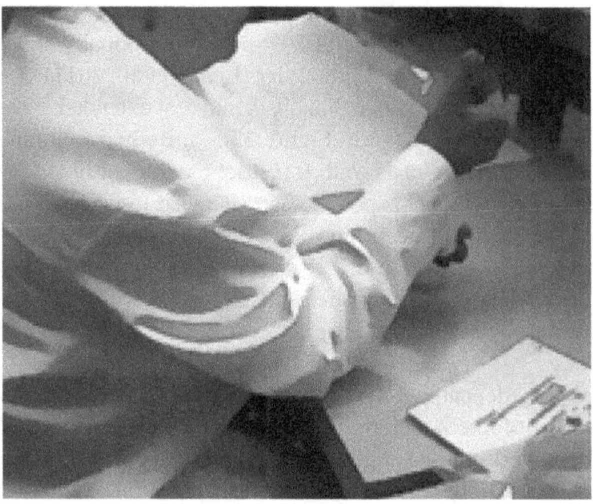

Figure 15.39 Elbow resting on bench top (Adapted photo by Ostrom)

top, as shown in Figure 15.39. Resting the elbow on this hard surface can compress the ulnar nerve in the arm against the medial epicondyle causing cubital tunnel syndrome (Erickson & Woodward, 2001). Similarly, many workers will rest their forearm against the edge of the bench top, often again creating contact stress that can restrict blood flow or cause inflammation.

Other Personal and Contributing Risk Factors

Workload/Stress Although many laboratory workers work in a research and development environment, production laboratories are quite common. With constant deadlines needing to be maintained, stress can contribute to WMSDs. Also, along the same line, signs of WMSDs can occur relating to inadequate recovery time. In addition, some samples such as in a radiological laboratory can only be used for a short period of time due to ALARA considerations so the worker might be rushed to complete their task. This also might elevate the risk of symptoms.

Temperature Temperature can also be a factor. Some laboratories are temperature controlled because of the nature of the work being performed. This can reduce dexterity especially over long periods of time spent in colder temperatures. Temperature, similarly, can pose problems as a result of holding samples that come directly from a freezer, refrigerator, or heating device. All these can exacerbate ergonomic stressors.

Age Age is also a factor. According to McGowan (2011), sensory/motor perception, strength, and movement control, as well as other physiological changes, begin to occur at age 45. These factors can impact abilities such as grasping, reaching, and overall movement.

Physical Condition/Extracurricular Activities/Prior Injuries Ergonomic stressors can be worsened depending on a person's physical condition. Smoking, medical issues, and certain medications can restrict blood flow to the areas of the upper body, affecting the muscles and causing fatigue. Prior injuries to or extracurricular activities using the shoulder, arm, or hand can also be a significant personal risk factor that can contribute to further injury.

Recommendations

Thumb/Force The use of an electronic pipette greatly reduces the force needed by the thumb to complete one cycle of pipetting. By using electronic pipettes, the use of a plunger to aspirate and deliver liquids as well as blow out the pipette tip is completely eliminated. Instead, a light touch of a button will complete these tasks. Unfortunately, they do not eliminate the necessity of the thumb to eject the tip, but with the reduction in force needed for other parts of the pipetting cycle, the cumulative stress on the thumb is greatly decreased.

Costs for electronic pipettes are only around twice as expensive as manual ones, so the cost difference is minor when compared to the cost of an injury. If it is not feasible to use an electronic pipette, using a manual pipette with low spring pressure and short length of travel to reduce thumb force needed is recommended (California Department of Health, 2001).

Hand and Wrist Movement/Repetitive Motion It is important to use a pipette with a finger hook to reduce the time needed to grip the pipette. Finger hooks make it easy to rest the hand before, during, and after a pipette cycle (Erickson & Woodward, 2001). By relaxing the grip, possible injuries from inflammation of the tendons, such as carpal tunnel syndrome, can be decreased.

Keeping the wrist in a neutral posture is also important to reduce the prevalence of WMSDs. By taking the wrist out of neutral position, grip strength can be reduced by up to 42% (Kattel et al., 1996). This means that muscle use is not at full potential. Keeping the wrist neutral and choosing pipettes that do not require the wrist to be bent or tilted to aspirate or deliver liquids can substantially reduce the chance of injury.

In order to aid in muscle recovery and prevent ergonomic hazards, rotating pipetting tasks among several people or between laboratory tasks is suggested (Indiana University, 2014). Taking frequent breaks can also aid in preventing long periods of repetitive motion that occurs while pipetting.

Workstation/Awkward Postures Laboratory workers have a higher prevalence of hand and shoulder ailments than other workers (Bjorksten et al., 1994) (Figure 15.40). One way to reduce these conditions is to decrease the time spent with the arm extended at shoulder height. Shorter pipettes decrease the height of the hand, keeping the arms in a more neutral posture close to the body, effectively reducing the tension on the shoulder that leads to fatigue and discomfort.

Proper workstation surface height is similarly important to reduce WMSDs. If standing, the workstation should be at a height that allows the worker to keep arms close to the body to keep neutral posture as much as possible. Antifatigue mats can also help with static postures while standing since pipetting can be repetitive in nature. If sitting, chair height should be adjusted to maintain proper neutral posture and reduce the need to bend the neck or hunch the shoulders or back. If possible, alternate sitting and standing (Indiana University, 2014).

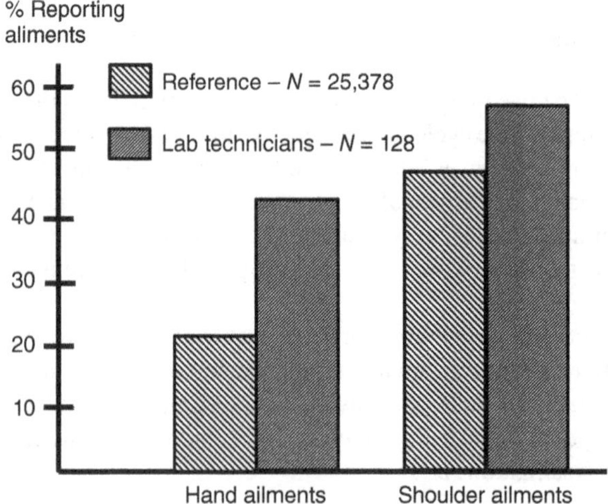

Figure 15.40 Hand and shoulder ailments among female laboratory workers and female state employees (Adapted from Bjorksten et al. (1994))

If working in a hood or other workstation that requires awkward postures, limiting the time or spreading the work throughout the day can aid in muscle recovery to reduce the chance of stress on the body. Keeping work at the edge of the work surface can lower the instances of having to reach and extend the elbow and wrist.

To reduce contact stress injuries to muscles and tendons around the elbow, the workstation should have a padded or curved edge. This will maintain good blood flow to the forearm, wrist, and hand. Elevating the hands should be limited in order to minimize bending the elbow resulting in awkward postures.

CONCLUSION

Using a pipette has many ergonomic issues that can become problematic if not addressed. Fortunately, researchers have noticed this and have been able to give laboratory workers new knowledge and many suggestions to thwart the damage caused by the poor ergonomic tool use in the workplace. With ergonomic programs becoming more commonplace, tasks can be more expertly monitored to promote safe and rigorous ergonomic work practices and work habits that can be used to mitigate a lifetime of risk factors associated with WMSDs. Proper tool use can make a considerable difference in the reduction of injuries in the workplace.

MEDICAL RECORDS CASE STUDY

CASE STUDY

A medical record contains, at a minimum, a 2-year medical history of the patient to include physical examination findings, treatment plans, operative records, consent forms, medication distribution, referral papers, admission and discharge records, and medical certificates. Care providers from different specialties within the hospital (dental, urgent care, and ophthalmology) request records, and then return them to the medical records department, who is responsible for their accuracy and completeness.

Medical records management requires a high level of detail. Inappropriate record keeping can result in medical claims, operative procedure complications, and alleged medical negligence. In addition, confidentiality laws and ethical principles protect the privacy of communication between patient and their doctor. These laws also protect the contents of the medical record that is maintained by a doctor and the hospital.

A general procedure for handling a medical record begins with the medical records department receiving a "pull" request generated from the appointment desk. Care providers are primarily responsible for documenting the patients' health history under their care; such records have a direct relationship with the management of the patients' care. Twice a day, the technicians query the database to obtain a list of records that are needed for the next day's appointment calendar. Technicians spend up to 2 h a day pulling medical records for either appointments or audit purposes and another 2 h filing them back into the system after they are verified and complete. Three of the seven shelves are within easy reach. Some technicians use step stools for the highest shelve (Figure 15.41).

Figure 15.41 Medical record storage

Each department (or specialty) retrieves the records before the shift begins and returns them at the end of the day; these records are not necessarily complete. Returned records are verified for completeness by the records department; pending laboratory or X-ray results and other documentation (patient sheets), not directly noted by the caregiver in the record, must be retrieved from a computer system, printed and manually filed into the record. This process is time consuming and accounts for a majority of the technicians' time. It results in batch sorting records on the technician's work surface. For example, sorting categories include completed, incoming, needs laboratory work. The limited size of the workstation surface requires batches and individual records be handled unnecessarily. In other words, the technicians move piles of records to make room for more urgent work.

Once the record is complete, it is filed back into/onto the shelves. Other responsibilities include assembling a new record and archiving. The department "pulls" approximately 5000 appointment records a month. In addition, a minimum of 200 records are "pulled" for audits and another 50 for other purposes (e.g., needs to be copied, patient relocating). Each record is handled a minimum of four times and an average of six times: 5250 records × 4 moves = 21,000 moves/month; 5250 records × 6 moves = 31,500 moves/month.

Ergonomics Hazards Handling Records Within the Storage System

Managing medical records in the current storage system requires repetitive, awkward postures while pulling or filing records that are above shoulder height and below elbow height. Sixty percent of the records are outside of the comfortable reach envelope for standing. Technicians also pull records while kneeling or standing on a step stool.

An automated storage system, regardless of the increase in storage space, will reduce reaching in awkward postures and provide a standing work surface. Having a work surface in front of the file shelves reduces records handling because the patient sheets can be filed in-place instead of bringing the record to the desk.

The recommendations were to install an automated record storage equipment shown in Figure 15.42.

The Interim recommendations included the following:

- Use rolling tool stools when accessing lowest shelves.
- Use rolling tables when managing records within the isle.
- Continue with archiving to decrease the volume of records and, when feasible, consolidate records to the center three shelves.
- Implement stretch and flex into the work day to oxygenate fatigued muscles and increase circulation.

HOSPITAL CASE STUDY

Ergonomic Risk Assessment for a Hospital

Introduction This report summarizes the ergonomic risk assessment conducted at a hospital. Two areas were observed in order to determine solutions to reduce ergonomic stressors: the labor and delivery patient rooms and the patient transport by ambulance gurney. This assessment is based upon interviews with nurses, corpsmen,

Figure 15.42 Automated record storage system

emergency medical technicians (EMTs), local fire department response personnel, and base safety personnel. In addition, the hospital ergonomic team identified the priority areas and suggested improvements.

The operations reviewed present opportunities to reduce the risk of WMSDs and improve safety, health, and productivity. MSDs are injuries and illnesses that affect muscles, nerves, tendons, ligaments, joints, spinal discs, skin, subcutaneous tissues, blood vessels, and bones. WMSDs are

- Musculoskeletal disorders to which the work environment and the performance of work contribute significantly or
- Musculoskeletal disorders that are aggravated or prolonged by work conditions.

LABOR AND DELIVERY WARD

Description of the Operation

Patients arrive at the emergency room entrance via ambulance or personal vehicle. Typically, for patients arriving via personal vehicle, one or two nurses help the patient out of the personal vehicle and into a wheelchair, then out of the wheelchair and onto an Obstetrics/Gynaecology stretcher for transport to the labor and delivery room.

For the labor and delivery procedure, the OB/GYN stretcher leg support collapses downward. As the leg support folds down, the nurse must pull on the seating area to slide the patient closer to the edge of the stretcher for care. In addition, leg supports (stirrups) are difficult to adjust and in the labor and delivery room, the limited space makes it difficult to care for the patient from all sides (see Figure 15.43).

Ergonomic Issue Description

The major ergonomic risk factors for the labor and delivery task are excessive lifting from transferring patients, and awkward postures and excessive force caused by operating the stretcher.

Excessive Lifting

As long ago as 1965, a study of the physical loads on nurses noted that an adult human being is awkward to lift or carry because it is not a rigid load and it has no handles. The study further noted that a patient lying in a bed is particularly hard to lift just by virtue of physical position relative to the lifter. Added to the physical burden is the mental stress of knowing that a human being can be severely damaged if dropped or handled incorrectly. The study observed that loads on nurses may be worse than those on industrial workers.

Half a century later, the situation of nurses remains unchanged. Manually lifting and transferring patients is a high-risk activity, both for the healthcare worker and for the patient. Of all occupations, nursing has one of the highest incidences of work-related back problems.

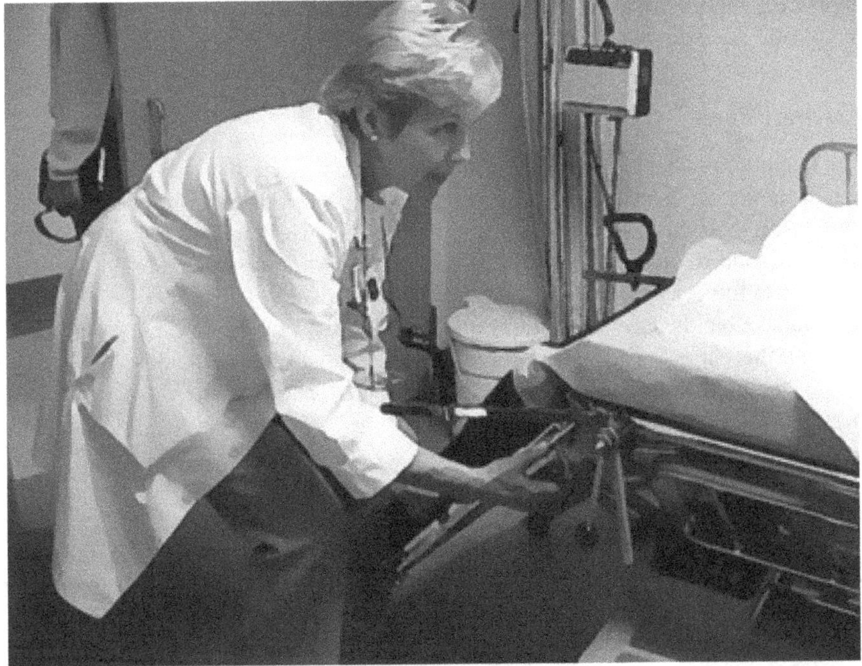

Figure 15.43 Nurse assumes awkward posture while exerting force to pull the patient

Recommendations

- Ensure that yearly back injury prevention training covers proper body mechanics, lifting techniques, stretching, and information on the use of patient transfer/handling equipment. Refer to Appendix A.
- Where feasible, use a mechanical assist patient transfer aids such as a portable lift to transfer patients from a personal vehicle to the stretcher or wheelchair. Use raise-to-stand units to help patients stand and walk. Refer to Appendix B.
- Where feasible, use manual assist patient transfer/handling aids such as gate belts to assist with patient movement and lateral transfer. Refer to Appendix B.
- Where feasible, transfer the patient directly onto the OB/GYN stretcher.
- Since OB patients are typically 30–60 lb heavier than other patients, the hazard abatement program will provide bariatric wheelchairs; these larger wheelchairs allow more room for patient transfer and comfort.

Awkward Posture

The patient rooms do not allow the nurse to easily care for the patient from all sides. The space restriction forces the nurses to bend forward or twist at the trunk while providing care. Depending on the type of care being administrated, the nurse may have to assume this posture for a few seconds to a few minutes. The muscles must apply considerably more contraction force to hold awkward postures, particularly if the position is maintained for more than a couple of seconds.

Excessive Force

The height of the hands affects the amount of force needed to push or pull an object. When the hands are slightly above waist height, a worker gets the most from the muscles. As the hands are moved lower or higher, the working posture becomes more awkward, and the muscles must exert more force. Nurses in the labor and delivery ward exert force while in an awkward posture to position the patient for treatment. The nurse's hands are well below waist height when pulling on the leg support to position the patient. Performing forceful exertions can irritate tendons, joints, and discs, leading to inflammation, fluid buildup, and constriction of blood vessels and nerves in the area.

RECOMMENDATION

Replace the dated OB/GYN stretcher. The facility is evaluating three OB/GYN stretchers with smaller dimensions to increase space in the patient room and reduce reaching. Features of the new model include lightweight construction; side rails; 8-in. locking casters; and easy glide controls that are operated from a standing, not stooped, posture to provide quick and easy patient positioning with reduced force. This stretcher also has integrated foot supports that increase patient comfort and provide ease of operation by the nurse.

PATIENT TRANSPORT

Injury Data

Documented shoulder and arm strain from attempting to arrest a falling gurney with patient when the equipment failed to lock into position.

Description of the Operation

Ambulance personnel are faced with an ever-changing, uncontrolled work environment in which patients are commonly moved from homes with narrow passageways to the ambulance for transportation to hospitals. Once at the hospital, the ambulance crew transports the patient from the ambulance to the emergency room, then transfers the patient to a short-term care ward stretcher for treatment.

Loading the Gurney

At the pickup site, with the patient on the gurney, the ambulance technicians position the gurney at the rear of the ambulance. One technician folds up the gurney's legs, disengages the wheel locks, and pushes the gurney into the ambulance (Photo 2). The second technician stands at the side of the gurney to help with guiding and

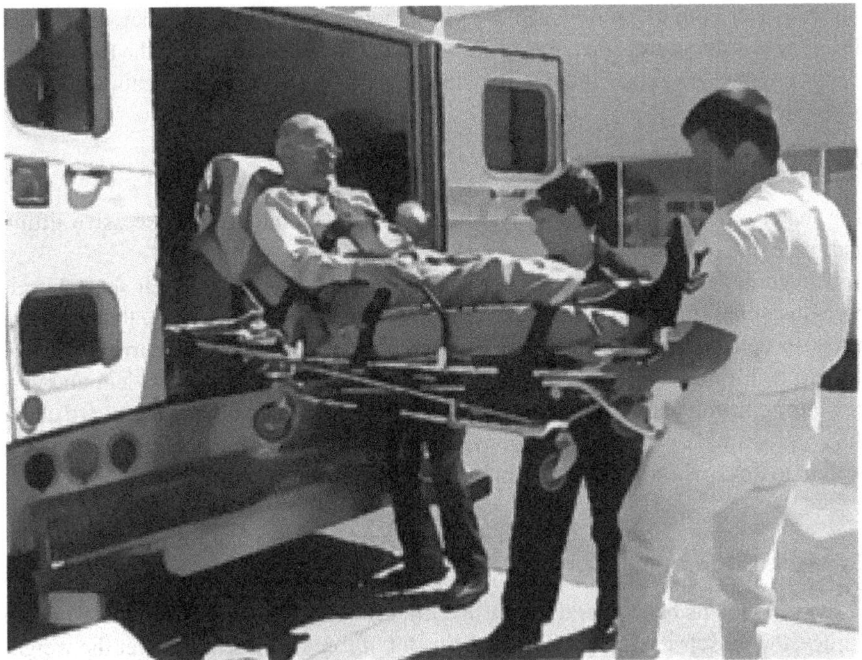

Figure 15.44 Patient transport

to reassure the patient. During this process, the technician at the end of the gurney bears the weight of both the gurney and the patient while pushing the gurney into the ambulance. The process is complicated because the patient's feet interfere with the gurney's base controls and handhold at the foot end, as shown in Figure 15.44.

Unloading the Gurney

The gurney is removed from the ambulance with one technician outside the ambulance at the foot of the gurney and the other technician inside the ambulance. The technician at the foot pulls the gurney out until the base drops and supports the gurney weight. Again, the patient's feet can obstruct the handhold and require the technician to bend, as shown in Photos 2 and 3. The gurney lacks standard safety equipment to arrest a fall from the ambulance in the event the base does not engage or technician footing is compromised.

Gurney Operation

The gurney is raised and lowered with one technician at each end, as shown in Figure 15.44. To lower the gurney, the technician at the foot of the gurney disengages a lock located beneath the patient's feet. The feet of a tall patient make it difficult for the technician to grasp the mechanism. Locks at the head and the foot of the gurney must be disengaged at the same time. Once the locks are disengaged, the crew bears the weight of both the gurney and the patient while lowering the gurney. If one end disengages without the other, the gurney can fall. To raise the gurney, the technicians disengage the locks and lift the gurney until they hear an audible indication.

Ergonomic Issue Description

The major ergonomic risk factor for the ambulance technicians is excessive lifting and force due to manual handling of the gurney.

The gurneys used at this facility have limited height positions. Their short wheelbase and small diameter wheels make them unstable on some terrain, thus creating a risk to both the technician and the patient if a gurney falls. The gurneys are not tension controlled; therefore, the technician bears the weight of the patient and gurney during lifting and lowering. The handles on the foot end of the gurney are under the cot, forcing the technician into awkward postures. The gurneys have a no catch system to arrest a gurney falling from the back of the ambulance.

Excessive Lifting

When technicians bend over to perform a lift, such as when raising and lowering the gurney, the muscles in the back must exert a lot of force to raise and lower the weight of the upper body. This causes the back muscles to fatigue more rapidly and puts pressure on the discs in the lower back. When technicians have to maintain awkward postures for more than a few seconds, their back muscles and discs experience the application of a large amount of static force. The problem becomes worse when either greater weight or greater distance is required.

If the weight of the load were to suddenly shift while being lifted, the resulting awkward posture, combined with the weight and distance of the load from the lower spine, could tear tendons, ligaments, and muscles.

Recommendations

- Ensure that yearly back injury prevention training covers proper body mechanics, lifting techniques, stretching, and information on the use of patient transfer/handling equipment.
- The staff are evaluating two stretchers with the following features: a catch mounted at the rear of the ambulance to arrest a fall; a minimum of five height positions; weight-sensitive tension control or lift assist; a single hand lever to set cot height and lock release; lightweight construction; 6-in.-diameter wheels with locks; foot-end lifting system; and four-point patient restraining harness.

KEY POINTS

Working in health care is a very stressful job, both from a psychological and physical perspective.

Use of the correct patient handling devices and techniques can help to reduce the potential for injuries associated with patient handling.

The use of medical devices also poses a risk to the user from an ergonomic perspective. Proper use of these devices and the proper selection of the device in the first place can help to reduce the potential for injury.

REVIEW QUESTIONS

1. What factors make lifting a patient so potentially hazardous to healthcare professionals?
2. How will the current trend of Americans becoming heavier impact the potential for a WMSD?
3. Surgeons use devices such as retractors and scalpels on a daily basis. What types of injuries might they develop?
4. Seating is important for healthcare professionals. What factors would be important in the design of these sorts of chairs?
5. The posture a healthcare professional has to attain to move a patient can be quite awkward. Besides the devices listed in this chapter, what other types of equipment might be important?
6. Seating is important for healthcare professionals. What factors would be important in the design of these sorts of chairs?
7. The posture a healthcare professional has to attain to move a patient can be quite awkward. Besides the devices listed in this chapter, what other types of equipment might be important?

REFERENCES

Asundi, K., Bach, J., and Rempel, D. (2005). Thumb Force and Muscle Loads Are Influenced by the Design of a Mechanical Pipette and by Pipetting Tasks. *Human Factors*, 47(1), 67–76.

Björksten, M., Almby, B., and Jansson, E. (1994). Hand and Shoulder Ailments Among Laboratory Technician Using Modern Plunger-Operated Pipettes. *Applied Ergonomics*, 25(2), 88–94.

Bureau of Labor Statistics. (2011). *USDL-11-16-12*. News Release.

California Department of Health. (2001). Lab Workers: Take the Pain Out of Pipetting. Retrieved from CDPH: http://www.cdph.ca.gov/programs/hesis/Documents/labwork.pdf.

Erickson, J. and Woodward, B. (2001). Smart Pipetting: Using Ergonomics to Prevent Injury. Retrieved from us.mt.com: http://us.mt.com/dam/RAININ/PDFs/ErgoPapers/pipetting-ergonomics-prevent-injury.pdf.

FDA. (2006). Hospital Bed System Dimensional and Assessment Guidance to Reduce Entrapment.

Fox, T. (1999). *A Pipette with Ergonomic Benefits*. American Biotechnology Laboratory.

Indiana University. (2014). Ergonomics in the Laboratory. Retrieved from ehs.iupui.edu: http://ehs.iupui.edu/ergonomics.asp?content=ergonomics-in-the-laboratory.

Kattel, B., Fredericks, T., Fernandez, J., and Lee, D. (1996). The Effect of Upper-Extremity Posture on Maximum Grip Strength. *International Journal of Industrial Ergonomics*, 18, 423–429.

Lichty, M., Janowitz, I., & Rempel, D. (2011). Ergonomic Evaluation of Ten Single-Channel Pipettes. *Work*, 39(2), 177–185.

Mayo Clinic. (2012). Diseases and Conditions: De Quervain's Tenosynovitis. Retrieved from Mayoclinic.org: http://www.mayoclinic.org/diseases-conditions/de-quervains-tenosynovitis/basics/definition/con-20027238.

McGowan, B. (2011). *Ergonomics and Aging Population.* Retrieved from http://www.humantech.com/blog/ergonomics-and-the-aging-population/

Occupational Safety and Health Administration, John L. Henshaw, Assistant Secretary, OSHA 3182-3R, 2009.

Workplace Health Promotion. (n.d.). Retrieved 2015, from CDC: http://www.cdc.gov/workplacehealthpromotion/implementation/topics/disorders.html.

ADDITIONAL SOURCE

Guidelines for Nursing Homes Ergonomics for the Prevention of Musculoskeletal Disorders, U.S. Department of Labor, Elaine L. Chao, Secretary.

16

CASE STUDIES

The following section includes a number of ergonomic case studies where hazards were recognized and evaluated and recommendations made to reduce the extent of the exposure.

REPAIR PROCESS

The repair process consists of the following steps:

- Disassembly
- Send out for cleaning
- Painting process
- Reassembly
- Release mechanism test
- Load test
- Quality check
- Storage.

These steps and the related ergonomic issues are discussed below.

Occupational Ergonomics: A Practical Approach, First Edition.
Theresa Stack, Lee T. Ostrom and Cheryl A. Wilhelmsen.

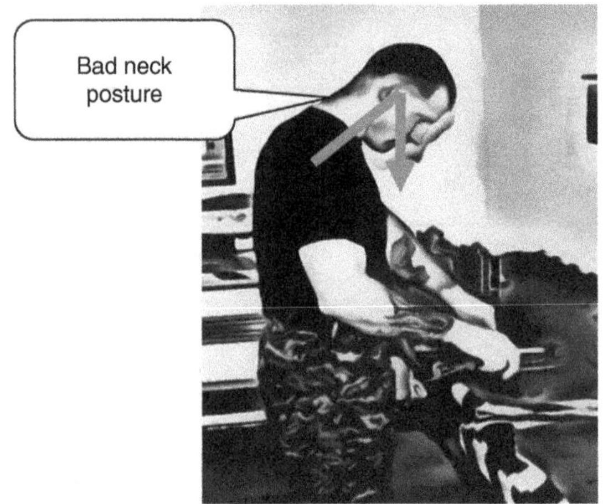

Figure 16.1 Disassembly task – 1 (Graphic by Lee Ostrom)

Disassembly

The disassembly task takes approximately 30 min/unit. The steps to perform the disassembly are as follows:

- Remove unit from storage rack.
- Place into support jig on workbench.
- Disassemble rack.

Figure 16.1 shows a technician preparing to perform the disassembly. Note that his elbows are at approximately 135° angle. This task would be considered light work, so the angle of the elbow joint should be no more than 100–110°. Figure 16.2 shows the typical posture of a technician while disassembling the device. Notice that he has his legs wide apart so that he can get low enough to work on the device without bending his back.

Simple hand tools are used to take the device apart. These are shown in Figure 16.3.

Cleaning and Painting

Once the device is disassembled, it is sent out for cleaning. There is a 1-day turnaround time for this step. The device is then painted. There is a half-day curing time for the paint. A cleaning device was purchased but has not been installed. The new device can clean five racks at a time and will save considerable turnaround time in this process.

Reassembly

The device is reassembled and this process takes approximately 45 min. The device is placed on a supportive rack during parts of the process. Figure 16.4 shows this

Figure 16.2 Disassembly task – 2 (Graphic by Lee Ostrom)

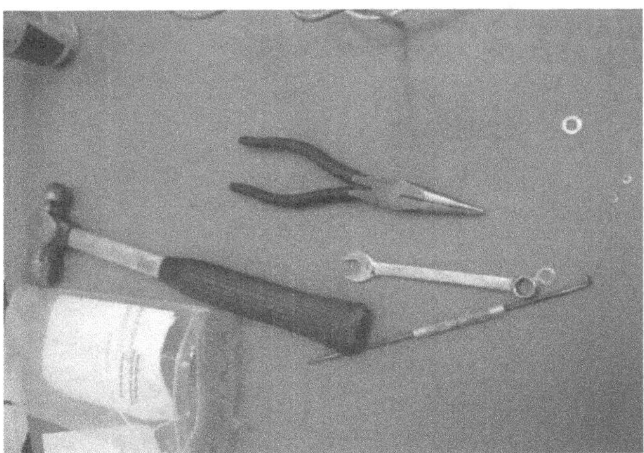

Figure 16.3 Hand tools

step. The reassembly process is considered easier because it requires less force to accomplish. The same workstation is used for reassembly as that of disassembly.

Mechanism Test

This step is accomplished by a technician using electronic test equipment. It is a simple step, taking a few minutes. However, as Figure 16.5 shows, the technician is

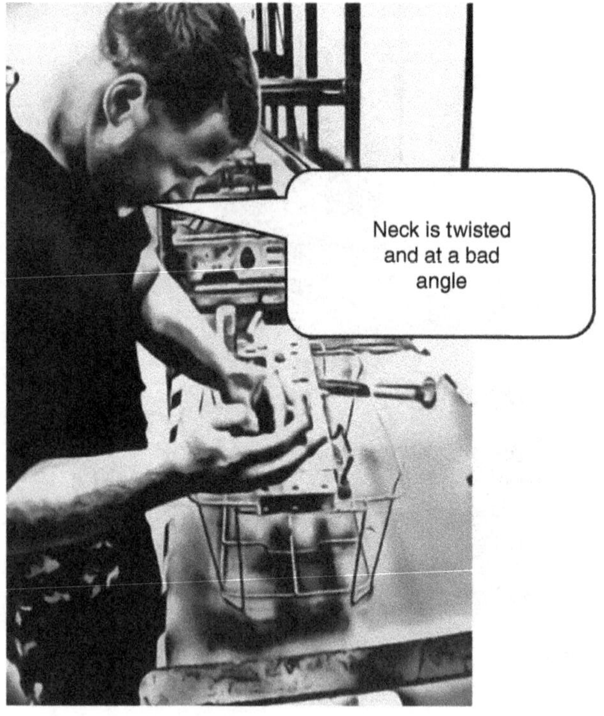

Figure 16.4 Reassembly (Graphic by Lee Ostrom)

Figure 16.5 Release mechanism test (Graphic by Lee Ostrom)

slightly bent forward while he performs this test. The table height is approximately 38 in.

Load Test

The most difficult step in this process is the load test. The steps involved in the load test are as follows:

- The device rack is placed on a steel load test machine at a height of 32 in.
- The safety cage is 46 in. high.
- Technician has to climb into the cage to place the device on the tester.
- Technician has to move in and out of the cage 13 times to accomplish one test.
- Technician is bent over at the waist 90° each time he enters the cage.

When the device is placed in the cage, the technician has to kneel down to make the initial connections. The technician's knee is placed onto the horizontal cross assembly, and the lower leg is compressed on that sharp cross member. This process causes the compression on the soft tissues and causes fatigue to the opposite leg. Figure 16.6 shows this step.

In the process of performing this test, the technician enters the cage multiple times to check if the device is connected appropriately. Figure 16.7 shows the posture the technician must attain to do this step.

Next the technician closes the cage and begins the test by using a hand-pumping hydraulic pump. The technician kneels on the floor to pump the handle (Figure 16.8).

The technician enters the cage up to four more times to complete the test. Figure 16.9 shows the posture he must assume to accomplish this.

Quality Check and Storage

Finally, a quality check is performed on the reassembled device, and the device is stored on racks as shown in Figure 16.10.

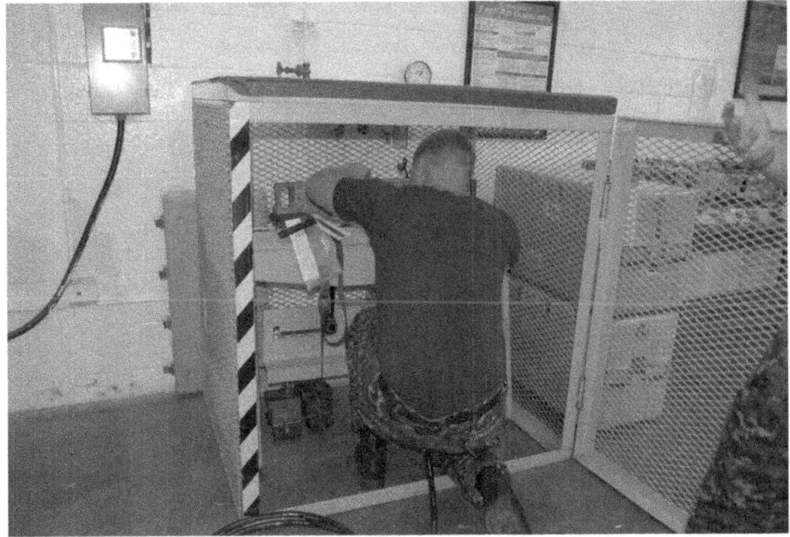

Figure 16.6 Connecting device to test machine (Graphic by Lee Ostrom)

Figure 16.7 Checking connection to test equipment (Graphic by Lee Ostrom)

Figure 16.8 Hand-pumping hydraulic pump (Graphic by Lee Ostrom)

Figure 16.9 Entering cage (Graphic by Lee Ostrom)

Figure 16.10 Storage racks (Graphic by Lee Ostrom)

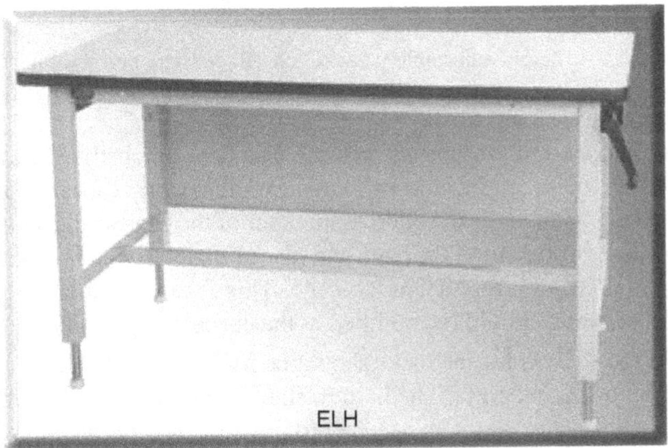

Figure 16.11 Manual adjustable height work bench (Photo with permission from Pro-Line)

Recommendations

1. An adjustable height fixture should be developed so that the device can be maintained at an appropriate work height for each technician. If this cannot be provided, then an adjustable height work bench should be provided. There are a wide variety of choices in adjustable height work benches, ranging from simple hand-crank models as shown in Figure 16.11 to electrically adjustable models as shown in Figure 16.12. The cost of these benches ranges from $800 on up.
2. Redesign the load test cage. There are several options to redesigning the cage. The first is to build a cage of 7 ft tall and 4 ft wide so the technician can enter

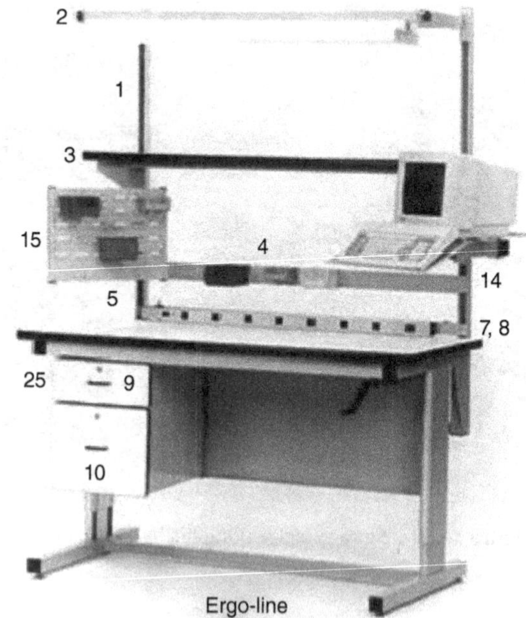

Figure 16.12 Electric height adjustable work bench (Photo with permission from Pro-Line)

and exit the cage in an upright posture and have unobstructed access to the test article. The second would be to add lockable roller casters to the load test machine and remove the lower cross member of the cage so the tech can roll the test machine in and out of the cage and place the device onto the machine from an unrestricted posture outside the cage. This is an extremely ergonomically stressful task and should be modified at the earliest opportunity.

3. The storage shelves for the racks should be modified so that the top shelf is no more than 60 in. in height. The bottom shelf should be at approximately 30 in. and the middle shelf can be placed approximately in between the two. This will reduce the ergonomic stress of placing the devices in and out of storage.
4. The cleaning device that was purchased should be placed in service. This will better optimize the process.

ERGONOMIC RECOMMENDATIONS FOR BACKPACK WELDER APPARATUS

Introduction

In a review of the design of a backpack welding/cladding system from an ergonomic perspective, there were only two pieces of physical dimensions of the proposed backpack welding/cladding system found when reviewing the information. This was that it would weigh between 40 (18 kg) and 60 lb (27.2 kg) and be approximately 29 in. (73.7 cm) in diameter. The exact design of the backpack was not available to review. Therefore, in this regard, ergonomic guidelines were provided from a variety of sources for the design of the backpack.

Weight of Load

There are not good guidelines for weight of the load that can be carried on the shoulders. Mil Std 1472F states that for a mixed population of males and females the maximum weight should be 42 lb (19 kg) and weights should only be carried 33 ft (10 m) (Mil Std 1472F, 1999). For a male-only population, the weight can be up to 82 lb (37.2 kg), with carry distances being again 33 ft. However, military backpacks infantry soldiers carry are much heavier. A NATO commissioned study puts the maximum weight for backpacks for a strong person at 95 lb (43 kg). The question is, of course, "what is a strong person?" (Ros).

Snook's tables (Liberty Mutual Insurance Company) for carrying loads in front of a person show that 90% of the male population can carry 41 lb (18.6 kg) 28 ft (8.5 m) every 5 min (hand height 33 in.) and 57% of the female population can carry 40 lb (18 kg) every 5 min (hand height 31 in.). The tables also show that 69% of the male population, and less than 10% of the female population would find an approximately 60-lb (27.2 kg) load acceptable (same hand heights). From all these sources, it appears that a 40-lb backpack would be acceptable to 90% of the male population and over 50% of the female population. A 60-lb backpack would limit the population that could perform this task.

Strap and Belt Configuration

Wide, padded straps should be used that distribute the weight over the widest area possible. Though we could not find good guidance, a survey of quality mountaineering backpacks shows the approximate width to be 2.5 in. (6.3 cm). A sternum strap helps to prevent the straps from slipping off the shoulders. The load must be symmetrical or divided equally on each shoulder (Mil Std 1472F, 1999; Healthworks Medical Group, 2015).

Most modern mountaineering backpacks are also contoured to the back.

A hip belt should be provided. The weight of the load should be distributed one-third on the shoulders and two-thirds on the hips (Ros).

Center of Mass

Most all studies found concerning the biomechanics of loads concern lifting and carrying in front of a person and not carrying in the back. There are some on children's book bags, but these are not applicable to this application. However, the center of mass of the backpack should be as close to the body as possible. The further the center of mass is away from the user's back, the greater the movement there would be to try to topple the person backward. Table 16.1 shows these movements. Note that these movements are calculated from the back and not from the core. If calculated from the center of the user's core, the movement would be even higher.

Figure 16.13 from an Aarn backpack advertisement shows the comparison of a large load in which the center of mass is solely on the back and one in which the

TABLE 16.1 Center of Mass and Moments

Center of Mass from Body (in.)	40 lb Weight (ft-lb)	60 lb Weight (ft-lb)
6	20	30
12	40	60
18	60	90
24	80	120

Figure 16.13 Comparison of the center of mass from a standard backpack and an Aarn backpack (Permission from Aarn)

load is spread from back to front. Note that the posture of the person with a standard backpack is inclined forward. The user would be even more inclined forward if the center of mass of this backpack was farther back. A hip belt might help counter the effect of the center of mass not being very close to the body. However, it will not eliminate it.

Using a backpack weighing 40 lb (18 kg) with the center of mass 12 or more inches (30.5 cm) from the user would possibly cause a smaller individual great difficulty navigating stairs and small hatchways. In fact, there are no guidelines for how much weight a person can carry up- and downstairs. Keep in mind that if the backpack is 29 in. in diameter (73.7 cm) and the user is large, the combined measurement from the front of the person to the back of the backpack could be 45 or more inches (114 cm). The hatchway openings would have to be large enough to allow access. Also, it might actually be difficult to descend certain staircases face forward if the slope of the stairs is at a high angle. The back of the backpack might snag on the stair risers as the person descends them. There also might be a situation where an emergency develops, and a person has the backpack on. The person must be able to exit the emergency area with

Figure 16.14 Aarn bodypack and standard backpack (aarnpacks, 2015)

the backpack on. Also, the straps should allow the person to drop the backpack easily (quick release) (Figure 16.14).

The guidance we can provide for the center of mass is keep it as close to the body as possible. Also, a backpack 29 in. in diameter (73.7 cm) might be problematic for negotiating small hatchways and stairways.

Lifting and Lowering the Backpack

We recommend that the backpack be positioned on a stand so the user can back into the backpack in a standing posture and then snap on the straps and hip belt. The same setup should be positioned where the device is used. This would eliminate the need to pick up the backpack and swing it into position on the back. Another option would be to have a second person always help the user put the backpack on and remove it.

Summary of Guidance and Recommendations

1. The backpack should not be more than 40 lb.
2. The shoulder straps should be 2.5 in. wide and padded. The backpack should be contoured to the back.

3. There should be a hip belt and sternum strap.
4. The weight of the backpack should be divided one-third on the shoulders and two-thirds on the hips.
5. The load must be symmetrical.
6. The center of mass of the backpack should be as close to the body as possible.
7. The diameter of the backpack must be small enough to fit through a hatchway with a large individual.
8. The straps should be quick release.
9. A stand should be used to help the person put on and take off the backpack, and/or a second person should always help the user put on and take off the backpack.

BEAD-BLASTING OPERATION CASE STUDY

Purpose and Introduction

The purpose of this case study is to document the findings from an ergonomic assessment of the aircraft bead-blasting facility. Also, to present alternative potential solutions to alleviate the musculoskeletal stressors associated with these sorts of decoating operation.

This decoating operation uses modern, safer plastic beads and recovers and reuses the decoating material. The impurities and spent beads are captured in a bag house and disposed of appropriately.

The workers who are tasked with performing this task are motivated and wish to have the task optimized. Several of the workers have experienced musculoskeletal injuries. The most common injury report is carpel tunnel syndrome (CTS). The workers are required to wear heavy coveralls with a disposable synthetic suit over the top. They also wear vinyl gloves, a cotton gauntlet type glove over the vinyl gloves, and a supplied air hood with apron and hearing protection. The supplied air hood can provide cooled or heated air, depending on the time of year. Figures 16.15 and 16.16 show workers preparing for the decoating task.

The air hoods the workers use sit on the head via head band webbing. On the day the assessment was conducted, the workers did not wear knee pads. The workers might perform this task from between 2 h a day up to 6 h a day. This is the actual time the workers spend suited up and spraying decoating material. The task constraints are listed below:

- The workers have to suit up in the protective gear.
- The workers have to have the ability to get away from a broken hose.
- The task requires that the workers articulate their wrists in several deviated postures to direct the spray of decoating beads under the edges of floor beams and angled/bent structural members.
- Potentially all surfaces of the aircraft must be decoated.

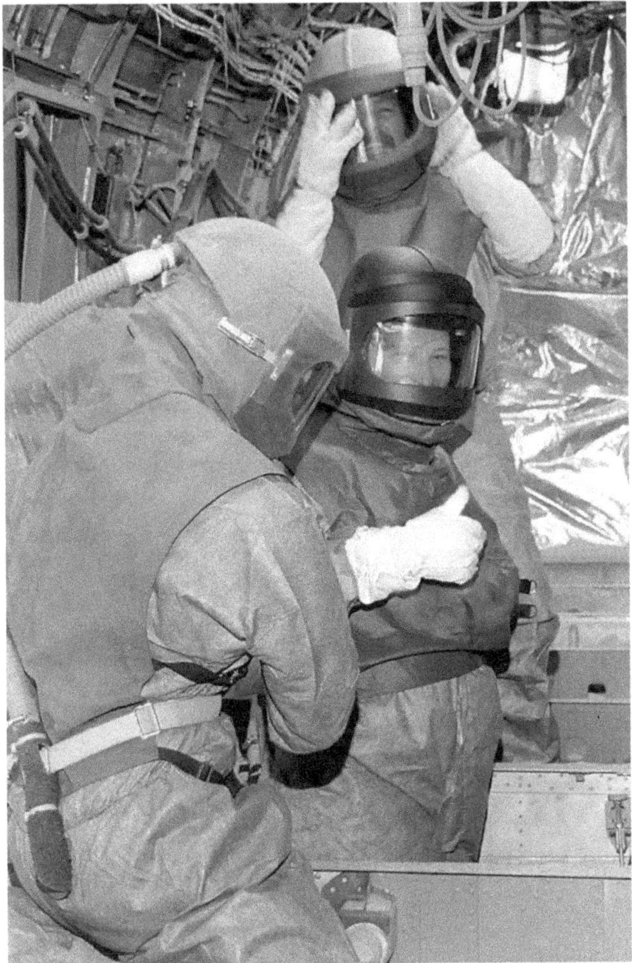

Figure 16.15 Decoating (Graphic by Lee Ostrom)

Ergonomic Assessment Procedures

The ergonomic assessment of the decoating was conducted in the following manner:

- Introduction to the tasks
- Observation of the workers
- Interviews with the workers
- Ergonomist suited up and participated in the decoating operation
- Evaluated alternative potential solutions.

Findings

Figures 16.17–16.22 show aspects of the decoating task. This task has all the attributes of a very stressful task. It requires the workers to wear several layers of protective

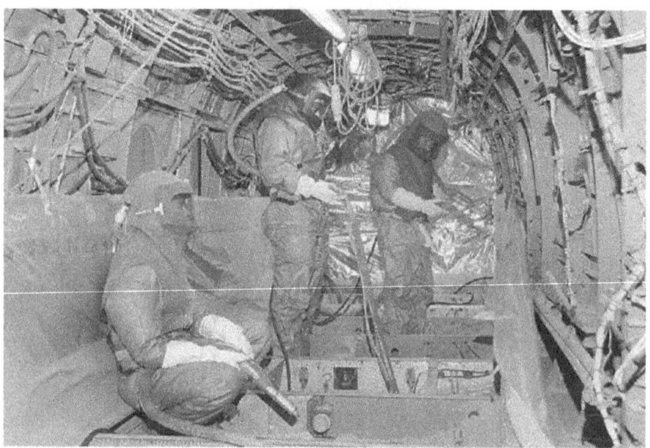

Figure 16.16 Decoating 2 (Graphic by Lee Ostrom)

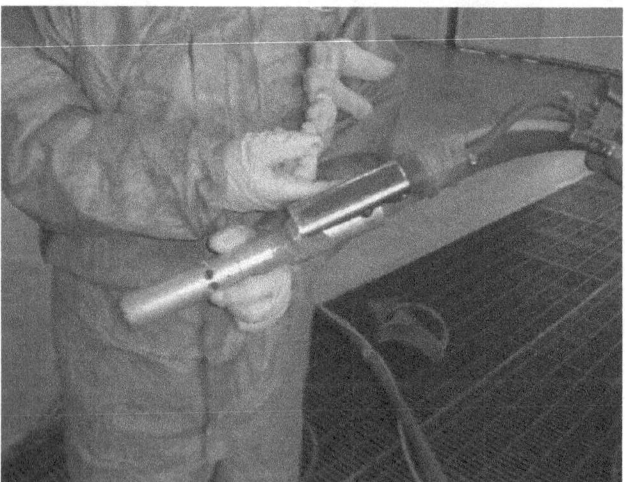

Figure 16.17 Nozzle (Graphic by Lee Ostrom)

gear. The task requires the workers to use the equipment with their wrists in deviated postures, and the task requires the workers to attain several stressful postures. The workers have to grip the nozzle and depress a dead-man lever to allow the stream of decoating beads to exit the nozzle. However, the actual grip on the lever, once it is depressed, is relatively light. Though, a grip has to be maintained on the lever or it will stop the stream. The grip required on the nozzle barrel to hold and direct it to the surface to be decoated varies from a relatively high level to a relatively low level depending on where the stream is being directed. For instance, the worker can brace the hose against their shoulder and maintain a lighter grip if they are spraying directly in front of them. However, this posture is not possible if they are directing the spray alongside of a floor beam. Also, the nozzle is a smooth and somewhat polished

Figure 16.18 Standing posture (Graphic by Lee Ostrom)

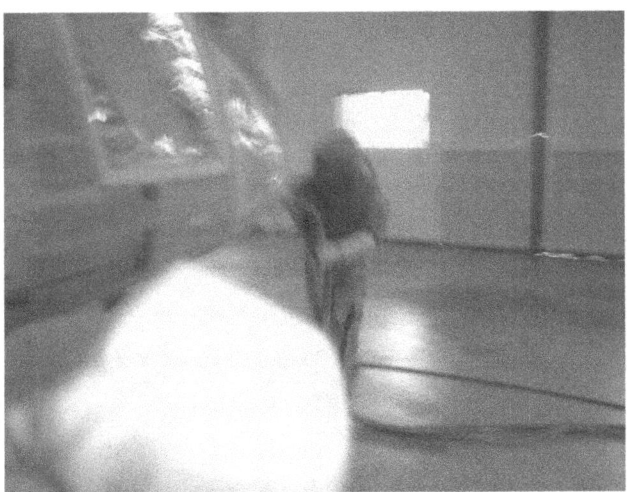

Figure 16.19 Standing posture 2 (Graphic by Lee Ostrom)

finish. Therefore, a stronger grip has to be maintained on it to prevent the nozzle from slipping through the hand (Figures 16.17–16.22).

The ergonomist who performed the assessment found that they started with their right hand, shifted to their left hand when their left hand became fatigued, then used both hands.

Since the air hood helmet sits on the head via head band webbing, it tends to want to slip off the head in any posture, other than standing/kneeling/sitting straight up. The task requires all postures, so a part of the effort of this task is trying to keep the hood in place so one can see the part they are decoating.

Figure 16.20 Kneeling posture (Graphic by Lee Ostrom)

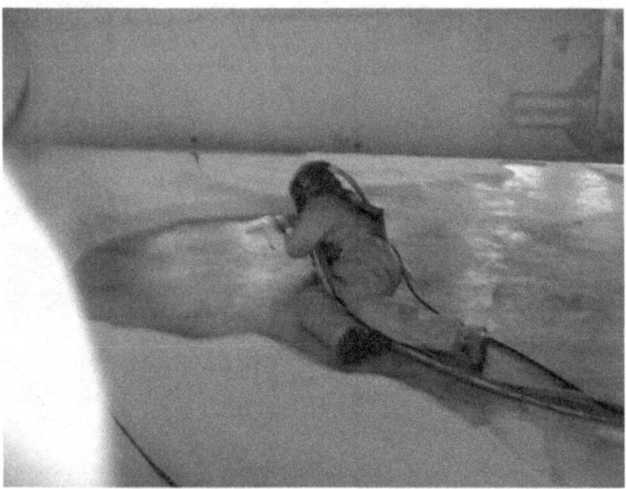

Figure 16.21 Another required posture (Graphic by Lee Ostrom)

The easiest way to perform the task is in an upright, standing posture. However, this is not possible because all postures are required.

Evaluation of Solutions and Recommendations

Table 16.2 defines the different levels of controls for hazards related to ergonomics.

Using this table as a guide, the possible controls that can be used to reduce the stress related to this task are discussed below:

Figure 16.22 Aircraft to ground clearance (Graphic by Lee Ostrom)

TABLE 16.2 Levels of Hazard Control

1. **Elimination** – A redesign or procedural change that eliminates exposure to an ergonomics risk factor; for example, using a remotely operated soil compactor to eliminate vibration exposure
2. **Engineering controls** – A physical change to the workplace; for example, lowering the unload height of a conveyor
3. **Substitution** – An approach that uses tools/materials/equipment with lower risk; for example, replacing an impact wrench with a lower-vibration model
4. **Administrative** – This approach is used when none of the above is practical to implement. Administrative controls are procedures and practices that limit exposure by control or manipulation of work schedule or the manner in which work is performed. Administrative controls reduce the exposure to ergonomic stressors and thus reduce the cumulative dose to any one worker. If you are unable to alter the job or workplace to reduce the physical stressors, administrative controls can be used to reduce the *strain and stress* on the work force. Administrative controls are most effective when used in combination with other control methods; for example, requiring two people to perform a lift

- Supplying workers with a better fitting helmet
- Improve gripping:
  - Modifying the nozzles so they have a better gripping surface
  - Purchasing nozzles with a better gripping surface
  - Finding a glove with a better gripping surface.
- Parking the aircraft on the grating, using jacks to elevate the aircraft, and using a creeper to better access the underside of the aircraft
- Using coveralls with integral kneepads.

The most thorough, long-term hazard reduction strategy modifications to the facility should be considered. It is speculated that a considerable time savings would occur with reduced processing and clean-up times associated with facility modifications.

In the event all other recommendations are not implemented, then a worker rotation schedule should be developed.

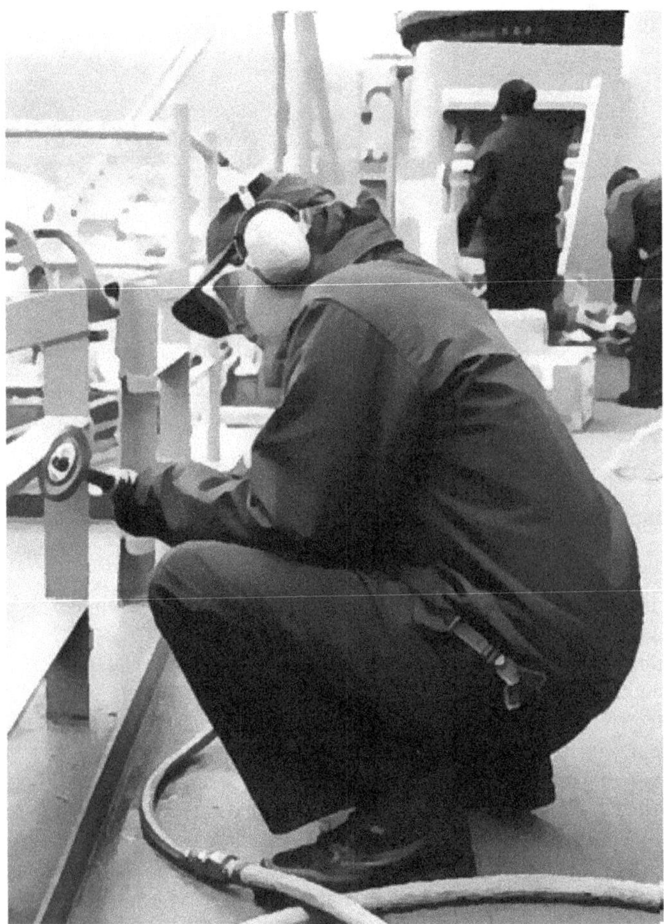

Figure 16.23 Sailor grinding a handrail

PROTECTION FROM BAD VIBRATIONS

This case study is adapted from a US Navy Case Study on Hand Arm Vibration (US Navy public website, Navy, 2015).

Some potentially serious occupational hazards in Navy workplaces, such as noise-induced hearing loss and heat stress, are well known, heavily reported, and well documented (Figure 16.23). However, certain other workplace hazards, some of which can produce serious, irreversible, and unsuspected diseases, are not as widely recognized. One such hazard is hand arm vibration, which can cause Raynaud's syndrome or hand arm vibration syndrome (HAVS).

Raynaud's syndrome/HAVS is a medical condition that can lead to permanent disability. HAVS is caused by people's hands being exposed to chronic vibration, which damages the nerves, blood vessels, and bones. Exposure to cold temperatures also increases the probability of acquiring HAVS and the likelihood of exhibiting symptoms.

TABLE 16.3 Tasks and the Vibrating Tools Associated with the Tasks

Potential HAV Exposures and Tasks Relevant to US Navy		
Task	Type of Tools	Remarks
Dismantlement of ships, particularly submarines	Electric and pneumatic cutting tools, grinders, and electric saws	The presence of hazardous materials often prevents use of torch cutting to dismantle vessels. This forces the use of hand tools to cut metal sections. Significant vibration exposures have been associated with this work because of duration, tool size, and substrate and work postures
Paint removal/surface preparation	Hand grinders, needle guns, hydroblast nozzles, and abrasive blast nozzles	Heavy metal lead exposure can also affect peripheral nerve conduction and may have an additive neurological effect
Preparation of welding surfaces – precleaning or smoothing after welding	Hand grinders	
Foundry cleaning departments. Removal of burrs and projections on newly cast work	Grinders, chippers	Improved quality control can reduce need for finishing and cleaning. Silica exposure also may be an issue
Sheet metal and fiberglass work	Hand grinders, orbital sanders, and polishers	
Road repair	Jackhammers	Much of the noise and vibration are associated with air-blow off and escape post tool impact. Devices controlling exhaust and recoil control exposure with minimal or no effect on productivity
Forestry (chain saw use)	Chain saws	Tool maintenance increases safety while reducing vibration. Cold is an additional hazard to hands

Most occurrences of HAVS affecting Navy personnel involve workers who use gasoline, pneumatic, hydraulic, or electric vibratory tools, such as grinders used for surface preparation, or rivet guns and bucking bars for airframes maintenance. These tools are common in Navy shipyards, aircraft-maintenance shops, and other environments such as construction sites and foundries. Table 16.3 lists tasks in the US Navy and the vibrating tools associated with the tasks.

Historically, there has been inconsistent and often limited progress in eradicating or even recognizing the HAVS problem. Puget Sound Naval Shipyard & Intermediate

Maintenance Facility (PSNS & IMF) in Bremerton, WA, was one of the first Navy activities to look at issues involving HAV. Early in the 1990s many of the jobs at PSNS & IMF that used handheld power tools were labor intensive. The safety office evaluated some of these jobs by slow motion videotaping workers using power tools, since vibration measuring instruments were not available at the time. As a follow-on effort, PSNS & IMF made a considerable effort to evaluate the ergonomics of handheld tools and the benefits of antivibration gloves (Figure 16.24).

Mindful of the need for further study, the Defense Safety Oversight Council (DSOC) initiated a project in 2007 to address the root causes of HAVS. The Council collaborated with the General Services Administration (GSA) and the National Institute for Occupational Health and Safety (NIOSH) to provide guidelines for low-vibration and other ergonomics characteristics in procurement criteria for new power hand tools. A concurrent effort worked to identify and incorporate International Organization for Standardization (ISO) 10819 and American National Standards Institute (ANSI) S2.73 certified vibration absorbing gloves into the federal procurement process.

A working group with Department of Defense (DoD)/GSA/NIOSH and US Coast Guard members was formed. The Navy was recognized as a leader within DoD in identification of HAVS having a focused effort within several fleet concentration areas. PSNS & IMF, Naval Base San Diego, and the Navy Fleet Readiness Center, East (Cherry Point, NC), provided leadership and technical support in their areas of expertise for this project. Procurement criteria for vibration absorbing gloves, low-vibration tools, and third-party certification guidelines were developed.

As a result of the DSOC project in September 2009, several lower vibration tools were introduced into the federal supply system (Figure 16.25).

Figure 16.24 Builder from Naval Mobile Construction Battalion Four Zero breaks apart the asphalt with a jackhammer on a road repair project

Figure 16.25 Operations specialist wears certified antivibration gloves while using a needle gun to chip paint off a bulkhead

With input from Navy subject matter experts, GSA is continuing to incorporate low-vibration and other ergonomics characteristics into procurement criteria for new and updated power hand tools.

Collaboration with the Navy Clothing and Textile Research Facility in Natick, MA; the Defense Logistics Agency; and support from the office of the Secretary of Defense for Manpower, Personnel and Readiness (see OSD MPR Memo of December 15, 2009; Prevention of Vibration-Induced Hand and Arm Injury) resulted in the introduction of certified vibration absorbing gloves into the federal supply system.

Only full-finger protected gloves are tested since HAVS always begins at the fingertips and moves toward the palm (Finger-exposed "half-finger" gloves are not recommended). Using certified antivibration gloves alone will not solve the HAV problem, and the Navy recommends that the gloves be used in combination with low-vibration tools such as the ones listed above, worker education, and appropriate work practices.

The Navy, in conjunction with the US Army Center for Health Promotion and Preventive Medicine, has also developed guidelines for workers and supervisors on the use of low-vibration tools and antivibration gloves to protect Navy workers from hand arm vibration exposures as shown below:

Guidelines to protect Navy workers from harm vibration exposures:

➢ Workers and their supervisors should ensure use of appropriate work practices and protective equipment. These include the following:

- Use of certified ANSI S2.73/-ISO 10819 (third-party tested) vibration absorbing gloves. (Many models are now available within the federal supply system.)
- Use power tools with reduced-vibration characteristics.
- Keep fingers, hands, and the body warm.

- Do not smoke. (Nicotine in tobacco constricts the blood vessels and can reduce circulation in the fingers.)
- Let the tool do the work, grasping it as lightly as possible, consistent with safe work practices.
- Keep tools well maintained.
- For pneumatic tools, keep the cold exhaust air away from fingers and hands.
- Take breaks from tool use for at least 10 min/h to allow circulation to recover.
- Wear hearing protective equipment as appropriate. (Most operations producing significant hand arm vibration are also noisy.)
- Have vibration exposure evaluated by a professional if they feel they are exposed to high levels of vibration.
- If signs and symptoms of HAVS appear, seek medical help.

➢ Work with your supply points of contact and process managers (engineers, shop supervisors, and technical authorities) to specify and order suitable low-vibration tools and certified antivibration gloves. **The continued and expanded availability of these products will depend on user demand.**

➢ Report concerns and worker complaints to the appropriate industrial hygiene and occupational health professionals through their safety office. Specialized assistance such as that provided by the Navy and Marine Corps Public Health Center may be beneficial.

➢ Review the process specification and technical manuals. If they feel that low-vibration tools and/or antivibration gloves might be considered for the relevant processes, use the comment sheet, typically on the last page of nearly every DoD/Navy technical manual to describe potential issues and concerns.

The Navy faces the continual challenge of finding better and improved vibration-reducing materials and technologies that meet ANSI/ISO guidelines and standards and can be incorporated into ships and shore facility designs during the acquisition process. Because Navy leadership is concerned about the safety and health of its military and civilian workers, they are working hard to address HAVS as an under-recognized occupational health problem through acquisition of safe, cost-effective, and performance-improving designs and equipment.

HEAVY CABLE HANDLING

Introduction

Ship–to-shore cable handling is common on cruise ships, commercial vessels, and navy ships.

There are two ends of the cable, the ship end and the shore end. Before a ship arrives in port, electricians perform additional tests in conjunction with the ship's crew to again ensure the ship-to-shore cables and outlet assemblies meet the minimum requirements for delivering power to the vessel.

After testing, the head of the ship-to-shore cable is installed into the outlet assembly. Several other cables may also be installed into the outlet assembly, depending on the power requirements of the particular ship. In some instances, one outlet assembly might be used to supply part of the power and another outlet assembly might be used to supply the remaining power. The shore end of the cable is heavy, weighing approximately 25 lb with each foot of the attached cable weighing approximately 12 lb.

In general, the electricians do not handle the ship end of the cable, unless they are performing a specific test. The ship end weighs approximately 44 lb.

The ship's crew normally removes both the ship end and shore end of the cables when they are readying to leave port. In many cases, the ship's crew leaves the cables in a disorganized pile.

The discussion section of this report addresses these tasks in additional detail and describes the hazards associated and methods to abate the hazards.

Approach

To obtain an understanding of actual operations, the following was included in the site visit:

- Observing the process and procedures
- Meetings with task performers, a supervisor, and an industrial hygienist
- Identifying, evaluating, and discussing the physical workplace risk factors and possible ergonomic solutions.

The ergonomic assessment report was prepared by the following:

- Evaluating the severity of the exposure to the physical workplace risk factors
- Identifying ergonomic solutions for controlling the hazards that fit within the constraints of the tasks
- Performing cost-benefit trade-off analyses for proposed solutions
- Providing recommendations for the most advantageous control measures as well as the related cost impacts and ergonomic benefits for each.

Hazard Control Options

Ergonomic recommendations developed to mitigate the identified physical workplace risk factors were evaluated against each other in a cost-benefit trade-off analysis.

There are four basic approaches to controlling risk from eliminating the hazard to using administrative controls. Table 16.4 includes a full description of the hazard control options in order of highest to lowest priority.

Testing and Installing Ship-to-Shore Cables

A day before a ship arrives in port, each ship-to-shore cable that will be used to supply power is tested to ensure it meets minimum standards. The test that is used is called a Megger Test. One electrician is usually required to perform the test. This test entails testing each of the leads to ensure there are no faults in the system. To do the tests,

TABLE 16.4 Levels of Hazard Control for Implementing Ergonomic Improvements

Hazard Control Hierarchy
1. **Elimination** – A redesign or procedural change that eliminates exposure to an ergonomic risk hazard; for example, using a remotely operated soil compactor to eliminate vibration exposure
2. **Engineering controls** – A physical change to the workplace; for example, lowering the unload height of a conveyor
3. **Substitution** – An approach that uses tools/material/equipment with lower risk; for example, replacing an impact wrench with a lower-vibration model
4. **Administrative** – This approach is used when none of the above can be used or are impractical to implement. Administrative controls are procedures and practices that limit exposure by control or manipulation of work schedule or the manner in which work is performed. Administrative controls reduce the exposure to ergonomic stressors and thus reduce the cumulative dose to any one worker. If you are unable to alter the job or workplace to reduce the physical stressors, administrative controls can be used to reduce the strain and stress on the work force. Administrative controls are most effective when used in combination with other control methods; for example, requiring two people to perform a lift

the electrician has to remove the cover on the shore end of a cable and ensure it is dry. Then the electrician maneuvers the shore end of the cable to a point where it is accessible to perform the test. The shore end weighs approximately 25 lb and each foot of cable lifted weighs approximately 12 lb. The amount of weight lifted by the electrician varies depending on how much cable must be moved or lifted with the shore end. In addition to the weight of the head and cable, there are frictional forces the electrician must overcome related to the cable sliding along the surface of the pier or the quay wall. If the cable is left in a pile, there may also be frictional forces related to disentanglement of the cable. Ships can enter the port at any time of the year, and, during winter months, the cables can be frozen to the deck of the pier or the quay wall.

Once a ship arrives in port, an electrician meets the boat and again performs tests to ensure the cables meet the minimum standards required. The shore end of the cable is then connected to the appropriate outlet on the outlet assembly. The heights of the outlets and the assemblies themselves vary from pier to pier. There are two types of outlet assemblies. Figure 16.26 shows the type that is standard on most piers. This style of outlet assembly uses the Shore end connector. Figure 16.27 shows the newer type of ship-to-shore cable receptacle. The newer style outlet assembly uses a Camlock style fitting for each lead and not the shore end. The Camlock style reduces many of the physical risk factors found with the shore end connector because of the lighter weight cables. Figure 16.28 shows the height range of the outlets on the outlet assemblies.

The weights of lift for the shore end to the various heights were measured using a handheld load cell. When the shore end was lifted 1 ft from the deck, the reading was approximately 37 lb. At the height of the racks (31 in.) the weight of lift was approximately 55 lb. At 60 in. the weight of the cable was between 75 and 85 lb. The measurements taken did not account for any additional difficulties, including freeze,

Figure 16.26 Standard outlet assembly (Graphic by Lee Ostrom)

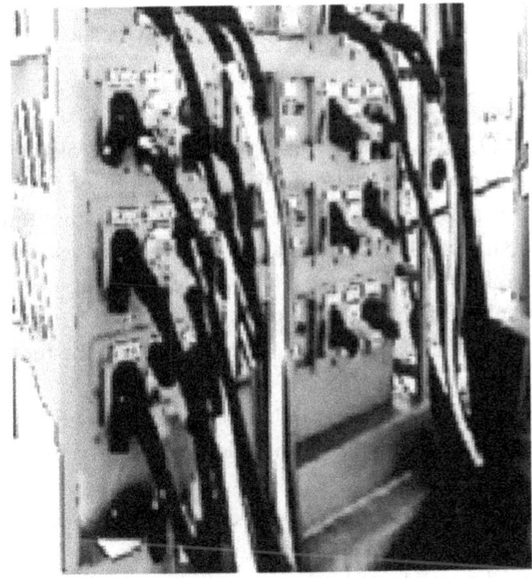

Figure 16.27 Newer style outlet assembly (Graphic by Lee Ostrom)

friction, or tangled cables, but only assessed the weight of the cable under ideal conditions.

Two of the electricians interviewed said that the sailors always help with the connection of the shore end to the outlet assembly, whereas one former electrician who performed this task said he did it alone.

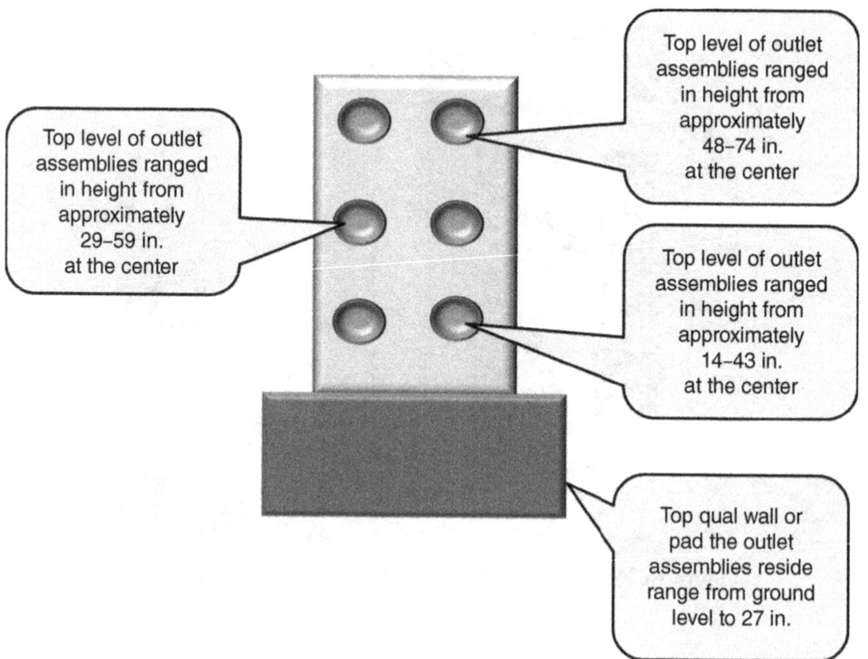

Figure 16.28 Outlet assembly measurements (Graphic by Lee Ostrom)

The MIL-STD 1472G states that the maximum weight of lift for a uniform load of 18 in. wide by 12 in. deep and 18 in. high to a height of 5 ft is 56 lb. The maximum weight of lift is to be reduced if the weight of load is not uniform and/or there are obstacles. The ship-to-shore cables are not a uniform package, and the weight increases as the height of lift increases. The cable handling operation does include negotiation of obstacles. Many of the heights of lift exceed 5 ft and they can be above 6 ft or more. Therefore, the maximum weight of lift should be reduced by as much as 66% to help ensure injuries do not occur.

This task clearly has a high potential to cause injury. Injuries that could occur from lifting the cables to those heights include potential for harm to the back, neck, shoulder, and lower extremities.

While connecting the Camlock-style lead connectors, the electricians have to turn the connector 180° to the left, then insert the connector, and then seat it 180° to the right, while holding the weight of lead and possible some length of the main cable.

In each current configuration, the most stressful part of this job is holding the weight of the cable and seating the connector. The possible injuries include those to the wrist, back, and shoulder. The current practice is to use two electricians to perform the task that does reduce some of the stress associated with this task but does not reduce the risk of injury.

The ship end weighs approximately 34 lb without the cap, and the cap weighs approximately 12 lb. The electricians have to sometimes persuade the cap to come loose from the ship end using a hammer.

Figure 16.29 Truck mounted crane (Graphic by Lee Ostrom)

The stressful aspects of this task are that the ship end is heavy (approximately 44 lb assembled) and the electricians have to be on their knees to perform the task. The potential injuries are primarily back, shoulder, and wrist injuries. The risk of injury is lower for this task because it is performed very infrequently.

The recommended engineering controls that should be implemented for this sort of issues are mount electric actuated jib cranes on the existing electricians' service trucks to manipulate the cables. The crane should have a capacity of at least 200 lb. Figure 16.29 shows this type of crane. The example depicted below has a capacity of 700 lb in the extended configuration.

Other devices considered for moving the cable included a fork lift or a walk-behind forklift. However, these devices might not be effective on an icy pier and would require a higher level of maintenance.

AN ERGONOMIC IMPROVEMENT ALSO PROVIDES IMPROVED FALL PROTECTION

Naval Submarine Base Kings Bay is located adjacent to the town of St Marys in Camden County, Georgia, in southeastern Georgia, and not far from Jacksonville, Florida. Kings Bay Trident Refit Facility (TRIREFFAC) is the largest tenant command at Kings Bay and has quietly and efficiently kept a significant portion of the United

States Fleet Ballistic Missile submarines at sea since 1985. TRIREFFAC provides quality industrial-level and logistics support for the incremental overhaul, modernization, and repair of Trident Submarines. It also furnishes global submarine supplies and spare parts support. In addition, TRIREFFAC provides maintenance and support services to other submarines, regional maintenance customers, and other activities as requested.

TRIREFFAC machinists were subjected to potential injuries due to ergonomic hazards while operationally testing and repairing pumps that circulate water on the Trident Submarines. These pumps support diving and surfacing operations. While being inspected and/or repaired, the pumps are placed on the test stand by a crane. The workers are required to access the entire pump (top, bottom, and sides) and were required to stand on temporary staging that was not conducive for allowing them to perform work in optimal ergonomic postures. Workers reached and extended their bodies at times as much as 4 ft, to access bolt threads, wiring, seals, and other components. They were also required to twist and bend their bodies into awkward postures for extended periods of time to perform repairs and tests. These awkward postures resulted in a higher potential for an injury. At times, workers have been observed standing on rails of the temporary staging as well as piping to access components of pumps as they disassemble/reassemble them. This was not only an ergonomic issue, but a fall protection issue as well (Figures 16.30 and 16.31). The shapes of the pumps are similar to a tower and pot belly that require the necessity for a versatile staging configuration.

These tasks are performed by three to five workers at a time and the same employees work the tasks until the entire pump repair is complete. The task duration is normally 3 days and performed 30–40 times per year.

A Mishap Prevention and Hazard Abatement (MP/HA) project was developed to provide an ergonomic solution to this issue. The project was submitted through Naval Facilities Engineering Command (NAVFAC) for ergonomic funding in March 2008

Figure 16.30 Staging

Figure 16.31 Staging for pumps

for $31,000 to design and purchase customized staging for permanent access to the pumps to eliminate the stressors created by overexertion on the workers. The Navy's MP/HA Program Team assessed the pump test stand, and the potential solution to the ergonomic issues was identified. TRIREFFAC developed a proposed design for permanent staging for the pump stand area. The design was reviewed and approved by the Building 4026 Safety Committee and Facilities Representative. Meetings were held with the potential vendor of the permanent staging, and the Navy MP/HA Team reviewed the final design of the staging. The newly acquired staging was installed by TRIREFFAC personnel (Figure 16.32).

Figure 16.32 Final staging

The completed project allows workers to work in ergonomic neutral postures, provides access to pumps in a manner that reduces the potential for injury, and reduces the potential for a life-threatening fall.

REFERENCES

Aarnpacks (2015). Retrieved from aarnpacks: http://www.aarnpacks.com. Accessed December 22, 2015.

Healthworks Medical Group (2015). Healthy Tips. Retrieved from UShealthworks: http://www.ushealthworks.com/HealthyTips_Ergonomics_BackPack.html. Accessed December 22, 2015.

Liberty Mutual Insurance Company (1978). Manual Materials Handling.

Mil Std 1472F (1999). Department of Defense Design Criteria Standard.

Ros, H. (2005). Prevention of Low Back Complaints. Ministry of Defense HDP/DMG/AMG, RTO-MP-HFM-124.

US Navy public web site, Navy (2015). http://www.public.navy.mil/comnavsafecen/Documents/SuccessStories/SuccessStories2/149_HAV.pdf. Accessed December 22, 2015

ADDITIONAL SOURCES

Ergonomic Guidelines for Manual Materials Handling (2007). DHHS (NIOSH) Publication No. 2007-131.

Gillette, J. C., Stevermer, C. A., Meardon, S. A., Derrick, T. R., & Schwab, C. V. (2009). Upper Extremity and Lower Back Moments During Carrying Tasks in Farm Children, *Journal of Applied Biomechanics*, 25, 149–155.

MIL-HDBK-759C (1995). Handbook for Human Engineering Design Guidelines.

APPENDIX A

EXERCISES

ERGONOMICS LABORATORY EXPERIMENT MANUAL MATERIAL HANDLING

Objectives

- Analyze manual material handling task and make recommendations to reduce exposure to the physical workplace risk factors.
- Use ergonomic principles to recommend engineering and/or administrative changes to reduce or eliminate.

Apparatus

- Liberty mutual manual material handling tables and/or
- Manual materials handling checklist

Procedure: Evaluate the following task and make recommendations for intervention strategies based on the results. When making recommendations, use a tiered a tiered approach covering administrative and engineering controls if feasible.

Occupational Ergonomics: A Practical Approach, First Edition.
Theresa Stack, Lee T. Ostrom and Cheryl A. Wilhelmsen.

Figure A.1 Super sacks weight in excess of 50 lb each are lifted into the truck without use of the lift gate

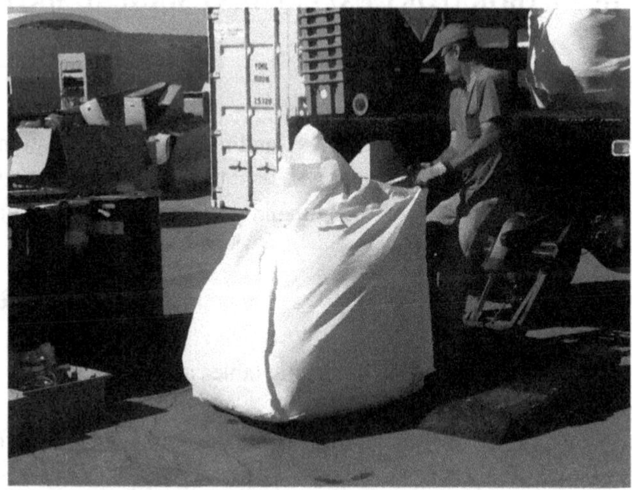

Figure A.2 Super sacks weight in excess of 50 lb each are dragged across the staging area

Work Description

Two hazardous waste technicians work in excess of 50 h a week performing this task and other tasks such as waste water pump out. They are on emergency response call alternating weekends.

Figure A.3 (a) Over packs are lifted via fork truck but moved manually once on the truck. (b) Truck is packed to capacity

Packaged hazardous waste (tightly packed) is manually lifted onto the back of a rack truck. Larger bulk hazardous waste is dragged from the staging area onto the back of the lift gate (along with the drums). After the lift gate elevates, the technician drags the bulk waste and drums onto the truck (Figures A.1 and A.2).

Other items such as the bulk containment are lifted with a fork truck and placed onto the truck (Figure A.3).

Figure A.4 All items are off-loaded by hand except for the full drums

Figure A.5 Items are staged on pallets by hazard category after off loading

After replacing the back racks on the truck, the technician maneuvers the waste products and stacks empty drums onto the truck to limit the amount of return trips to the facility.

Upon returning to the storage/staging site, the bags are off-loaded onto pallets. All the transfer work from the truck to the storage site (onto pallets) is done manually (Figures A.4 and A.5), except for the drums that are handled with a drum dolly. Other items are handled in a similar way as shown in Figure A.6.

Figure A.6 Many batteries are recycled

ERGONOMICS LABORATORY EXPERIMENT THRESHOLD LIMIT VALUE FOR LIFTING (ACGIH TLV FOR LIFTING), MANUAL MATERIAL HANDLING AND RULA/REBA

Objectives

- Analyze laundry operation and determine:
  - The TLV for lifting for each similar exposure group
  - The RULA or REBA score for each similar exposure group
  - Areas of improvement with regard to manual material handling
- Use ergonomic principles to recommend changes to reduce or eliminate the exposure if found above the action level

Apparatus

- Photographics
- ACGIH TLV tables, RULA/REBA score sheets, Liberty Mutual Manual Materials Handling website

Procedure: Evaluate the following task and make recommendations for intervention strategies based on the results. When making recommendations a tiered approach covering administrative and engineering controls if feasible. For example:

1. Description: Describe tasks and state variables evaluated and assumptions applied.
2. Evaluation: State the methods used, assumptions, and results of the evaluation

3. Interpret these results.
4. Make at least one recommendation to improve each task found above the recommended exposure limit.

- A: Distance from midpoint between inner ankle bones and the load
- B: Lifting tasks should not be started at a horizontal reach distance more than 80 cm (32 in.) to the midpoint between the inner ankle bones (Figure A.1)
- C: Routine lifting tasks should not be conducted from starting heights greater than 30 cm (12 in.) above the shoulder or more than 180 cm above floor level (Figure A.1)
- D: Routine lifting tasks should not be performed for shaded table entries marked "No known safe limit for repetitive lifting." While the available evidence does not permit identification of safe weight limits in the shaded regions, professional judgment may be used to determine if infrequent lifts of light weights may be safe
- E: Anatomical landmark for knuckle height assumes the worker is standing erect with arms hanging at the sides
- F: Reproduced with permission

See Figures A.7–A.19.

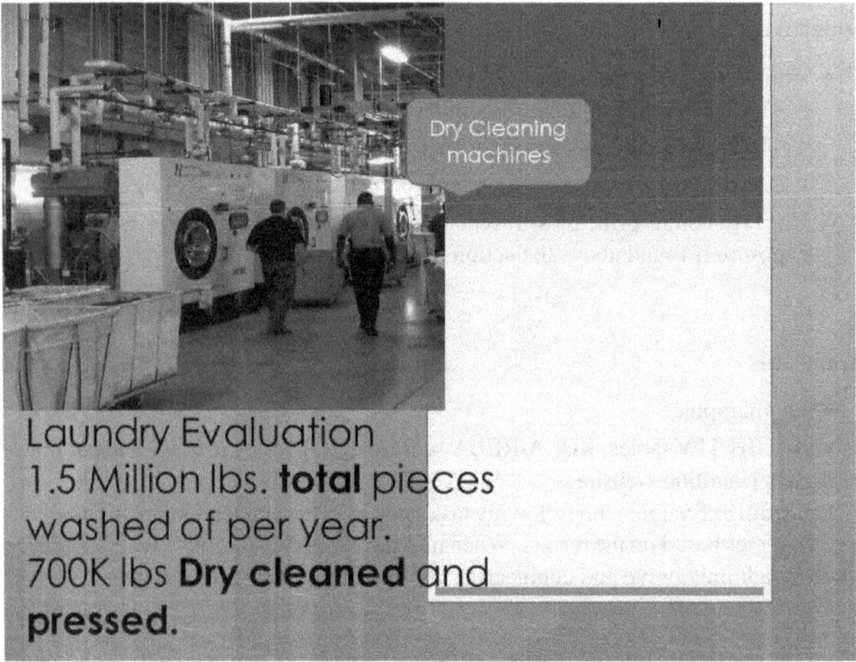

Figure A.7 Laundry evaluation. Laundry processes 1.5 million lb of total laundry per year with 700 thousand dry-cleaned

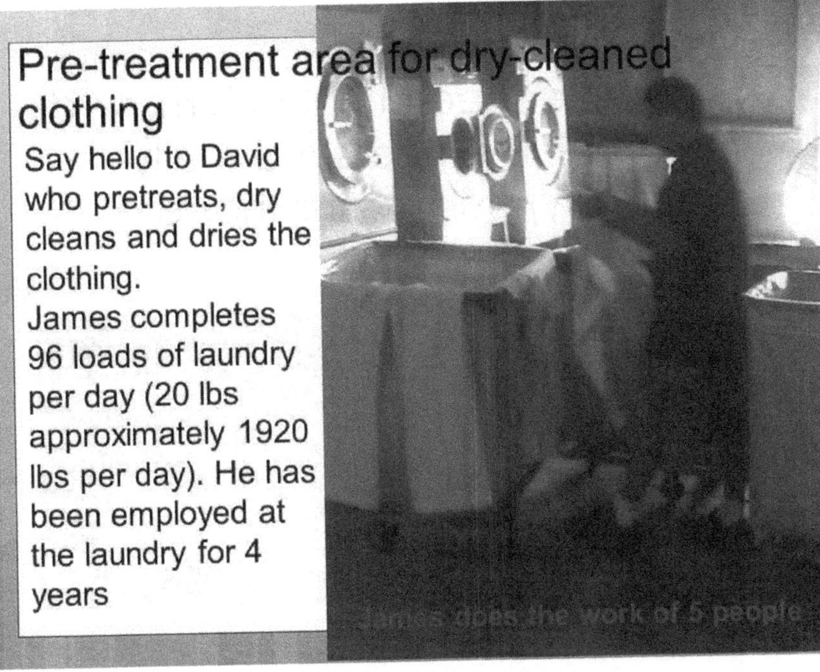

Figure A.8 Clothing requiring dry cleaning is first pretreated

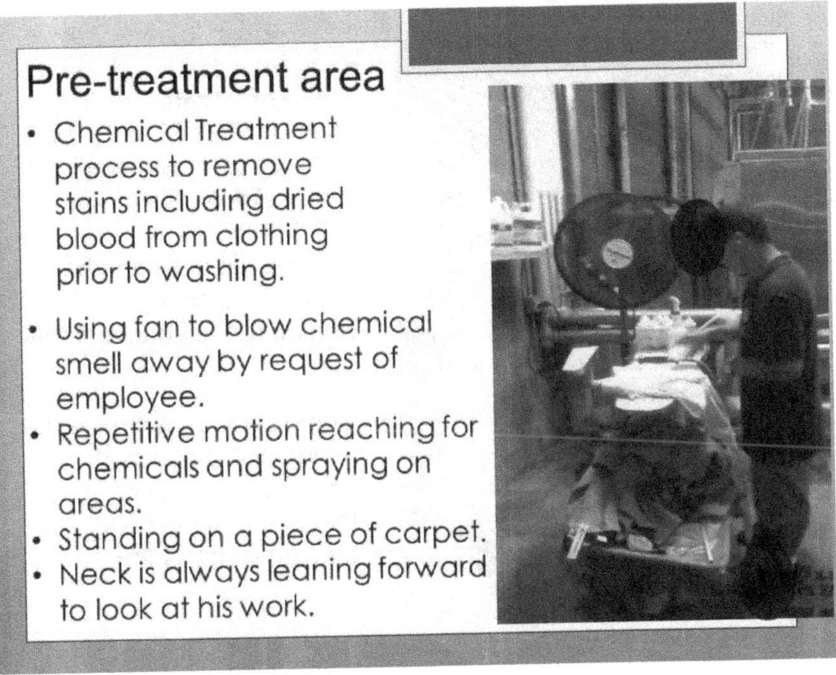

Figure A.9 Pretreating is time consuming

All dry-cleaned clothing is moved through this process manually (wash/dry) using carts that are 26" deep.

Figure A.10 Pretreated laundry is washed

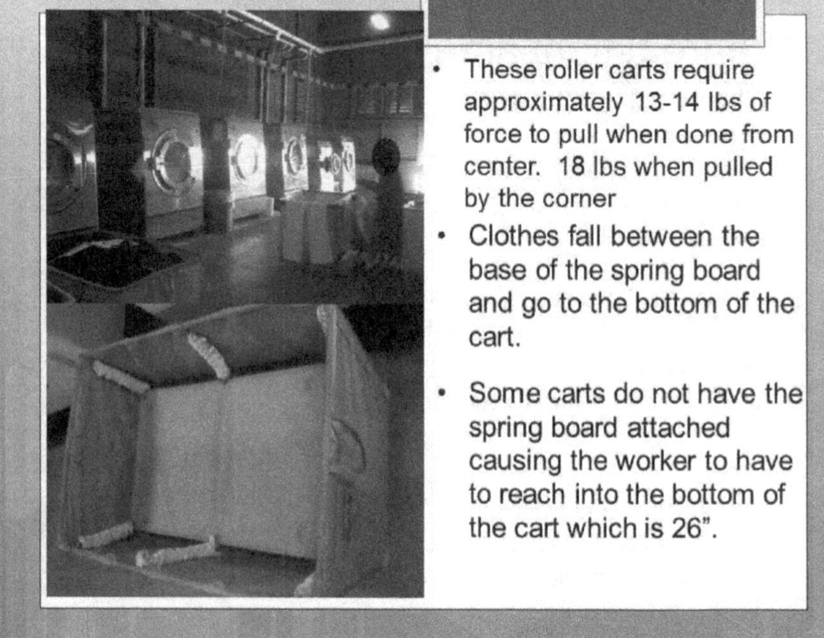

These roller carts require approximately 13-14 lbs of force to pull when done from center. 18 lbs when pulled by the corner
Clothes fall between the base of the spring board and go to the bottom of the cart.
Some carts do not have the spring board attached causing the worker to have to reach into the bottom of the cart which is 26".

Figure A.11 Laundry is transported through the entire facility via spring-loaded carts

Say hello to Michael.
He run the dry cleaning side of the laundering procedure.
He performs this task about 15-20 times a day for about 4 hours he sorts approximately 96 LOADs per day of dry-cleaned laundry per day; 20 lbs to a load. Henry has been doing this job for about 5 years.

Figure A.12 After wishing laundry is sorted

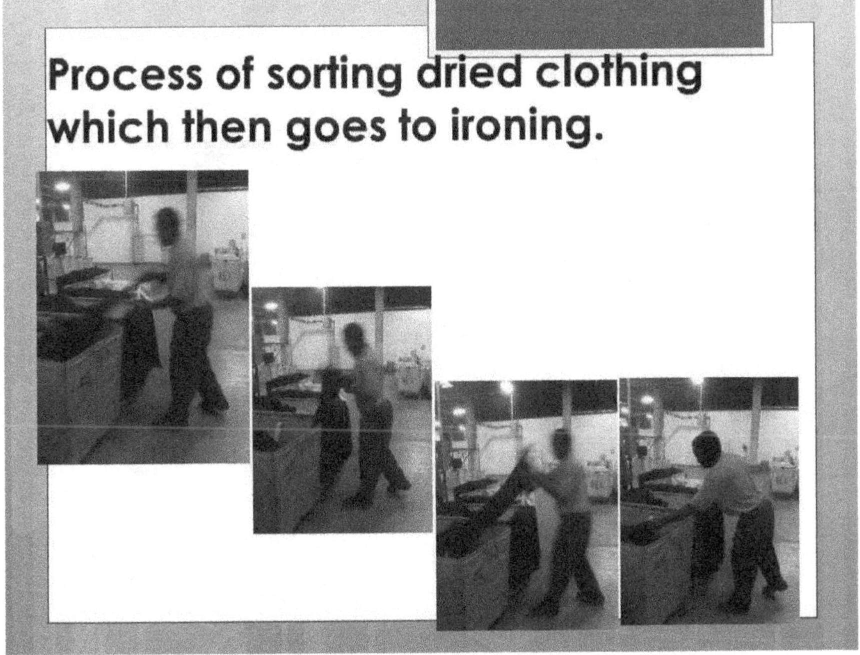

Figure A.13 Sorting is highly repetitive

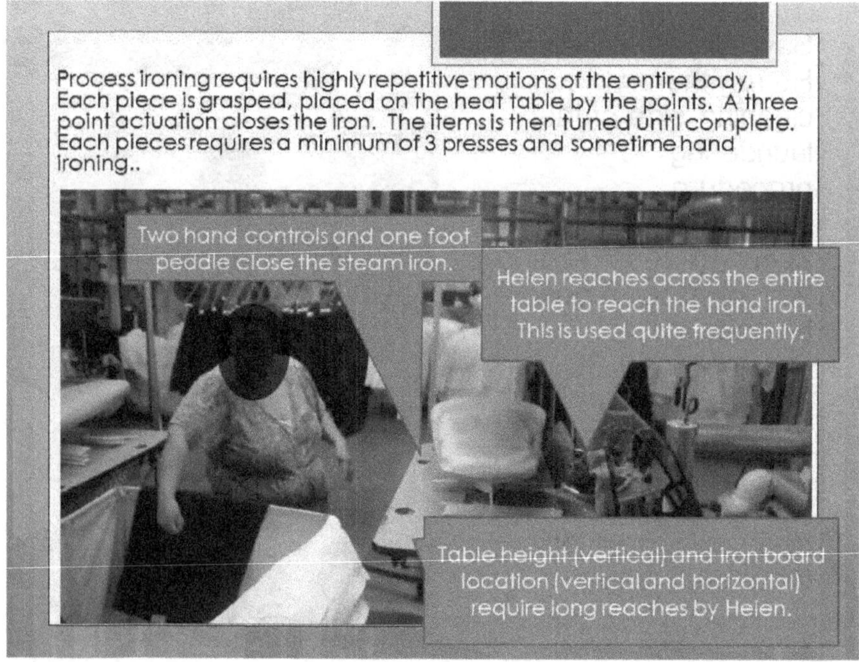

Figure A.14 Each item is then pressed; pressing takes multiple steps due to folding each piece in its pleats and then hanging

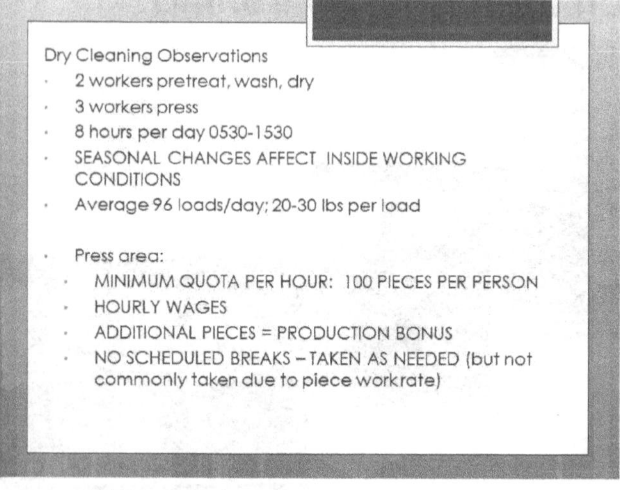

Figure A.15 Press work is highly repetitive

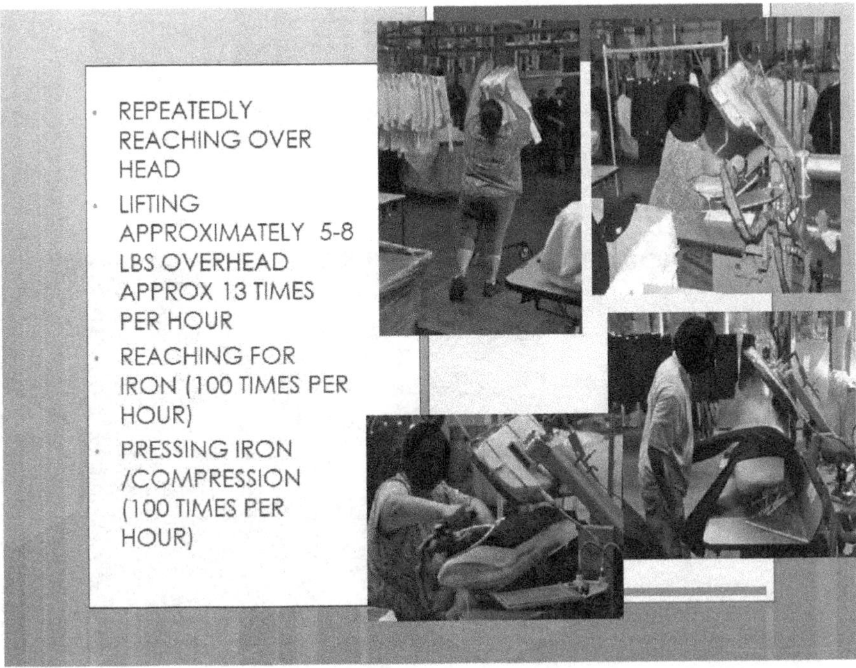

Figure A.16 Press work is highly repetitive, worker tends to avoid breaks due to the compensation method

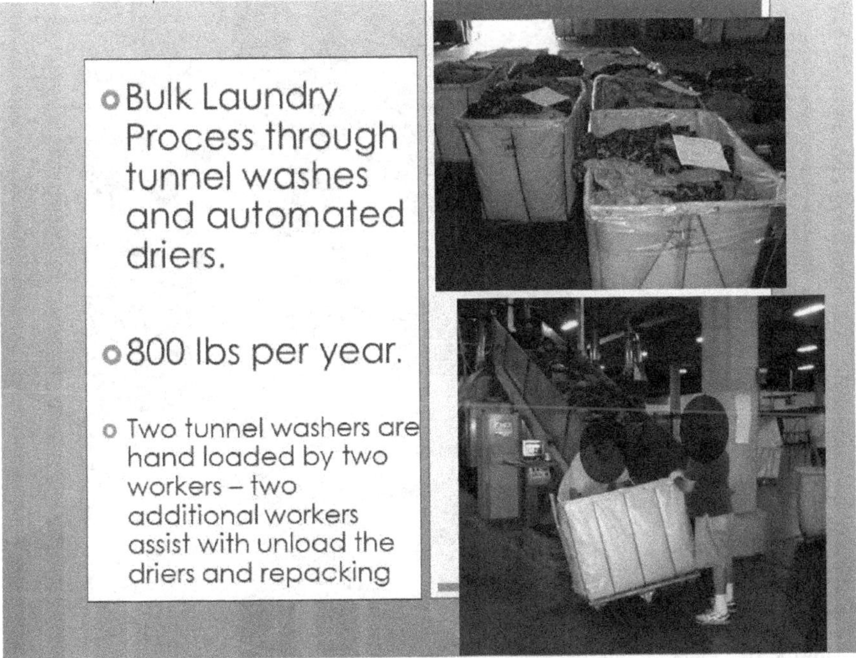

Figure A.17 Bulk laundry bags weight up to 50 lb and are loaded on two-tunnel washer conveyors manually by two people

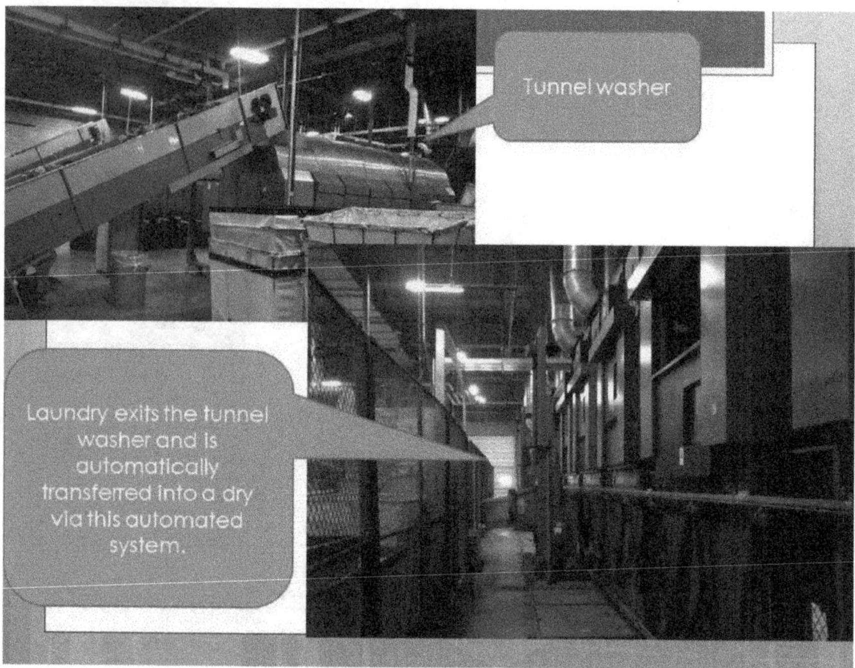

Figure A.18 Laundry is automatically transferred from the washer into dries

Figure A.19 Laundry exits the driers through shoots at the bottom. Workers poke at the laundry with a stick when it becomes stuck in the shoot, they then re-bag the bulk laundry

APPENDIX B

GUIDES

ERGONOMIC GUIDE FOR WELDERS

Foreword

This guide is for managers and supervisors at worksites performing welding tasks. It provides a brief background on ergonomics and offers suggestions for improving the workplace to reduce the risk of work-related musculoskeletal injuries.

A healthy workforce helps ensure the highest level of quality and productivity. Welding work contributes to a large percentage of injuries in the United States. Many of these injuries involve the musculoskeletal system and include strains or sprains to the lower back, shoulders, and upper limbs. They can result in pain, disability, medical treatment, financial stress, and a change in the quality of life for those affected with them.

INTRODUCTION – WHAT IS ERGONOMICS?

Ergonomics has several meanings. The first is literal. Derived from two Greek words: *Ergo* = work; and *Nomos* = laws, ergonomics literally means *the laws of work*.

The practical meaning of Ergonomics is fitting the task and work environment to the human. Ergonomists try to design tasks and workplaces within the capability of the human. There are many examples of tasks, work environments, and even products, which did not take human capabilities into consideration.

Occupational Ergonomics: A Practical Approach, First Edition.
Theresa Stack, Lee T. Ostrom and Cheryl A. Wilhelmsen.

For example, in your home, kitchen countertops are usually about 36 in. in height. Is this a good height for everyone? No, it is usually too low for men and may or may not be correct for women. The height was selected because it was the average elbow height for women in the 1930s. If countertops were designed for the average male, the height would be 43 in. In a perfect ergonomic world, countertop height would be adjustable or there would be work surfaces at the proper height for your spouse, your children, and yourself in kitchens.

In the design of workplaces, human body size and capabilities must be considered. Injuries are more likely if task demands and the environment exceed the capabilities of the human. There are times when all ergonomics risk factors cannot be removed and the worker is exposed. Refer to Figure B.1, where the worker is in an awkward posture. The goal is to limit the amount of time the person is in those less than adequate situations.

Think about how well one's workplace accommodates them. Can they reach the items they use most without straining? Is it comfortable to lift the materials they need to move around? Are there handles to grasp the items or are they cumbersome and awkward? Can they change posture when they perform their work or are they in a static posture all day? These are just a few of the items that may indicate a mismatch between the workplace and the worker.

Most of the tasks welders perform are dictated by the design of the item being worked. In many cases, the materials are big, heavy, and might be covered with dirt, rust, and/or grime. However, numerous job aids such as fixtures, jigs, and part holders available to aid welders in doing their jobs and reduce the risk of injury. In addition, special chairs and/or creepers are available that provide body support while performing tasks so that an awkward posture is avoided. These types of ergonomic solutions will be presented in the following sections.

Figure B.1 Welding with the hands in front of the body. *Ergonomics Stressors: Compression: Leaning on a hard surface and Awkward and Static Posture: Holding the arms away from the body for long durations* (Photo adapted from Ergonomics Guide for Welders, unpublished Navy Occupational Safety and Health Working Group, Ergonomics Task Action Team with technical support provided by the Certified Professional Ergonomists at General Dynamics Information Technology (GDIT), San Diego, CA, 2010)

Physical Workplace Risk Factors

The many possible causes of injury are not limited to one industry or to specific occupations but result from a pattern of usage. Common causes of injury are as follows:

- Repetitive gripping/twisting
- Repetitive reaching
- Repetitive moving
- Static postures
- Lack of rest to overcome fatigue.

These causes fall into six major physical workplace risk factors:

- Force to perform the task
- Highly repetitive tasks
- Poor, awkward, or static postures
- Pressure points or compression
- Vibration
- Duration.

When present with sufficient frequency, magnitude, or in combination, these risk factors may cause work-related musculoskeletal disorders (WMSDs). Musculoskeletal disorders (MSDs) are injuries and illnesses that affect muscles, nerves, tendons, ligaments, joints, spinal discs, skin, subcutaneous tissues, blood vessels, and bones. WMSDs are:

- MSDs to which the work environment and the performance of work contribute significantly, or
- MSDs aggravated or prolonged by work conditions.

In addition, environmental conditions such as working in temperature extremes may contribute to the development of WMSDs. Personal risk factors such as physical conditioning, preexisting health problems, gender, age, work technique, hobbies, and organizational factors (e.g., job autonomy, quotas, deadlines) may also contribute to, but do not cause, development of WMSDs.

WORK-RELATED MUSCULOSKELETAL DISORDERS

Many injuries can develop when there is a mismatch between the capabilities of the workforce and the demands of the task. These injuries are generally called **work-related musculoskeletal disorders** or WMSDs. These have also been called cumulative trauma disorders (CTDs) or repetitive stress injuries (RSIs).

In general, these conditions develop because of microtraumas that occur to the body over time. Consider lower back vertebral disc degeneration. The vertebral disc is made of flexible cartilage and contains a semiliquid gel. The cartilage is in the form of rings. When a person performs lifts beyond his/her capability, these rings can degrade. If the person continues performing such lifts, the disc can rupture. When it ruptures, it bulges out and can place pressure on a spinal nerve causing severe pain. Common WMSDs for welders include the following:

Back injuries – From strains and sprains to degradation of the vertebral discs

Bursitis – Inflammation of a sac-like bodily cavity, containing a viscous lubricating fluid located between a tendon and a bone or at points of friction between moving structures (i.e., inflammation of a bursa)

Carpal tunnel syndrome – A complex disorder that starts with the inflammation of the tendon sheaths in the wrist and progresses into the degradation of median nerve

Tendonitis – Inflammation of the tendons

Tenosynovitis – Inflammation of tendon sheath

Thoracic outlet syndrome – A disorder in which blood vessels and nerves in the upper shoulder region are compressed and cause pain. This condition is sometimes caused by chronic postures associated with overhead work

Trigger finger – Tendons in the fingers "lock down" due to injury to the tendons

If you are experiencing signs/symptoms, please contact your physician or occupational medicine clinic for an evaluation.

Common Physical Workplace Risk Factors for Welders

Welding tasks may expose workers to physical workplace risk factors (or ergonomics stressors). If tasks are performed repeatedly over long periods of time, they can lead to fatigue, discomfort, and injury. The main physical workplace risk factors (ergonomics stressors) associated with the development of a work-related musculoskeletal injury in welding tasks include the following:

- Awkward body postures
- Heavy lifting
- Static position.

Applying proper ergonomic principles can reduce the potential for these and other WMSDs.

Nature of Welding

Ask any professional welder and they will say that welding is a highly skilled profession that requires years to perfect. Welding is a precise task that requires the welder to maintain static postures for relatively long periods of time. In almost all cases, welding in the field requires the welder to adapt to the workplace, rather than adapting the workplace to the welder. This is because metal is heavy, and it is easier to have the welder assume an awkward posture, than move a ship. Welding also is hot work and generates metal fumes that can contain many relatively harmful metals.

The use of proven ergonomic principles can improve the way a particular task is performed, thereby reducing welder exposure to risk factors. This generally translates to a healthier workforce, improved morale, greater productivity, and increased product quality. The Navy literally floats on quality welds.

Implementing Solutions

The following section on ergonomic solutions for welders describes changes to equipment, work practices, and procedures (administrative controls) that can address ergonomics-related risk factors, help control costs, and reduce employee turnover. These changes may also increase employee productivity and efficiency because they eliminate unnecessary movements and reduce heavy manual work.

The recommended solutions in the following pages are not intended to cover all ergonomic challenges, nor is it expected that all of these solutions are applicable to each and every welding environment. It is recognized that implementing engineering solutions may present certain challenges, which includes work that is performed outdoors and in cramped spaces.

However, welding personnel are encouraged to use the examples in this document as a starting point for developing innovative solutions tailored to the specific ergonomic challenges. The solutions have been categorized according to the equipment type.

TYPES OF ERGONOMICS IMPROVEMENT

See Tables B 1–B 9.

TRAINING AND ACTION PLAN

Training

A key element in implementing solutions for ergonomics stressors is training; refer the Ergonomics Program chapter for more detail.

Action Plan for Implementing Solutions to Reduce Workplace Risk Factors

- Step 1: Look for clues
  - Observe work activities
    - Risk factors
    - Worker fatigue
    - Tool/equipment modifications
    - Increased absenteeism
    - Decreased production
    - Bottlenecks/missed deadlines

TABLE B 1 Site-Wide Ergonomic Improvements for Welding Operations

Equipment such as gas cylinder transportation carts, sheet metal carts and the gas safety/guardian transportation cart developed by the USAF houses cylinders, leads, and hoses (right); can reduce hazards associated with pushing, pulling, carrying and lifting, Figure B.2

Advantage:

- Reduces pushing, pulling, lifting, and carrying forces
- Safer, saves time
- Carts, trucks, dollies can be customized

Points to remember:

- Wheels should be appropriate for the flooring conditions
- Larger diameter wheels reduce push forces
- Wheels should be well maintained
- Ensure proper load capacity for equipment being moved
- Pushing is preferred over pulling

Figure B.2 Gas safety/guardian transportation cart developed by the USAF houses cylinders, leads, and hoses

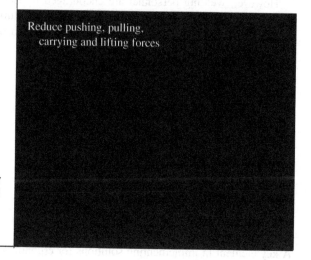

Reduce pushing, pulling,
carrying and lifting forces

- Talk to workers (form ergonomics action teams or designate an ergonomic point of contact)
- Use assessment tools
  - Risk factor physical check list or manual material handling checklist both found in Appendix C
- Step 2: Prioritize job for improvements
  - Consider
    - Frequency and severity of the risk factors
    - Frequency and severity of complaints
    - Injuries

Description: Safety and health personnel recognized the ergonomics hazards associated with moving and carrying sheet metal and round stock and unnecessary preparation work due to outdoor storage conditions

Project summary and advantages:

- A vertical index system virtually eliminates the manual handling storage and retrieval tasks. The storage system houses all the raw materials and delivers it to the personnel around elbow height. Elbow height handling of material typically results in less back bending, Figure B.3
- The index system delivers the sheet metal quickly. Improves sheet metal quality and reduces handling time and unnecessary finish work because storage will be exclusively indoors and in one unified location, Figure B.4
- Saves time and effort when completing a project due to less preparation work through product specific storage and quicker product access time, Figure B.5
- Round stock system has security controls to limit access to level one stock and incorporates a grabbing claw and overhead hoist
- Systems have inventory control functions to track stock in real time

Figure B.3 Vertical sheet metal storage system used at New London, CT

Figure B.4 Vertical sheet metal storage system. Before index unit, round stock retrieval was time consuming and required multiple pieces to be moved in order to locate the correct piece

TABLE B 2 (*Continued*)

Points to remember:

- Material index systems exist for material as small as medical supplies to as large as sheet metal
- Jib cranes and clamps can be incorporated

Figure B.5 Vertical index unit for round stock has security access codes for level one stock and incorporates a jib crane at New London, CT

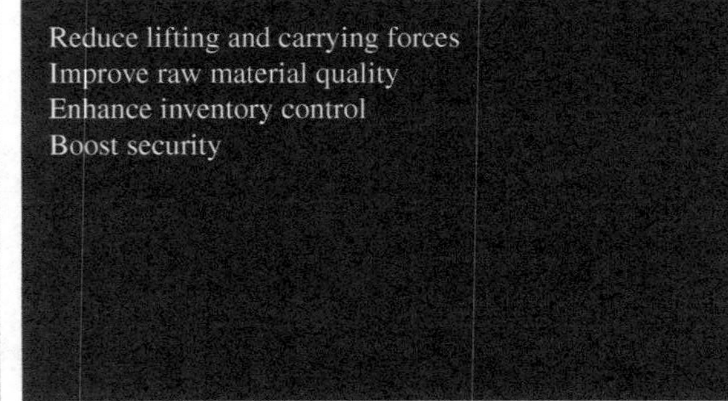
Reduce lifting and carrying forces
Improve raw material quality
Enhance inventory control
Boost security

TABLE B 3 Grippers and Handles for Moving Raw Materials and Finished Work

Description: Overhead handling equipment from engine hoists to boom cranes can be used to move heavy equipment to reduce the frequency of heavy lifting

Advantages:

- Eliminates heavy lifting and carrying
- Allows heavy or awkward materials, tools or equipment to be moved without carrying as seen with the suction manipulator in Figure B.6
- Saves time and effort
- Provides a neutral lifting posture of the hands and torso as seen with the grabbing clamp in Figure B.7

Points to remember:

- Different handling system and grippers are available for many situations

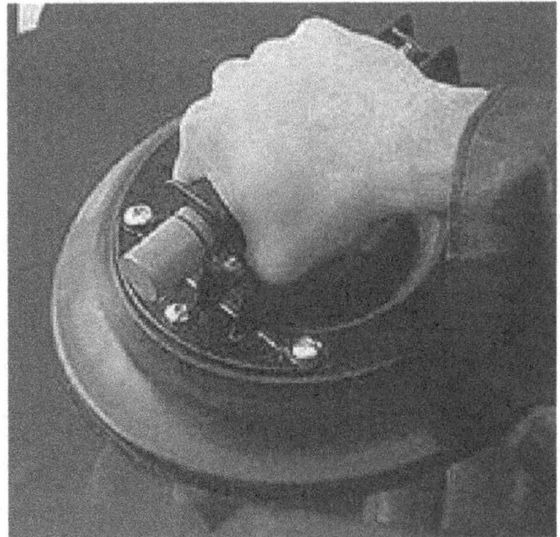

Figure B.6 Suction manipulator used to move sheet metal

Figure B.7 Grabbing clamp

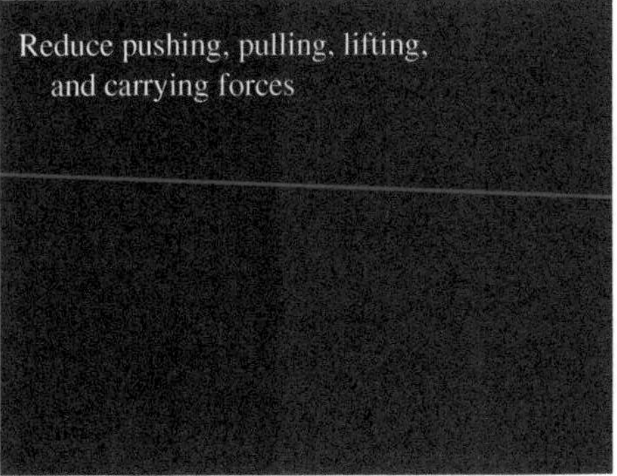
Reduce pushing, pulling, lifting, and carrying forces

TABLE B 4 Overhead Handling Equipment

Description: Overhead handling equipment for lifting and moving large pieces

Advantages:

- Eliminates heavy lifting and carrying
- Allows heavy or awkward materials to be moved, stored and loaded for distribution without carrying, Figure B.8
- Saves time and effort
- Reduces lifting, pushing, and pulling forces

Points to remember:

- Different handling systems and grippers are available for many situations
- Ensures system/device is rated for the load weight

Figure B.8 Manual lifting and carrying of fabricated work piece. Possible solution, an overhead crane eliminates hazards, saves time and effort, and protects finished product (Photo adapted from Ergonomics Guide for Welders, unpublished Navy Occupational Safety and Health Working Group, Ergonomics Task Action Team with technical support provided by the Certified Professional Ergonomists at General Dynamics Information Technology (GDIT), San Diego, CA, 2010)

Reduce lifting, pushing, and pulling forces

  - Workers' ideas
  - Time frame for making improvements
  - Difficulty in making improvements
- Step 3: Make improvements
  - Improve the fit between task demands and worker capabilities
    - Talk to various employees
    - Contact other industries

TABLE B 5 Material Positioning Equipment – Simple Solution

Description: In the earlier photos you can see the workers are in a very awkward posture. Simply elevating the work piece significantly reduces the stresses on the worker's back

Advantages:

- Reduced awkward neck and back posture, Figures B.9 and B.10
- Safer
- Low cost

Points to remember:

- Solutions do not need to be costly to be effective

Figure B.9 Before: Welder's neck and back are in an awkward posture while he kneels on a hard surface

Figure B.10 After: Simple trailer jack lifts work piece and improves posture at Fleet Readiness Center East (Photos adapted from Ergonomics Guide for Welders, unpublished Navy Occupational Safety and Health Working Group, Ergonomics Task Action Team with technical support provided by the Certified Professional Ergonomists at General Dynamics Information Technology (GDIT), San Diego, CA, 2010)

Reduce awkward body positions
Decrease contact stress on the lower body from kneeling on hard and cold surfaces

TABLE B 6 People Positioning Equipment

Description: Angle-adjustable creeper

Advantages:

Height- and or angle-adjustable creepers and tool stools allow the welders to get closer to their work while supporting some of the body weight, Figures B.11 and B.12

Points to remember:

- Ensure materials used for these devices are flame resistant or retardant
- Maintain casters for ease of movement

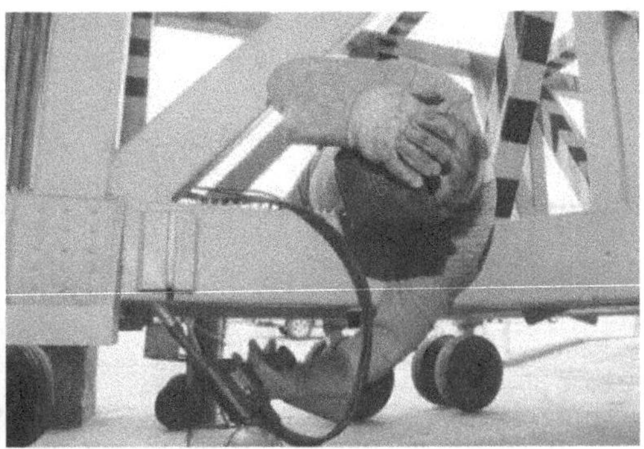

Figure B.11 Before: Welder is in awkward posture to reach beneath work piece

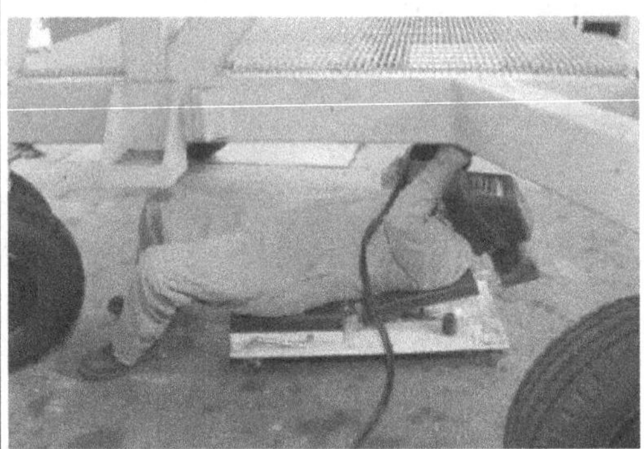

Figure B.12 After: Welder is in a supported poster and can easily see the work piece (Photos adapted from Ergonomics Guide for Welders, unpublished Navy Occupational Safety and Health Working Group, Ergonomics Task Action Team with technical support provided by the Certified Professional Ergonomists at General Dynamics Information Technology (GDIT), San Diego, CA, 2010)

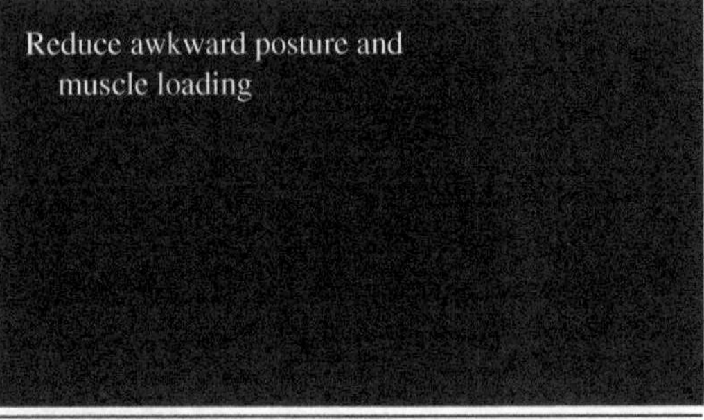

Reduce awkward posture and muscle loading

TABLE B 6 (*Continued*)

Description: Positions and supports welder

Advantages:

- Height- and/or angle-adjustable creepers allow the welders to get closer to their work while supporting some of the body weight
- Lean forward-type welding stool relieves some static back loading

Points to remember:

- Ensure materials used for these devices are flame resistant or retardant

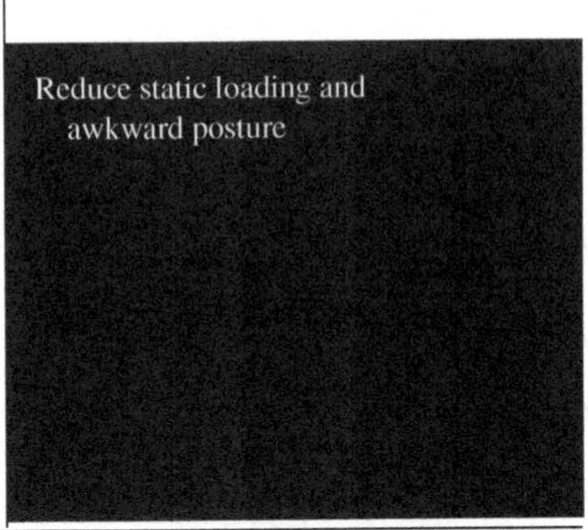

TABLE B 7 Welding Helmet with Auto-darkening Lens

Description: Lens automatically darkens when spark ignites

Advantages:

- Greatly reduces repetitive forward neck impact to close traditional helmet, Figure B.13
- Lenses are interchangeable
- Models incorporate sensitivity or delay control

Points to remember:

- Ensure shade, delay, and sensitivity features match working environment and welding type

Figure B.13 Auto-darkening welding lens

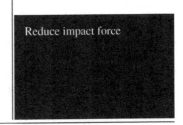

- Consult ergonomics experts
- Use Internet resources (see references and resources section)

- Step 4: Follow-up:
  - Has each improvement reduced or eliminated the risk factors, fatigue, discomfort symptoms, or injury reporting?
  - Has each improvement been accepted by the workers?

TABLE B 8 Material Positioning Equipment Positioning Equipment

Description: An anvil or positioning device is useful to hold smaller components during welding or material at an appropriate level

Advantages:

- The anvil can be positioned to reduce awkward postures of the neck, back, hands, and wrists by positioning the work piece instead of the worker

Work surfaces that are easy to raise or to lower allow employees to work in neutral postures

Points to remember:

- An adjustable work table, work chair, or standing platforms may be necessary to accommodate different size work pieces
- Surfaces must be able to support the weight of the object or material
- Welding tables come in a variety of sizes with fixtures for small or large parts and can be joined together for larger work pieces

TABLE B 9 Robotic Welders, Electromagnetic, and Air Presses

Description: Remote operation welding

Advantages:

- Reduced fatigue associated with welding in awkward or constrained postures
- More efficiency
- Greater consistency of quality welds
- Increased repeatability
- Reduced production costs

Points to remember:

- Not applicable for all welding applications

Description: Portable electromagnetic and air presses hold parts in place during welding

Advantages:

- Reduces awkward postures and forceful exertions associated with manually clamping parts prior to weld operations
- Can reduce job completion time

Points to remember:

- Useful for long seams

- Have any improvements created new risks or other problems?
- Have any improvements impacted production or quality?
- Are implemented improvements supported by training?

ADDITIONAL SOURCES

OSHA Guidelines for Shipyards; OSHA 3341-03N (2008). Ergonomics for the Prevention of Musculoskeletal Disorders. http://www.osha.gov/dsg/guidance/shipyard-guidelines.html.

National Institute for Occupational Safety and Health: Publication Number 2007-131: Ergonomics Guidelines for Manual Material Handling. http://www.cdc.gov/niosh/docs/2007-131/.

Navy Safety and Occupation Heath (SOH) Program Manual (OPNAVINST 5100.23G). Chapter 23 Ergonomics Program, www.navfac.navy.mil/safety (Select Hazard Abatement). Chapter 23 Ergonomics Program and Chapter 12 Hazard Abatement.

Department of Defense Instruction Safety and Occupation Health Program (DoDI) 6055.1. Enclosure (6) DoD Ergonomic Program Requirements and Procedures, http://www.dtic.mil/whs/directives/corres/pdf/605501p.pdf.

APPENDIX C

TOOLS

RAPID ENTIRE BODY ASSESSMENT TABLES

See Tables C 1–C 3.

RAPID UPPER LIMB ASSESSMENT TABLES

See Tables C 4–C 6.

Hand Tool Selection or Comparison Tool

Date	Evaluator
Type of tool	ID number/storage location
Department	Job/task
Purpose	
Description	
Manufacture/model	

Occupational Ergonomics: A Practical Approach, First Edition.
Theresa Stack, Lee T. Ostrom and Cheryl A. Wilhelmsen.

TABLE C 1 REBA Table A: Neck, Leg, and Trunk Score

		Neck Score											
		1				2				3			
	Legs	1	2	3	4	1	2	3	4	1	2	3	4
Trunk Posture Score	1	1	2	3	4	1	2	3	4	3	3	5	6
	2	2	3	4	5	3	4	5	6	4	5	6	7
	3	2	4	5	6	4	5	6	7	5	6	7	8
	4	3	5	6	7	5	6	7	8	6	7	8	9
	5	4	6	7	8	6	7	8	9	7	8	9	9

Source: Adapted from McAtamney & Corlett (1993) and Hignett & McAtamney (2000).

TABLE C 2 REBA Table B: Lower Arm, Wrist, and Upper Arm Score

		Lower Arm					
		1			2		
	Wrist	1	2	3	1	2	3
Upper arm score	1	1	2	2	1	2	3
	2	1	2	3	2	3	4
	3	3	4	5	4	5	5
	4	4	5	5	5	6	7
	5	6	7	8	7	8	8
	6	7	8	8	8	9	9

Source: Adapted from McAtamney & Corlett (1993) and Hignett & McAtamney (2000).

TABLE C 3 REBA Table C: Final Evaluation Score

Score A Score from Table A + Load/Force Score	Score B (Table B Value + Coupling Score)											
	1	2	3	4	5	6	7	8	9	10	11	12
1	1	1	1	2	3	3	4	5	6	7	7	7
2	1	2	2	3	4	4	5	6	6	7	7	8
3	2	3	3	3	4	5	6	7	7	8	8	8
4	3	4	4	4	5	6	7	8	8	9	9	9
5	4	4	4	5	6	7	8	8	9	9	9	9
6	6	6	6	7	8	8	9	9	10	10	10	10
7	7	7	7	8	9	9	9	10	10	11	11	11
8	8	8	8	9	10	10	10	10	10	11	11	11
9	9	9	9	10	10	10	11	11	11	12	12	12
10	10	10	10	11	11	11	11	12	12	12	12	12
11	11	11	11	11	12	12	12	12	12	12	12	12
12	12	12	12	12	12	12	12	12	12	12	12	12

Source: Adapted from McAtamney & Corlett (1993) and Hignett & McAtamney (2000).

TABLE C 4 RULA Table A: Upper Limb Score

Table A		Wrist Posture Score							
		1		2		3		4	
Upper Arm	Lower Arm	Wrist Twist		Wrist Twist		Wrist Twist		Wrist Twist	
		1	2	1	2	1	2	1	2
1	1	1	2	2	2	2	3	3	3
	2	2	2	2	2	3	3	3	3
	3	2	3	3	3	3	3	4	4
2	1	2	3	3	3	3	4	4	4
	2	3	3	3	3	3	4	4	4
	3	3	4	4	4	4	4	5	5
3	1	3	3	4	4	4	4	5	5
	2	3	4	4	4	4	4	5	5
	3	4	4	4	4	4	5	5	5
4	1	4	4	4	4	4	5	5	5
	2	4	4	4	4	4	5	5	5
	3	4	4	4	5	5	5	6	6
5	1	5	5	5	5	5	6	6	7
	2	5	6	6	6	6	7	7	7
	3	6	6	6	7	7	7	7	8
6	1	7	7	7	7	7	8	8	9
	2	8	8	8	8	8	9	9	9
	3	9	9	9	9	9	9	9	9

Source: Adapted from McAtamney & Corlett (1993) and Hignett & McAtamney (2000).

TABLE C 5 RULA Table B: Trunk, Leg, and Neck Score

Neck Posture Score	Trunk Score											
	1		2		3		4		5		6	
	Legs		Legs		Legs		Legs		Legs		Legs	
	1	2	1	2	1	2	1	2	1	2	1	2
1	1	3	2	3	3	4	5	5	6	6	7	7
2	2	3	2	3	4	5	5	5	6	7	7	7
3	3	3	3	4	4	5	5	6	6	7	7	7
4	5	5	5	6	6	7	7	7	7	7	8	8
5	7	7	7	7	7	8	8	8	8	8	8	8
6	8	8	8	8	8	8	8	9	9	9	9	9

Source: Adapted from McAtamney & Corlett (1993) and Hignett & McAtamney (2000).

TABLE C 6 RULA Table C: Final Score

Table C		Neck, Trunk, and Leg Score						
		1	2	3	4	5	6	7+
Wrist and Arm Score	1	1	2	3	3	4	5	5
	2	2	2	3	4	4	5	5
	3	3	3	3	4	4	5	6
	4	3	3	3	4	5	6	6
	5	4	4	4	5	6	7	7
	6	4	4	5	6	6	7	7
	7	5	5	6	6	7	7	7
	8+	5	5	6	7	7	7	7

Source: Adapted from McAtamney & Corlett (1993) and Hignett & McAtamney (2000).

	Tool A	Tool B	Optimum
Weight (lb)			Counterbalanced >5 lb power >1 lb precision
Speed/torque (single, dual, variable)			
Specifications (RPM, motor size)			
Direction (one-way, reversible)			
Torque			Air shut-off, pulse, clutch, limiting device
Noise level (dBA) (idle/under load)			Noise < 85 dBA
Vibration level (m/s^2) (idle/under load)			See ACGIH guidelines
Comfort during use (poor, good, excellent)			
Grip diameter			
Handle insulation			No slip, nonconductive frequency dampening material
Grip angle			70–80° pistol
Grip length			See tool survey tables
Trigger (1–4 fingers)			Finger strip, levers
Exhaust pathway (rear, front, side)			Adjustable diverter, away from hands/wrist suspended hoses
Center of gravity (appropriate)			
Quality of job (poor, good, excellent)			
Cost			
Overall rating (poor, good, excellent)			

Tool Evaluation Checklist	Yes	No
Basic principles		
1. Does the tool perform the desired function effectively?		
2. Does the tool match the size and strength of the operator?		
3. Can the tool be used without undue fatigue?		
4. Does the tool provide sensory feedback?		
5. Are the tool capital and maintenance costs reasonable?		
Anatomical concerns		
1. If force is required, can the tool be grasped in a power grip?		
2. Can the tool be used without shoulder abduction?		
3. Can the tool be used with the wrist straight?		
4. Does the tool handle have large contact surface to distribute forces?		
5. Can the tool be used comfortably by everyone?		
Handles and grips		
1. For power or precision use, is the tool grip within permitted range per tables?		
2. For power grips, can the handle be grasped with the thumb and fingers slightly overlapped?		
3. Is the grip cross section circular?		
4. Is the grip length within permitted range per tables?		
5. Is the grip surface finely textured and slightly compressible?		
1. Does the handle surface have a nonslip material?		
2. Is the handle nonconductive and stain-free?		
3. For power/pistol grip uses, does the tool have a grip angle 70–80°?		
4. Can two-handle tools be operated with less than 90 N of grip force?		
Power tool consideration		
1. Are triggers activation forces within permitted range per tables?		
2. For repetitive use, is a finger strip trigger provided and at least 1 in. length?		

Tool Evaluation Checklist	Yes	No
3. Are less than 10,000 triggering actions required per shift?		
4. Is a reaction bar provided for torques exceeding 6 N-m for in-line tools		
a. 12 N-m for pistol grip tools?		
b. 50 N m for right-angled tools?		
5. Does the tool create less than 85 dBA for a full day of noise exposure?		
6. Is the vibration exposure within the ACGIH recommendations?		
7. Is there room for 2–3 fingers on the trigger for power tools, and 4 fingers if the tool is balanced (not thumb operated)?		
8. Do handles have vibration-absorbing materials?		
Miscellaneous and general considerations		
1. For general use, is the weight of the tool less than 2.3 kg? (5 lb)		
2. For precision tasks, is the weight of the tool <4 kg? (1 lb)		
3. For extended use, is the tool suspended or equipped with a microbreak strap?		
4. Is the tool balanced (i.e., center of gravity is aligned with hand grip)?		
5. Does the tool have stops to limit closure and prevent pinching?		
6. Can you operate the tool with either/both hands?		
7. Does the tool have smooth rounded edges, without finger grooves?		
8. Does the tool protection against the generation of heat or cold exposure? (Handle temp should not exceed 65–95°.)		

COMPUTER WORKSTATION CHECKLIST

The computer workstation checklist is one method available for performing computer workstation assessments. The checklist is designed to be printed as two double-sided pages (front and back). The first sheet (page 1 and 2) is an educational guide for the employee and is meant to be left at the workstation for the employee's reference. At the beginning of the assessment, the evaluator should define ergonomics and explain

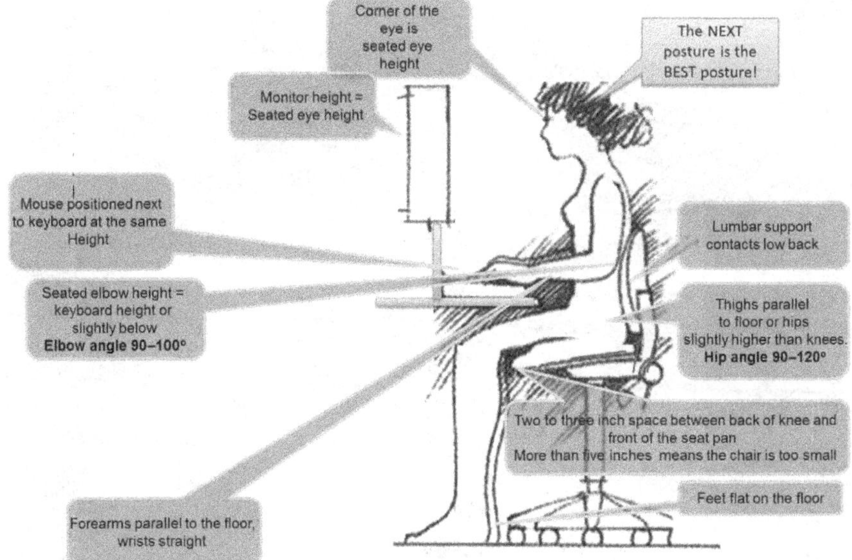
Corner of the eye is seated eye height
The NEXT posture is the BEST posture!
Monitor height = Seated eye height
Mouse positioned next to keyboard at the same Height
Lumbar support contacts low back
Seated elbow height = keyboard height or slightly below
Elbow angle 90–100°
Thighs parallel to floor or hips slightly higher than knees.
Hip angle 90–120°
Two to three inch space between back of knee and front of the seat pan
More than five inches means the chair is too small
Feet flat on the floor
Forearms parallel to the floor, wrists straight

Figure C.1 Office at a glance

the proper-seated neutral posture for a computer workstation. The second page sheet (page 3 and 4) contains a checklist, which is intended to guide the evaluator and be kept for the evaluator's records. The checklist is designed to step the evaluator through the workstation evaluation. Answering an item on the checklist with a "NO" indicates a potential ergonomics program. Possible solutions to address the issue are in the far right column.

Thank you for participating in an ergonomic computer workstation assessment.

Ergonomics is the science of fitting the workplace to the worker to reduce the risk of injury. In order to reduce the risk of developing work-related musculoskeletal disorders (WMSDs), it is important to change posture frequently (Figure C.1). This will help prevent soft tissue WMSDs, such as carpal tunnel syndrome and tendonitis. The following illustration is a guide to setting up our own computer workstation. The neutral posture is the optimal body position, which provides the greatest strength and control and minimizes stress. Even a neutral posture can be fatiguing if held all day; therefore, microchanges in posture and stretching are recommended (see page 2) (Figure C.2).

For more information, or to report pain or discomfort, you feel is associated with the job, please contact their Supervisor or Safety Officer who can refer you as needed to the Industrial Hygienist and/or Occupational Health Provider/Clinic.

The information in the figure accommodates 90% of the population, special considerations may be necessary.

Tip: Taking 20 s microbreaks throughout the day to refocus one's eyes will reduce fatigue at the end of the day. 20/20 rule: for every 20 min of work, rest the eyes 20 s (note: Page 1 and 2 are to be left with the employee).

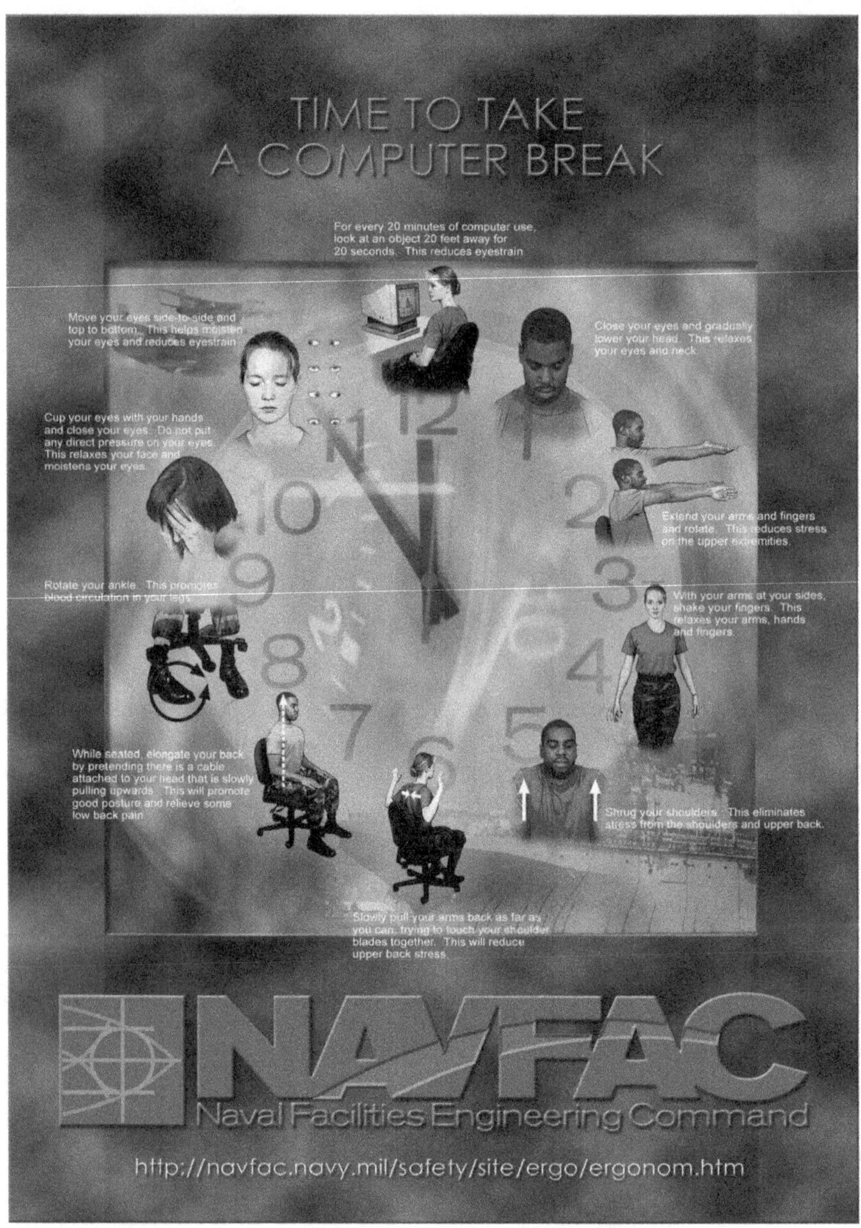

Figure C.2 Office stretching exercises – Department of Defense Ergonomics Working Group 2007 (Adapted from OPNAVINST 5100.23G Series Chapter 23 Appendix B)

COMPUTER WORKSTATION SURVEY TOOL

Evaluator:		Date:	
Employee name:	Title:	Location:	Email:
Time in current position:		Phone number:	
Percent of day (or hours per day) spent performing the following tasks:	Computer – keying:		Mouse, track ball:
Hours worked per week? Is workstation shared? Y/N	Telephone:		Writing: Other tasks:
Pain or discomfort, documented injuries, risk factors, etc.			

If the answer is **NO** to any of the following questions, there is a potential problem.

Y	N	Minimum and maximum recommendations accommodate 90% of the population Special considerations may be necessary for the extremes and users with special medical conditions	Possible solutions Circle if recommended
		Work chair **Seat height** Do the user's feet rest comfortably on the floor or a footrest with thighs parallel to the floor and hips at a height equal to or slightly above knee height? *Action*: If the workstation height is adjustable – adjust the chair so the user's feet rest comfortably on the floor/footrest. If the workstation is not adjustable raise the user to the keyboard height (refer to keyboard section) and use a footrest to encourage sitting back in the chair	Footrest
		Seat pan Does the seat pan support the thighs? The user should be able to fit two fingers between the backs of the knees and the edge of the seat. The seat pan should not be significantly shorter or longer than the length of the thighs. *Action:* Adjust seat pan and/or adjust backrest. (Fixed seat pan maximum length 16.9 in.)	Footrest Lumbar support Different chair

Y	N	Minimum and maximum recommendations accommodate 90% of the population Special considerations may be necessary for the extremes and users with special medical conditions	Possible solutions Circle if recommended
		Does the seat cushion have a rounded front edge?	Different chair
		Is the seat pan wider than the hip breadth of the user to allow space for movement and clothing? (Minimum inch)	Different chair
		Backrest	
		Does the backrest provide adequate lumbar support and buttocks clearance without interfering with the user's movement? The most pronounced part of the backrest should coincide with the middle of the user's lumbar area (small of the back) between 5.9 and 9.8 in. from the seat pan. *Action*: Adjust backrest	Lumbar support
		Is backrest wide and high enough to support the torso? (Minimum 14.2 in. W × 12.2 in. H)	Different chair
		Armrests	
		Do the armrests adjust to a height that is comfortable for the user and avoids hunched shoulders (armrests are too high) or slouching (armrests are too low) while allowing the user to get close enough to perform the task while sitting back in the chair? The user should not plant his/her elbows on the armrests while typing. Armrests should be soft and pliable. *Action:* Adjust armrests or remove armrests if they are not adjustable and interfere with the task	
		Do the armrests adjust to a width that comfortably fits the user's hips and allow the user to easily exit/enter a chair and perform his/her task? (Minimum separation 18 in.) *Action*: Adjust armrests or remove if necessary	
		Miscellaneous	
		Does the chair have a stable base supported by five legs with casters and swivel 360°?	Different chair
		Does the chair roll easily (casters appropriate for the floor surface)?	Chair mat Different casters

Y	N	Minimum and maximum recommendations accommodate 90% of the population Special considerations may be necessary for the extremes and users with special medical conditions	Possible solutions Circle if recommended
		Work surface	
		Is there adequate clearance beneath the workstation for the user to get close enough to the task, maintain freedom of movement, and not come into contact with obstructions such as table legs, filing cabinets, etc.? (Height clearance for legs minimum 25 in., depth clearance for knees minimum 17 in.) *Action*: Rearrange workstation, remove clutter/obstructions	Different work surface Raise or lower work surface
		Are the computer monitor and keyboard in alignment with (directly in front of) the user? *Action:* Rearrange workstation	
		Is the work surface with the keyboard positioned at seated elbow height? Seated elbow height is measured with the feet resting comfortably on the floor (or a footrest) and the back positioned against the backrest. The upper arms are close to the sides with elbows at a 90° angle. The seated elbow height is the distance from the floor to the bony protrusion on the elbow. *Action:* Adjust work surface, keyboard tray, or chair. If feasible, reposition a portion of the work surface used exclusively for computer tasks	Height-adjustable keyboard tray Leg lifters for desk Different work surface
		Is the mouse or other input device located at the same height as the keyboard (at elbow height) within close reach? When keying or using the mouse, the upper arms should be close to the body, elbows approximately 90° with forearms parallel to the floor and wrists straight	Mouse bridge or platform Keyboard tray Alternative or wireless pointing device
		Are frequently used support equipment/materials (telephone, documents, references) within 14–18 in. with occasionally used items within 22–26 in.?	

Y	N	Minimum and maximum recommendations accommodate 90% of the population Special considerations may be necessary for the extremes and users with special medical conditions	Possible solutions Circle if recommended
		Two-handed reach distances are shorter than single-handed reaches and reaches for items over 10 lb. should be performed standing. *Action:* rearrange workstation	
		Monitor	
		Is the monitor located about arm's length away from the user (minimum 15.7 in.)? Monitor distance depends on the user's eyesight and possible corrective vision use, and monitor depth. *Action:* Rearrange workstation	Suggest employee see personal eye care specialist Larger work surface
		Is the monitor height (measured from the top row of characters on the screen) at a height equal to or 20° below the user's seated eye height (measured from the corner of the eye when a person is looking straight ahead)? The monitor should be located so the user does not have to bend the neck back or forward to see clearly. *Action:* Elevate or lower monitor. If necessary, elevate chair and provide footrest	Monitor risers/arm
		Are the monitor images clear and stable, free of dust or glare (reflections)? *Action:* Turn off overhead lights, reposition blinds, or shield monitor to the side/top to assess glare. Rearrange workstation so that monitor is perpendicular to light source. Change lighting/blinds during the day to reduce glare	Add task lighting, reduce overhead lighting (removing bulbs), glare screen
		Accessories	
		Is the employee comfortable while receiving phone calls during the day, which require him/her to type or write while speaking	Telephone headset
		Does the employee type in a neutral posture without using the wrist rest? A wrist rest should be used for resting; the arms should float above the keyboard in a neutral posture (straight wrists) when typing. The keyboard should be flattened or at a negative tilt as close to the user as possible	

Y	N	Minimum and maximum recommendations accommodate 90% of the population Special considerations may be necessary for the extremes and users with special medical conditions	Possible solutions Circle if recommended
		If the worker references documents while typing, are they located in a holder next to, in front of, or at an equal distance to the monitor and not resting flat on the desk? Document position depends on eyesight, document and screen font, and task parameters	Rearrange workstation Document holder
		Is the worker able to get up from the computer on a regular basis (typing for less than 6 h a day)?	Sit/stand workstation
		Is the input device (mouse) appropriate for the task and is the user operating it with minimal force? Thumb-operated trackballs are typically not recommended for extended daily use	Alternative input device Input device sized to the user

LABORATORY CHECKLIST

This document aids in identifying physical workplace risk factors associated with laboratory environments.

How to Use the Checklist

Step One: Select a laboratory that has any of these characteristics: high musculoskeletal injury rate; workers have concerns about performing tasks, during a walk through during hazard recognition, or follow-up after an intervention.

Step Two: Advise the involved workers and supervisor that an ergonomic survey is being performed in the area.

Step Three: Complete the checklist for the task you have identified. Answer N/A if the question does not apply to the task. Include all meaningful comments for each area.

Step Four: *Each "NO" answer indicates a risk of injury or suboptimal condition.* For each "NO" answer, modify the workstation or task to result in a yes response. Gather input from the workers, supervisors, safety specialists, and occupational health professionals, to evaluate equipment before purchasing and before modifying the work areas or tasks. This process will help increase product acceptance, test product usability, and durability, and take advantage of worker experience.

How to Prevent Laboratory Injuries

Commitment and involvement of the entire workplace, from top management to line workers, are essential elements of a successful injury prevention program. The best approach to prevent injuries involves education, design, and hazard abatement.

Education is a key step in injury prevention. The workers should have a basic understanding of ergonomic principles and handling techniques, recognize risk factors and injury or illness symptoms. The workers need to have a clear procedure to report injury symptoms, risk factors, near misses, hazards, incidents, and accidents to their supervisor.

The design of the job itself (work/rest schedules, job rotation) and of the workstation (dimension/layout) have a direct impact on the risk of injury. In order to prevent injuries, these aspects may need modification.

There are additional costs incurred in redesigning or modifying a task. Considering the factors in the checklist before designing or modifying a workstation should help reduce costs. It is usually more costly to redesign and retrofit than to design it right the first time. Workers are a great source of creativity and innovation and should always be included in the process.

Evaluator:		Date:
Department name:	Location:	POC name: Phone number:
Job position evaluated:		Number of employees performing job:
Follow-up date:		Email address:
Job description/drawings		
Recommendations:		
Follow-up:		

Laboratory Benches/Work Surface	**Yes** ☐	**No** ☐
1. If the employee stands for 4 or more hours a day, is antifatigue matting supplied?	☐	☐
2. Is the height of the bench appropriate for the work that is performed? Typically, the work should be positioned at elbow height, slightly higher for precision work, slightly lower if applying force	☐	☐
3. Is there adequate leg and foot room so the worker's head, hips, and feet are aligned directly in front of the work? From the side, ears over shoulders, shoulders over hips	☐	☐
4. Are work areas free of contact stress points such as a bench top with sharp edges?	☐	☐

Laboratory Benches/Work Surface	Yes ☐	No ☐

See Figure C.3

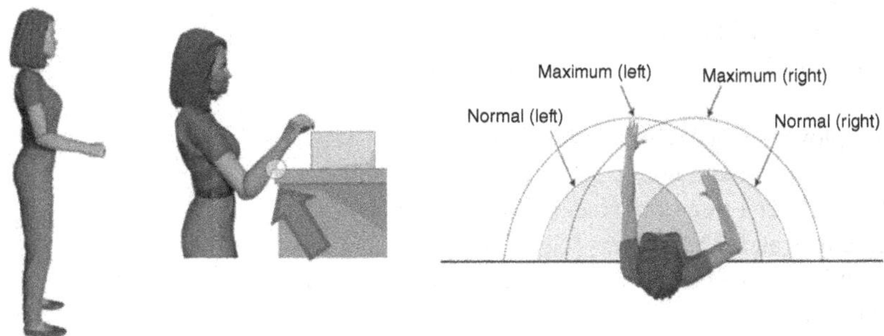

Figure C.3 Standing elbow eight, sharp edge on standing workstation with work above elbow height, reach envelope (Adapted from OPNAVINST 5100.23G Series Chapter 23 Appendix B Ergonomics Image Gallery)

5. Are frequently used items within easy reach, that is, within arm's length, allowing the worker to reach without bending the back?	☐	☐
6. Is the workstation at a height and distance that reduces arm/shoulder abduction as much as possible?	☐	☐
7. Are frequently accessed items stored between knuckle and shoulder height?	☐	☐

Laboratory Chairs	Yes ☐	No ☐
1. Can the laboratory chairs be adjusted to accommodate all employees who need to use the chair (i.e., do their feet reach the floor or foot support and do backs rest on the lumbar support? Reference Chair selection guide found in the Tools appendix	☐	☐
2. Are appropriate footrests or adjustable foot rings provided?	☐	☐
3. Do employees know how to correctly (or properly) adjust the chair?	☐	☐
4. Are chairs free of armrests if they interfere with work?	☐	☐

Microscope Use	Yes ☐	No ☐
1. Have employees been trained how to properly sit at a microscope workstation?	☐	☐
2 Do employees work with neutral shoulder postures (i.e., without rounded shoulders, without hunching, no elevated shoulders or elbows)?	☐	☐
3. Do employees work with neutral neck postures (neck flexion should not be >25°)?	☐	☐
4. Is the work area free of contact stress between sharp edges and the forearms? See Figure C.1	☐	☐
5. Is the microscope positioned at the edge of the workbench? See Figure C.4	☐	☐

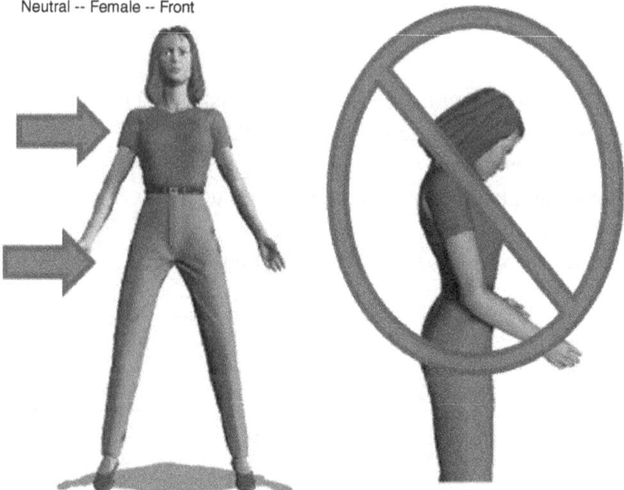

Figure C.4 Power zone between knuckles and shoulders, poor viewing posture (Ergonomics Image Gallery)

6. If armrests on the chair for the microscope use are provided, are they padded?	☐	☐
7. Is there sufficient legroom so the worker does not have to twist the torso?	☐	☐
8. Do employees rest their feet on the floor or an appropriate footrest (rather than the lab stool)?	☐	☐
9. Are microscope work breaks provided?	☐	☐

Pipette Use	Yes ☐	No ☐
1. Is manual pipette use limited to less than 4 h of cumulative or continuous activity a day?	☐	☐
2. Are electric pipettes provided where appropriate?	☐	☐
3. Are latch mode pipettes provided?	☐	☐
See Figure C.5		

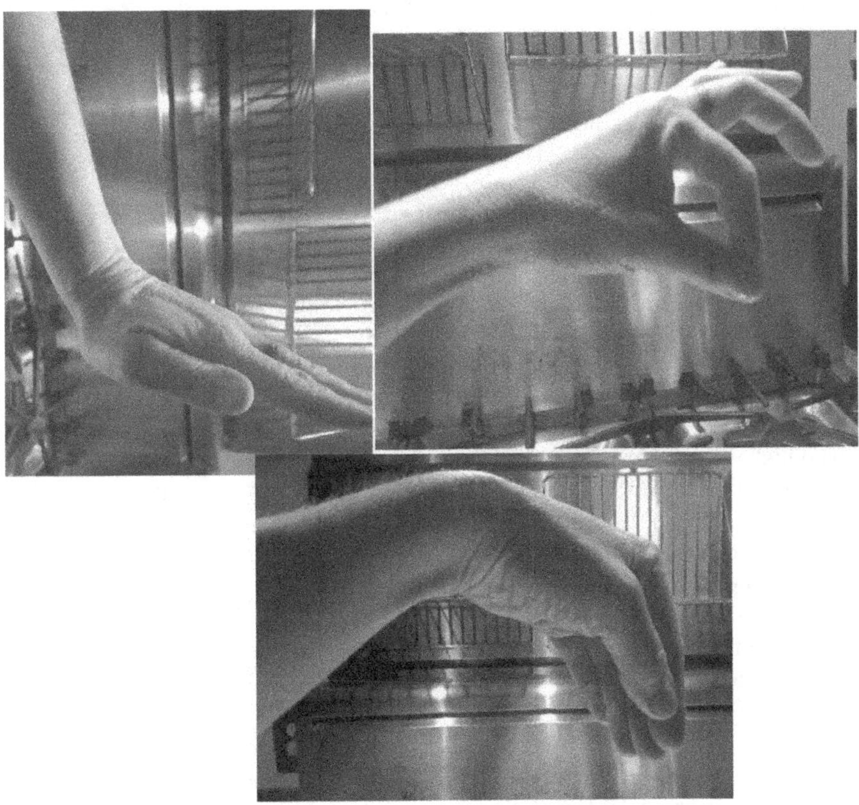

Figure C.5 Awkward wrist postures

4. Is the pipette designed to reduce contact stress with shape edges? Is the pipette handle oval shaped to reduce contact stress on sharp edges?	☐	☐
5. Have employees been trained how to properly operate the pipette (e.g., pickup tips, eject tips, program electronic pipettes)?	☐	☐
6. Are rest breaks provided?	☐	☐
7. Is the pipette electric or multichanneled to allow for computer activated multiple dispensing instead of finger activated dispensing?	☐	☐

Fine Motor Skills	Yes ☐	No ☐
1. Are vials with the fewest amounts of threads allowable used?	☐	☐
2. Are dissectors or micromanipulation tasks with forceps performed less than 5 h/week?	☐	☐
3. Are frequent microbreaks provided?	☐	☐
4. Is the work area free of contact stress between the forearms and the workbench?	☐	☐

Microtome/Cryostat	Yes ☐	No ☐
1. Do employees use neutral postures when operating the microtome or cryostat (without excessive wrist flexion or extension)?	☐	☐
2. Do employees repeat the same hand/arm/wrist motions with little or no variation for more than 2 h/day with the wrists bent in flexion (>30°) or extension (>45°) or ulnar deviation (>30°) combined with high, forceful exertions of the hands		
3. Do employees have access to automatic microtome/cryostat?	☐	☐
4. Is a fully adjustable chair provided that allows the feet to rest flat on the floor or footrest and back to lean against the lumbar support?	☐	☐
5. Is there sufficient legroom to prevent the worker from twisting the torso?	☐	☐

Laboratory Checklist adapted from the National Institutes of Health – National Institute of Environmental Health Sciences.

PHYSICAL RISK FACTOR CHECKLIST

This checklist is used for typical work activities that are a regular and foreseeable part of the job, occurring more than 1 day/week, and more frequently than 1 week/year.

The Physical Risk Factor Checklist is a tool used to identify ergonomics stressors in the workplace. For each category, determine whether the physical risk factors rate as a "caution" or "hazard" by placing a check (✓) in the appropriate box. Make a notation if a category is not applicable.

If a hazard exists, it must be reduced below the hazard level or to the degree technologically and economically feasible. Ensure workers exposed to ergonomics stressors at or above the "hazard" level have received general ergonomics training and provide a refresher of the ergonomics physical and contributing risk factors.

If the task rates a "caution", reevaluate at least yearly since changes in the work environment may create new ergonomics stressors.

Reduce to the lowest level feasible significant contributing physical risk factors and consider contributing personal risk factors in the evaluation. Contributing risk factors contribute to but do not cause work-related musculoskeletal disorders. Physical contributing risk factors may include temperature extremes, inadequate recovery time, and stress on the job. Personal contributing risk factors may include but are not limited to age, pregnancy, obesity, thyroid disorder, arthritis, diabetes, or preexisting injuries such as wrist/knee/ankle strain or fracture, back strain, trigger finger, and carpal tunnel syndrome. Professional judgment should be used in instances where personal or physical contributing factors are present. The risk of developing a work-related musculoskeletal disorder increases when ergonomics risk factors occur in combination.

Evaluator: Date:
Department name: Location: POC name: Phone number:
Job position evaluated: Number of employees performing job:
Follow-up date: Email address:
Job description/figures:
Recommendations: caution, hazard, neither, not applicable (circle one)
Follow-up:

Awkward Posture				
	Caution		Hazard	
1. Working with the hand(s) above the head, or the elbow(s) above the shoulders	□	More than 2 h total per day	□	More than 4 h total per day
2. Repeatedly raising the hand(s) above the head, or the elbow(s) above the shoulder(s) more than once per minute	□	N/A	□	More than 4 h total per day
3. Working with the neck bent (without support and without the ability to vary posture)	□	More than 30° for more than 2 h total per day	□	More than 30° for more than 4 h/day or More than 45° for more than 2 h total per day

Awkward Posture		
	Caution	Hazard
4. Working with the back bent forward (without support and without the ability to vary posture)	☐ More than 30° for more than 2 h total per day	☐ More than 30° for more than 4 h total per day, or More than 45° more than 2 h total per day

See Figure C.6

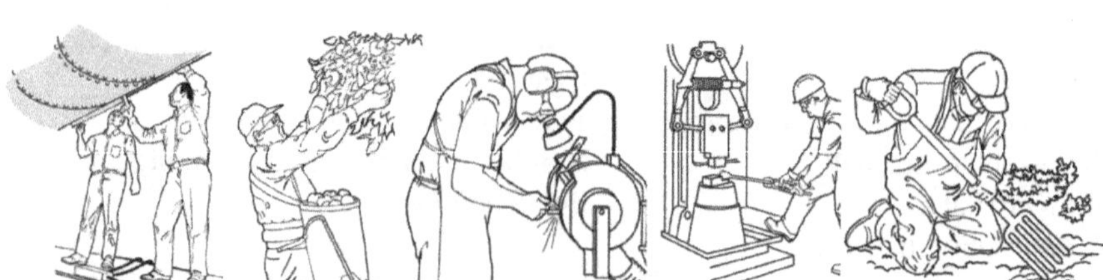

Figure C.6 Awkward postures (Adapted from OPNAVINST 5100.23G Series Chapter 23 Appendix B) and Public Domain Washington State Department of Labor and Industries Caution checklist)

	Caution	Hazard
5. Squatting	☐ More than 2 h total per day	☐ More than 4 h total per day
6. Kneeling	☐ More than 2 h total per day	☐ More than 4 h total per day
Moderate to High Hand-Arm Vibration		
7. Using impact or percussive type tools such as impact wrenches, carpet strippers, chain saws, percussive tools (jackhammers, scalers, riveting, or chipping hammers) or other tools that typically have high vibration levels	☐ More than 30 min total per day	☐ For exposures that exceed caution level of more than 30 min/day, perform

Awkward Posture				
		Caution		Hazard
				analysis using the Hand-Arm Vibration Analysis Tool Guide in the ACGIH TLV guide
8. Using grinders, sanders, jigsaws or other hand tools that typically have moderate vibration levels	□	More than 2 h total per day	□	For exposures that exceed caution level of more than 2 h/day perform analysis using the Hand-Arm Vibration Analysis Guide in the ACGIH TLV guide
High Hand Force				
9. Pinching an unsupported object(s) weighing 2 or more pounds per hand, or pinching with a force of 4 or more pounds per hand	□	More than 2 h/day (comparable to pinching half a ream of paper or the force required to open two wooden clothespins)	□	– More than 4 h/day with no other risk factors, or – More than 3 h/day with highly repetitive motion, or – More than 3 h/day with significant wrist deviation in flexion (>30°), extension (>45°), radial or ulnar deviation (>30°)

Awkward Posture				
		Caution		Hazard
10. Gripping an unsupported objects(s) weighing 10 or more pounds per hand, or gripping with a force of 10 or more pounds per hand	□	More than 2 h total per day (comparable to clamping light duty automotive jumper cables onto a battery)	□	– More than 4 h/day with no other risk factors, or – More than 3 h/day with highly repetitive motion, or – More than 3 h/day with significant wrist deviation in flexion (>30°), extension (>45°), flexion, radial or ulnar deviation (>30°)

See Figure C.7

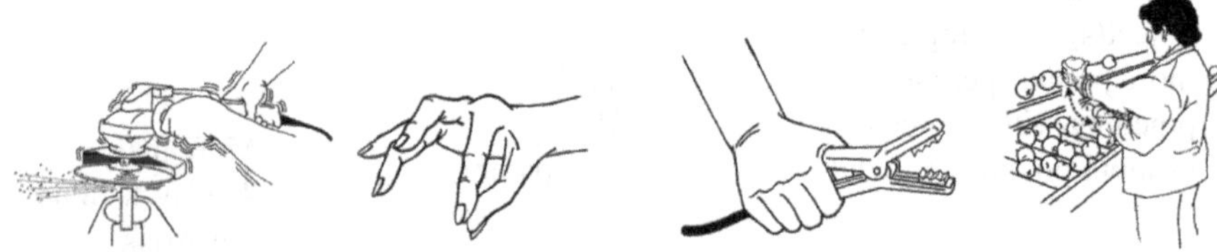

Figure C.7 Vibration, high hand forces, repetition (Adapted from OPNAVINST 5100.23G Series Chapter 23 Appendix B and Public Domain Washington State Department of Labor and Industries Caution checklist)

Highly Repetitive Motion				
11. Repeating the same motion with the neck, shoulders, elbows, wrists, or hands (excluding keying activities) with little or no variation every few seconds	□	More than 2 h total per day	□	– More than 6 h/day with no other risk factors, or

Awkward Posture				
		Caution		Hazard
				– More than 2 h/day with wrists bent in flexion (>30°), extension (>45°), radial or ulnar deviation (>30°) and high, forceful exertions of the hand(s)
Heavy, Frequent, or Awkward Lifting adapted from the MIL-STD 1472G				
12. Lifting objects	□	From the floor and place it on a surface not greater than 3 ft (36 in.) above the floor[a–c] Females 44 lb Males 87 lb	□	For exposures that exceed caution level perform lift analysis using the MIL-STD 1472F 5.9.11.3.1, ACGIH TLV guide, or NIOSH lifting equation
13. Lifting objects	□	From the floor and place it on a surface not greater than 5 ft (60 in.) above the floor[a–c] Females 37 lb Males 56 lb	□	For exposures that exceed caution level perform lift analysis using the MIL-STD 1472F 5.9.11.3.1, ACGIH TLV guide, or NIOSH lifting equation

Awkward Posture				
	Caution		Hazard	
14. Two people object lifting[d]	□	From the floor and place it on a surface not greater than 5 ft (60 in.) above the floor[a–c, e] Two females 74 lb Two males 112 lb One female and one male 74 lb From the floor and place it on a surface not greater than 3 ft (36 in.) above the floor[a–c, e] Two females 88 lb Two males 174 lb One female and one male 88 lb	□	For exposures that exceed caution level perform lift analysis using the MIL-STD 1472F 5.9.11.3.1
15. Carrying loads	□	Lifted from floor and carried (with one or two hands) 33 ft or less then placed on floor or another surface not greater than 3 ft (36 in.) above the floor[a–c, e–g] Females 42 lb Males 82 lb	□	For exposures that exceed caution level or a multiperson carry, perform lift analysis using the MIL-STD 1472F 5.9.11.3.5

Awkward Posture		
	Caution	Hazard
	Lifted from floor and carried (with one or two hands) 33 ft or less then placed another surface not greater than 5 ft (60 in.) above the floor[a–c, e] Females 37 lb Males 56 lb	
See Figure C.8		

Figure C.8 Problematic lifting (Adapted from OPNAVINST 5100.23G Series Chapter 23 Appendix B and Public Domain Washington State Department of Labor and Industries Caution checklist)

[a]Frequency is less than 1 lift/5 min or 20 lifts/8 h.
[b]Object size does not exceed 18 in. × 18 in. × 12 in. deep.
[c]If conditions 1 and 2 are not met, lifting recommendations are reduced. Use MIL-STD 1472F 5.9.11.3.1 for more information.
[d]Reference MIL-STD 1472F 5.9.11.3.1 for lifting teams greater than 2.
[e]Object weight is evenly distributed.
[f]If conditions 1 and 2 are not met, lifting recommendations are reduced. See MIL_STD 1472F 5.9.11.3.2 and MIL_STD 1472F 5.9.11.3.3, respectively, for more information.
[g]Reference MIL-STD 1472F 5.9.11.3.5 for carrying teams.

MANUAL MATERIAL HANDLING EVALUATION TOOL AND SOLUTION GUIDE

This document is an aid for the identification of risk factors associated with manual material handling (MMH) and will assist you in reducing these risk factors. Manual handling refers to tasks that require a person to lift, lower, push, pull, hold, or carry any object, animal, or person. This document focuses on the handling of objects and not the handling of persons or animals.

This checklist identification process assesses six factors: weight; posture and layout; frequency and duration; individual and object characteristics; and environmental factors. Potential solutions are given after each risk factor to reduce the stressors to the lowest amount feasible. The training module examples will aid in the development of an action plan to eliminate or reduce the risk of injury. The examples provided in the module and checklist corresponds to the six risk factors in the manual handling checklist.

How to Use the Checklist

Step One: Select a manual handling task that has any of these characteristics: high injury rate; workers have concerns about performing the task; or has been identified in the recognition phase of a risk assessment.

Step Two: Advise the involved workers and supervisor that they are performing the survey.

Step Three: Complete the manual handling checklist for the task(s) you have identified. Answer N/A if the question does not apply to the task. Include all meaningful comments for each factor. *Each "NO" answer indications a risk of injury or suboptimal condition.*

Step Four: For each "NO" answer, consult the potential solutions for examples of ways to reduce the risk of injury. Use these examples as a starting point for brainstorming solutions that can be implemented in the workplace. Gather input from the workers, safety specialists, occupational health professionals, and other personnel to evaluate equipment before purchasing. This process will increase product acceptance, test product usability, and durability, and take advantage of worker experience.

How to Prevent Manual Handling Injuries

Commitment and involvement of the entire workplace, from top management to line workers, are essential elements of a successful injury prevention program. The best approach to prevent manual handling injuries involves education and design as well as hazard abatement.

The design of the job itself (work/rest schedules, job rotation), the object being handled and the workstation (dimension/layout) have a direct impact on the risk of injury. In order to prevent injuries, these aspects may need modification.

There are additional costs incurred in redesigning or modifying a task. Considering the six factors in the checklist before designing or modifying a workstation should help reduce costs. It is usually more costly to redesign and retrofit than to design it right the first time. Workers are a great source of creativity and innovation and should always be included in the process. In addition, providing mechanical aids such as conveyors, floor cranes, carts, suspension tools, hoists, and lift tables can also reduce injuries and increase productivity.

Weight	Yes	No	Possible Solutions to Reduce the Risk of Injury
1. When standing, is object lifted less than 30 lb?	□	□	Reduce weight by changing the size or number of objects lifted (cumulatively through the day) Select or design objects that can be handled close to the body by changing shape/orientation Minimize the travel distance of the load Minimize the total weight handled each day Implement bulk handling (e.g., in place of twenty 10 lb bags, order in a single 200 lb bag and use an aid) Change from lifting or carrying to pushing; or from pulling to pushing Introduce lift teams Provide a mechanical aid like a cart or hoist
2. When seated, is object lifted less than 10 lb?	□	□	Stand when lifting object in excess of 10 lb

Other optimal conditions: Vertical distance the object travels is less than 10 in. (when work is performed at a stationary workstation)
Weight is evenly distributed between both hands
Horizontal distance between the person and the center of gravity of the object is less than 10 in.

Weight	Yes	No	Possible Solutions to Reduce the Risk of Injury

Notes/comments: See Figure C.9

Figure C.9 Keep load close when lifting (Ergonomics Image Gallery)

Posture and Layout			
3. Are objects handled between knuckle and shoulder height?	☐	☐	Minimize number of times the load is lifted below the knees or above the shoulders Adjust the height of the workstation to optimal working height Provide a mechanical aid like a hoist or cart Frequently used objects should be within easy reach, close to waist height if possible

Weight	Yes	No	Possible Solutions to Reduce the Risk of Injury
4. Are objects within arm's length, allowing the worker to reach them without bending the back?	□	□	Allow workers to use different postures and muscle groups through job rotation Provide a mechanical aid like a hoist or cart Frequently used objects should be within easy reach
5. Is the task performed in an open space, allowing worker to freely move feet and arms?	□	□	Allow workers to use different postures and muscle groups through job rotation Adjust the height of the workstation to optimal working height Provide a mechanical aid like a hoist or cart

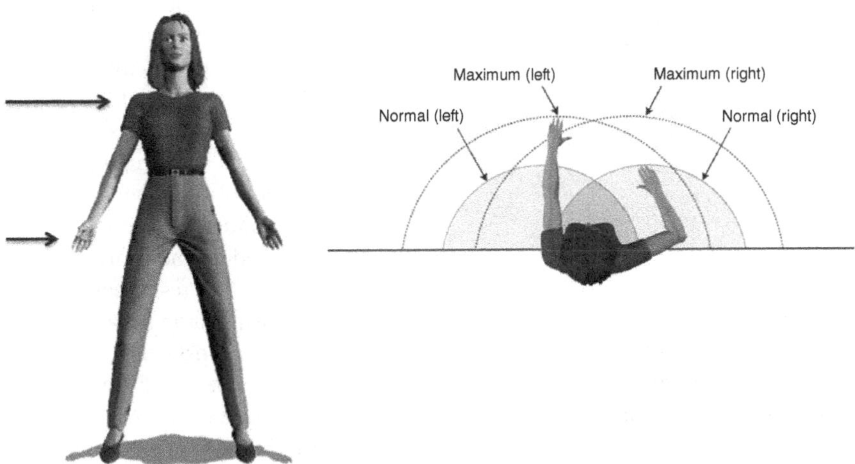

Figure C.10 Keep tasks between knuckles and shoulders an within arm's reach (Ergonomics Image Gallery used with license)

Posture and Layout			
6. Does the worker move without twisting the back during the handling process?	□	□	Use proper body mechanics – turn the feet rather than twisting Orient work toward the worker

Weight	Yes	No	Possible Solutions to Reduce the Risk of Injury
			Adjust the height of the workstation to optimal working height Provide a mechanical aid like a hoist or cart Align origin and destination of lift to avoid twisting
Other optimal conditions: Optimal work height is based on a worker's body dimension and type of work Allow the body to rest with sit/stand stool or footrest for standing work stations Provide antifatigue matting or shoe inserts for workers who stand for long durations Notes/comments:			
Frequency and Duration			
7. Does the worker perform the same motion less than once every 5 min?	□	□	Try using large muscle groups instead of small, for example, use the entire arm instead of just the hand Minimize the number of times a load is lifted Reduce the pace of the task Provide a mechanical aid Introduce an administrative control such as task rotation, varying the schedule and/or frequent and short rest breaks
8. Does the worker use different parts of the body every hour, giving muscle groups a chance to rest?	□	□	Try using large muscle groups instead of small, for example, use the entire arm instead of just the hand Minimize the number of times a load is lifted Reduce the pace of the task Provide a mechanical aid Introduce an administrative control such as task rotation, varying the schedule and/or frequent and short rest breaks

Weight	Yes	No	Possible Solutions to Reduce the Risk of Injury
Notes/Comments:			
Object Characteristics			
10. Is the object easy to handle, balanced, and stable?	☐	☐	Modify object – change shape, provide handles, use rigid containers, divide into smaller objects, move center of gravity closer to the body Balance and stabilize contents Provide a mechanical aid such as a cart or hoist Implement team lifting Use comfortable well-fitting gloves
11. Does the object provide a power grip handle in a neutral posture?	☐	☐	Provide a mechanical aid such as a cart or hoist Modify object
Notes/comments: See Figure C.11			

Figure C.11 Don't lift above shoulder height (Ergonomics Image Gallery used with license)

Weight	Yes	No	Possible Solutions to Reduce the Risk of Injury
Individual Characteristics	Yes	No	Comments/NA
12. Is the worker trained to perform the task, including safe handling principles?	□	□	Provide education in safe manual handling principles Provide clear, meaningful instruction for the task and evaluate worker's comprehension Ensure workers are physically fit to do the task
13. Does the worker's clothing and personal protective equipment allow for safe handling?	□	□	Provide education on proper selection and fit of PPE and clothing for environment Provide clear, meaningful instructions for the task and evaluate worker's comprehension Ensure workers are physically fit to do the task
Notes/comments:			
Environmental Factors			
14. Is the floor surface clean, nonslippery and even?	□	□	Keep work surface and floor clutter free; ensure good housekeeping Identify high traffic areas and flow direction using floor markers and overhead signs Use mirrors and other visual aids to help workers maneuver safely around corners and obstacles
15. Are temperature, humidity, lighting, noise, and airflow appropriate for the worker?	□	□	Optimal conditions: temperature 66–70°F, humidity 30–50%; lighting at least 200 lx Add lighting to improve visibility of objects without introducing glare Use appropriate clothing for temperature Notes/comments:

Education is a key step in injury prevention. The workers should have a basic understanding of ergonomic principles and handling techniques, recognize risk factors and injury or illness symptoms. The workers need to have a clear means of reporting injury symptoms, risk factors, near misses, hazards, incidents, and accidents to their supervisor.

Evaluator: Date:
Department name: Location: POC name: Phone number:
Job position evaluated: Number of employees performing job:
Follow-up date: Email address:
Job description/drawings
Recommendations:
Follow-up:

REFERENCES

McAtamney, L., & Corlett, E. N. (1993). RULA: A Survey Method for the Investigation of Work-Related Upper Limb Disorders. *Applied Ergonomics*, 91–99.

Hignett, S. and McAtamney, L. (2000) Rapid Entire Body Assessment: REBA, *Applied Ergonomics*, 31, 201–205.

GLOSSARY

Acceleration (*A*): Acceleration is a measure of how quickly speed changes with time. The measure of acceleration is expressed in units of measure (meters or feet per second) per second or meters/feet per second squared (m/s^2).

Administrative controls: Changes in work procedures such as written safety policies, rules, supervision, schedules, and training with the goal of reducing the duration, frequency, and severity of exposure to hazardous chemicals or situations.

Ambidextrous tools: Tools that can be used with either hand.

Amplitude : The amplitude of vibration is the magnitude of vibration. A vibrating object moves to a certain maximum distance on either side of its stationary position. Amplitude is the distance from the stationary position to the extreme position on either side and is measured in meters (m). The intensity of vibration depends on amplitude.

Anthropometry : Measurement of humans; physical anthropology refers to the measurement of the human individual for the purposes of understanding human physical variation.

Awkward postures: Postures that strain the neck, shoulders, elbows, wrists, hands, or back. Bending, stooping, twisting, and reaching are examples of awkward postures.

Borg scale : Ratings of perceived exertion scale. Borg developed two scales, an 11-point and a 15-point scale.

Occupational Ergonomics: A Practical Approach, First Edition.
Theresa Stack, Lee T. Ostrom and Cheryl A. Wilhelmsen.

Candela: Power emitted by a light source in a particular direction, weighted by the luminosity function (a standardized model of the sensitivity of the human eye to different wavelengths). A common candle emits light with a luminous intensity of roughly one candela. The candela is sometimes still called by the old name *candle,* such as *foot-candle*.

Carpel tunnel syndrome: A numbness and tingling in the hand and arm caused by a pinched nerve in the wrist.

Circumference: Distance around an object.

Contact pressure: Pressure from a hard surface, point, or edge on any part of the body.

Contrast : The ratio of the luminance of the object and the luminance of the background.

Cubital tunnel syndrome: Compression (pinching) of the ulnar nerve as it passes through the cubital tunnel in the arm (elbow).

Direct glare : One or more bright sources of light that shine directly into the eyes.

Double handle tools: Plier-like tools measured by handle length and grip span. Grip span is the distance between the thumb and fingers when the tool jaws are open and closed.

Engineering controls: Methods that are built into the design of a plant, equipment, or process to minimize the hazard.

Ergonomics specialists: Ergonomists are concerned with the safety and efficiency of equipment, systems and transportation. They use scientific information to ensure the health, comfort and protection of the people using them and, due to the nature of the work, can find themselves in a wide range of environments. Ergonomists match the capabilities and limitations of humans to tasks, workspaces and tools.

Eye strain: Condition that manifests itself through nonspecific symptoms such as fatigue, pain in or around the eyes, blurred vision, headache, and occasional double vision. Symptoms often occur after reading, computer work, or other close activities that involve tedious visual tasks.

Frequency (*F*): The frequency of a vibration, measured in Hertz (Hz), is simply the number of to and fro movements made in each second; 100 Hz would be 100 complete cycles in 1 s.

Glare : The sensation produced by luminance within the visual field that is sufficiently greater than the luminance to which the eyes are adapted to cause annoyance, discomfort, or loss of visual performance and visibility.

Hand-arm vibration (HAV): Transmitted from things in the hand to the hands and arms.

Hand-arm vibration syndrome : Vibration damages that may occur in the fingers, hands, and arms when working with vibrating tools or machinery.

Hazard mitigation : Any sustained action taken to reduce or eliminate the long-term risk to life and property from hazard events. It is an ongoing process that occurs before, during, and after disasters and serves to break the cycle of damage and repair in hazardous areas.

Hook grip: Characterized by a flat hand, curled fingers, and thumb used passively to stabilize the load, for example, auto steering wheels. Load is supported by fingers. This grip is most effective when the arms are down at the side of the body.

Illuminance : The amount of light falling on a work surface or task from ambient or local light.

Isoinertial: Derives from the words iso (same) and inertial (resistance), which in one terminology describes the primary concept of the isoinertial system, or expressing the same inertia in both the concentric phase and the eccentric.

LEAN 6 Sigma: Methodology that relies on a collaborative team effort to improve performance by systematically removing waste; combining lean manufacturing/lean enterprise and Six Sigma to eliminate the eight kinds of waste.

Lumen : A measure of the power of light perceived by the human eye.

Luminance : The amount of light leaving a surface; may be emitted or reflected.

Lux : A measure of the intensity, as perceived by the human eye. 1 Lux = 1 Lumen per square meter. Lux is a derived unit based on lumen, and lumen is a derived unit based on candela.

Mock-up: A model or replica of a machine or structure, used for instructional or experimental purposes.

Musculoskeletal disorder (MSD): Injuries or pain in the body's joints, ligaments, muscles, nerves, tendons, and structures that support limbs, neck, and back. MSDs are degenerative diseases and inflammatory conditions that cause pain and impair normal activities.

Naturalistic observation techniques: A study method that involves covertly or overtly watching subjects' behaviors in their natural environment, without intervention. Naturalistic observation is a common research method in behavioral sciences such as sociology and psychology.

Oblique grip: A variant of power grasp characterized by gripping across the surface of an object, that is, carrying a tray with handles – lift up and power grip ends.

Percentile: The *x*th percentile indicates that *x* percent of the population has the same value or less than that value for a given measurement.

Pinch grip: The hand grip that provides control for precision and accuracy. The tool is gripped between the thumb and the fingertips.

Power grip: The hand grip that provides maximum hand power for high hand force tasks. All the fingers wrap around the handle.

Power zone: Area between knuckles and shoulders. Also, can be expanded from knees to shoulders.

Proactive ergonomics: Emphasizes the prevention of work-related musculoskeletal disorders through recognizing, anticipating, and reducing risk factors in the planning stages of new systems of work or workplaces.

Psychophysics : The quantitative branch of the study of perception.

Rapid entire body assessment (REBA): An ergonomic assessment technique that is based on a biomechanical methodology.

Rapid upper body assessment (RULA): An ergonomic assessment technique that is based on a biomechanical methodology.

Ratings of perceived exertion (RPE): A psychophysical method for measuring physical stress.

Raynaud's syndrome or phenomenon: A condition (loss of blood circulation) to the fingers and hand. A more modern term for Raynaud's syndrome is Hand Arm Vibration Syndrome (HAVS).

Reflected glare: Caused by light reflected from an object or objects that an observer is viewing.

Resonance : Every object tends to vibrate at one particular frequency called the natural frequency. The measure of natural frequency depends on the composition of the object, its size, structure, weight, and shape. If we apply a vibrating force on the object with its frequency equal to the natural frequency, it is a resonance condition. A vibrating machine transfers the maximum amount of energy to an object when the machine vibrates at the object's resonant frequency.

Risk factors: Any attribute, characteristic, or exposure of an individual that increases the likelihood of developing a disease or injury.

Single-handed tools: Tube-like tools measured by handle length and diameter. Diameter is the length of the straight line through the center of the handle.

Speed/velocity (*V*): The speed of a vibrating object varies from zero to a maximum during each cycle of vibration. It moves fastest as it passes through its natural stationary position to an extreme position. The vibrating object slows down as it approaches the extreme, where it stops and then moves in the opposite direction through the stationary position toward the other extreme. Speed of vibration is expressed in units of measure per second (m/s). For instance, 1 foot/s or 1 m/s.

Tendonitis: A condition in which the tissue connecting muscle to bone becomes inflamed.

Threshold limit values: The maximum average airborne concentration of a hazardous material to which healthy adult workers can be exposed during an 8-h workday and 40-h workweek – over a working lifetime – without experiencing significant adverse health effects.

Vibration: The oscillating, reciprocating, or other periodic motion of a rigid or elastic body or medium forced from a position or state of equilibrium.

Wave length (λ): The measure of peak to peak of one vibration cycle.

White finger : Also known as HAVS or Raynaud's syndrome.

Whole-body vibration (WBV): Which is transmitted by mobile or fixed machines where the operator is standing or seated. Trucks, aircraft, heavy equipment, and boats are examples of things that cause whole-body vibration.

ACRONYMS

ACGIH: American Conference of Governmental Industrial Hygienists
ACL: anterior cruciate ligament
AIB: artillery instructional battery
AL: action limit
ALARA: as low as (is) reasonably achievable
AND: airplane nose down
ANU: airplane nose up
ATP: adenosine triphosphate
BCPE: Board of Certification for Professional Ergonomists
CASS: continuing analysis and surveillance system
CDC: Centers for Disease Control
CTDs: cumulative trauma disorders
CTS: carpal tunnel syndrome
CVR: cockpit voice recorder
GDIT: general dynamics information technology
HAL: hand-arm activity levels
HAV: hand-arm vibration
ISO: International Organization for Standardization
JHA: job hazard analysis

Occupational Ergonomics: A Practical Approach, First Edition.
Theresa Stack, Lee T. Ostrom and Cheryl A. Wilhelmsen.

JSA: job safety analysis
LI: lifting index
MCB: marine corps base
MCL: medial collateral ligament
MF: multifidus
MMH: manual materials handling
MPHA: Mishap Prevention/Hazard Abatement
MSDs: musculoskeletal disorders
NIOSH: National Institute of Occupational Safety and Health
NPF: normalized peak force
NSAIDs: nonsteroidal anti-inflammatory drugs
OA: osteoarthritis
PCL: posterior cruciate ligament
P_f: peak hand force
PFM: pelvic floor muscles
PFPS: patellofemoral pain syndrome
PPE: personal protective equipment
REBA: rapid entire body assessment
RMT: repetitive motion trauma
RPE: ratings of perceived exertion
RSIs: repetitive strain injuries
RULA: rapid upper limb assessment
RWL: recommended weight limit
STAN: sum total aft and nose
STEXs: sand table exercises
TA: transversus abdominis
TBS: The Basic School
TDGs: tactical decision games
TLV: threshold limit value
TOS: thoracic outlet syndrome
TRIREFFAC: trident refit facility
WBGT: wet bulb globe temperature
WBV: whole-body vibration
WMSDs: work-related musculoskeletal disorders
WPLG: work practices lifting guide

INDEX

Occupational Ergonomics: A Practical Approach, First Edition.
Theresa Stack, Lee T. Ostrom and Cheryl A. Wilhelmsen.